Manual of Structural Kinesiology

R. T. Floyd EdD, ATC, CSCS

Director of Athletic Training and Sports Medicine
Professor of Physical Education and Athletic Training
Chair, Department of Physical Education and
 Athletic Training
 The University of West Alabama
 (formerly Livingston University)
 Livingston, Alabama

SIXTEENTH EDITION

McGraw Hill

Boston   Burr Ridge, IL   Dubuque, IA   Madison, WI   New York   San Francisco   St. Louis
Bangkok   Bogotá   Caracas   Kuala Lumpur   Lisbon   London   Madrid   Mexico City
Milan   Montreal   New Delhi   Santiago   Seoul   Singapore   Sydney   Taipei   Toronto

The McGraw·Hill Companies

Mc Graw Hill **Higher Education**

Published by McGraw-Hill, an imprint of The McGraw-Hill Companies, Inc., 1221 Avenue of the Americas, New York, NY 10020. Copyright © 2007. All rights reserved. No part of this publication may be reproduced or distributed in any form or by any means, or stored in a database or retrieval system, without the prior written consent of The McGraw-Hill Companies, Inc., including, but not limited to, in any network or other electronic storage or transmission, or broadcast for distance learning.

This book is printed on acid-free paper.

1 2 3 4 5 6 7 8 9 0 WCK/WCK 0 9 8 7 6

ISBN-13   978-0-07-110655-9
ISBN-10   0-07-110655-3

Some ancillaries, including electronic and print components, may not be available to customers outside the United States.

www.mhhe.com

Contents

Preface

In revising this edition I have attempted to further refine the chapters for consistency and to improve the overall completeness and accuracy of the text. As with previous revisions, I have attempted to maintain the successful presentation approach the late Dr. Clem Thompson established from 1961 through 1989. I first used this book as an undergraduate and later in my teaching over the years. Having developed great respect for this text and Dr. Thompson's style, it is my intention to continue to preserve the effectiveness of this time-honored text, while adding material pertinent to the professions working with today's ever-growing physically active population. Hopefully, I have maintained a clear, concise, and simple presentation method supplemented with applicable information gained through my career experiences.

This text, now in its 60th year, has undergone many revisions over the years. My goal continues to be making the material as applicable as possible to everyday physical activity and to make it more understandable and easier to use for the student and professional. With this in mind, several tables and illustrations have been added while many of the previous ones have been revised and updated. While reading this text, I challenge kinesiology students and professionals to immediately apply the content to physical activities with which they are individually familiar. I hope that the student will simultaneously palpate his or her own moving joints and contracting muscles to gain application. Concurrently, I encourage students to palpate the joints and muscles of fellow students to gain a better appreciation of the wide range of normal anatomy and, when possible, appreciate the variation from normal found in injured and pathological musculoskeletal anatomy. Additionally, with the tremendous growth of information and media available via the Internet and other technological means, I encourage careful and continuous exploration of these resources.

Although these resources are helpful, they must be reviewed with a critical eye, as all information should be.

Audience

This text is designed for students in an undergraduate structural kinesiology course after completing courses in human anatomy and physiology. While primarily utilized in physical education, exercise science, athletic training, physical therapy, and massage therapy curriculums, it is often used as a continuing reference by other clinicians and educators in addressing musculoskeletal concerns of the physically active. Applied kinesiologists, athletic trainers, athletic coaches, physical educators, physical therapists, health club instructors, strength and conditioning specialists, personal trainers, massage therapists, physicians, and others who are responsible for improving and maintaining the muscular strength, endurance, flexibility, and overall health of individuals will benefit from this text.

With the ever-continuing growth in the number of participants of all ages in a spectrum of physical activity, it is imperative that medical, health, and education professionals involved in providing instruction and information to the physically active be correct and accountable for the teachings that they provide. The variety of exercise machines, techniques, strengthening and flexibility programs, and training programs is continuously expanding and changing, but the musculoskeletal system is constant in its design and architecture. Regardless of the goals sought or the approaches used in exercise activity, the human body is the basic ingredient and must be thoroughly understood and considered to maximize performance capabilities and minimize undesirable results. Most advances in exercise science continue to result from a better understanding of the body and how it functions. I believe that an individual in this field can never learn enough about the structure and function of the human body.

Those who are charged with the responsibility of providing instruction and consultation to the physically active will find this text a helpful and valuable resource in their never-ending quest for knowledge and understanding of human movement.

New to this edition

Major changes to this edition include the replacement and revision of several tables and illustrations in every chapter as well as the addition of others in most chapters. Content relative to understanding the structure and function of bones, joints, muscles, and nerves and their individual and combined contribution to human movement has been expanded. Special efforts have been made to enhance the explanation of muscle palpations.

Chapter 1, "Foundations of Structural Kinesiology," retains the foundations related to bones and joints with several additions regarding terminology, bony features, bone growth, bone markings, bone properties, and joint movements. Also added to this chapter is a table of joint action icons that will be utilized in the individual joint chapters to enable the reader to more quickly identify the actions of a particular muscle. Chapter 2, "Neuromuscular Fundamentals," contains the previous edition's information on neuromuscular function with greater detail regarding muscle nomenclature, muscle shape and fiber arrangement, muscle tissue properties, determination of muscle action, muscle contraction, neural control of voluntary movement, proprioception, and kinesthesis. Chapter 3, "Basic Biomechanical Factors and Concepts," has undergone some expansion and revision related to the types of machines found in the body and mechanical advantage with new illustrations including mathematical explanations. In the eight chapters that address specific body parts, tables have been revised to more clearly detail the origin, insertion, action, plane of motion, palpation, and innervation for each muscle. All palpations have been carefully reviewed, revised, and expanded with a greater emphasis placed on combining movement with palpation in most cases. Margin icons identifying the joint actions caused by each muscle have been added to the respective pages related to the muscles. Chapter 8, "Muscular Analysis of Upper Extremity Exercises," has been revised to include a greater discussion and explanation of kinetic chain activity and now contains the content related to conditioning that was previously in Chapter 13.

Several definitions in the Glossary have been revised, and over 40 new terms have been added to this edition. The listings of Web sites included at the end of each chapter have been updated and significantly expanded. Several questions and exercises have been added at the end of each chapter to complement the existing ones, which have also been revised. Finally, a fourth appendix has been added to provide the etymology of most terms used in this text.

Ancillaries

Image Presentation CD-ROM
ISBN 0-07-312323-4

The Image Presentation CD-ROM is an electronic library of visual resources. The CD-ROM comprises images from the text displayed in PowerPoint, which allows the user to view, sort, search, use, and print catalog images. It also includes a complete, ready-to-use PowerPoint presentation, which allows users to play chapter-specific slide shows.

Online Learning Center

www.mhhe.com/floyd16e
The Online Learning Center to accompany this text offers a number of additional resources for both students and instructors. Visit this Web site to find useful materials such as these:

For the instructor:
- Downloadable PowerPoint presentations
- Interactive Web activities
- Chapter-specific interactive animations
- Links to professional resources
- PowerWeb access

For the student:
- Self-scoring chapter quizzes
- Anatomy flashcards and crossword puzzles for learning key terms and their definitions
- Chapter-specific interactive animations
- Learning objectives
- Interactive labeling activities
- PowerWeb access

PowerWeb

PowerWeb is a reservoir of course-specific articles and current events. Students can access PowerWeb articles to take a self-scoring quiz, complete an interactive exercise, click through an interactive glossary, or check the daily news. An expert in each discipline analyzes the day's news to show students how it relates to their field of study. PowerWeb articles are available on the Online Learning Center that accompanies this text.

Acknowledgments

The seven reviewers have provided numerous comments, ideas, and suggestions that I very much appreciate. These reviews have been an extremely helpful guide in this revision, and their suggestions have been incorporated to the extent possible when appropriate. These reviewers are:

Eadric Bressel
Utah State University–Logan

James Johnson
Smith College

Gary Kamen
University of Massachusetts–Amherst

June Lindle
Cincinnati State Technical and Community College

Edith Smith
Troy University

Ben Velasquez
University of Southern Mississippi

Sean Wright
Tusculum College

I would like to especially thank Stephanie Bura, MAT, ATC; Randianne Sears, MAT; Brad Montgomery, MAT, ATC; and the kinesiology/athletic training students of The University of West Alabama for their advice and input throughout this revision. Their assistance and suggestions have been most helpful. I appreciate Elizabeth Swann, PhD, ATC, for her assistance with the critical thinking questions. I also acknowledge Mr. John Hood and Mrs. Lisa Floyd of Birmingham and Livingston, Alabama, respectively, for the fine photographs. Special thanks to Mrs. Linda Kimbrough of Birmingham, Alabama, for her superb illustrations and insight. I appreciate the models for the photographs: Mr. Fred Knighten, Mr. Darrell Locket, Miss Kala Peak, Mr. Donny Powe, Mrs. Zina Pruitt, Mr. Jay Sears, Mrs. Randianne Sears, Mr. Marcus Shapiro, and Mr. David Whitaker. My thanks also go to Lynda Huenefeld and the McGraw-Hill staff who have been most helpful in their assistance and suggestions in preparing the manuscript for publication.

R. T. Floyd

About the Author

R. T. Floyd is in his thirty-second year of providing athletic training services for the University of West Alabama. Currently, he serves as the Director of Athletic Training and Sports Medicine for the UWA Athletic Training and Sports Medicine Center, Program Director for UWA's CAAHEP accredited athletic training education program, and as a professor in the Department of Physical Education and Athletic Training, which he chairs. He has taught numerous courses in physical education and athletic training including kinesiology at both the undergraduate and graduate levels since 1980.

Floyd has maintained an active career throughout his years in the profession. He was recently reelected to serve on the National Athletic Trainers' Association (NATA) Board of Directors to represent District IX, the Southeast Athletic Trainers' Association (SEATA). He served two years as the NATA District IX Chair on the NATA Research and Education Foundation Board before being elected to his current position as the Member Development Chair on the Board. Previously, he served as the District IX representative to the NATA Educational Multimedia Committee from 1988 to 2002. He has served as the Convention Site Selection Chair for District IX from 1986 to 2004 and has directed the annual SEATA Competencies in Athletic Training Student Workshop since 1997. He has also served as a NATA BOC examiner for well over a decade and has served as a Joint Review Committee on Educational Programs in Athletic Training site visitor several times. Floyd was recently appointed to the advisory board of the Collegiate Sports Medicine Foundation. He has provided over a hundred professional presentations at the local, state, regional, and national level and has also had several articles and videos published related to the practical aspects of athletic training. He began authoring the *Manual of Structural Kinesiology* in 1992 with the twelfth edition after the passing of Dr. Clem W. Thompson, who authored the fourth through the eleventh editions.

Floyd is a certified member of the National Athletic Trainers' Association, a Certified Strength & Conditioning Specialist, and a Certified Personal Trainer in the National Strength and Conditioning Association. He is also a Certified Athletic Equipment Manager in the Athletic Equipment Managers' Association, a member of the American College of Sports Medicine, the American Orthopaedic Society for Sports Medicine, the American Osteopathic Academy of Sports Medicine, the American Sports Medicine Fellowship Society, the American Alliance for Health, Physical Education, Recreation and Dance, and the Sports Physical Therapy Section. Additionally, he is licensed in Alabama as an Athletic Trainer and an Emergency Medical Technician.

Floyd was presented the Most Distinguished Athletic Trainer Award by the NATA in 2003 and received the organization's Athletic Trainer Service Award in 1996. He received the District IX Award for Outstanding Contribution to the field of Athletic Training by SEATA in 1990 and also received the organization's highest award, the Award of Merit, in 2001. He was named to Who's Who Among America's Teachers in 1996, 2000, 2004, and 2005. In 2001, he was inducted into the Honor Society of Phi Kappa Phi and the University of West Alabama Athletic Hall of Fame. He was inducted into the Alabama Athletic Trainers Association Hall of Fame in May 2004.

To

my family,
Lisa, Robert Thomas, Jeanna, Rebecca, and Kate
who understand, support, and allow me to
pursue my profession

and to my parents,
Ruby and George Franklin,
who taught me the importance of a strong work ethic
with quality results

R.T.F.

Foundations of Structural Kinesiology

Objectives

- To review the anatomy of the skeletal system

- To review and understand the terminology used to describe body part locations, reference positions, and anatomical directions

- To review the planes of motion and their respective axes of rotation in relation to human movement

- To describe and understand the various types of bones and joints in the human body and their characteristics

- To describe and demonstrate the joint movements

Online Learning Center Resources

Visit *Manual of Structural Kinesiology's* **Online Learning Center** at **www.mhhe.com/floyd16e** for additional information and study material for this chapter, including:

- *Self-grading quizzes*
- *Anatomy flashcards*
- *Animations*

Structural kinesiology is the study of muscles, bones, and joints as they are involved in the science of movement. Bones vary in size and shape, which factors into the amount and type of movement that occurs between them at the joints. The types of joint vary in both structure and function. Muscles also vary greatly in size, shape, and structure from one part of the body to another.

More than 600 muscles are found in the human body. In this book, an emphasis is placed on the larger muscles that are primarily involved in movement of the joints. Details related to many of the small muscles located in the hands, feet, and spinal column are provided to a lesser degree.

Anatomists, athletic trainers, physical therapists, physicians, nurses, massage therapists, coaches, strength and conditioning specialists, performance enhancement specialists, personal trainers, physical educators, and others in health-related fields should have an adequate knowledge and understanding of all the large muscle groups, so they can teach others how to strengthen, improve, and maintain these parts of the human body. This knowledge forms the basis of the exercise programs that should be followed to strengthen and maintain all of the muscles. In most cases, exercises that involve the larger primary movers also involve the smaller muscles.

Fewer than 100 of the largest and most important muscles, primary movers, are considered in this text. Some small muscles in the human body, such as the multifidus, plantaris, scalenus, and serratus posterior, are omitted because they are exercised with other, larger primary movers. In addition, most small muscles of the hands and feet are not given the full attention provided to the larger muscles. Many small muscles of the spinal column are not considered in full detail.

Kinesiology students frequently become so engrossed in learning individual muscles that they lose sight of the total muscular system. They miss the "big picture"—that muscle groups move joints in given movements necessary for bodily movement and skilled performance. Although it is vital to learn the small details of muscle attachments, it is even more critical to be able to apply the information to real-life situations. Once the information can be applied in a useful manner, the specific details are usually much easier to understand and appreciate.

Reference positions

It is crucial for kinesiology students to begin with a reference point in order to better understand the musculoskeletal system, its planes of motion, joint classification, and joint movement terminology. Two reference positions may be used as a basis from which to describe joint movements. The **anatomical position** is the most widely used and accurate for all aspects of the body. Fig. 1.1 demonstrates this reference position, with the subject standing in an upright posture, facing straight ahead, feet parallel and close, and palms facing forward. The **fundamental position** is essentially the same as the anatomical position, except that the arms are at the sides and the palms are facing the body.

Anatomical directional terminology   FIGS. 1.1, 1.2, 1.3

It is important that we all are able to find our way around the human body. To an extent, we can think of this as similar to giving or receiving directions about how to get from one geographic location to another. Just as we use the terms *left, right, south, west, northeast,* etc. to describe geographic directions, we have terms such as *lateral, medial, inferior, anterior,* etc. to use for anatomi-

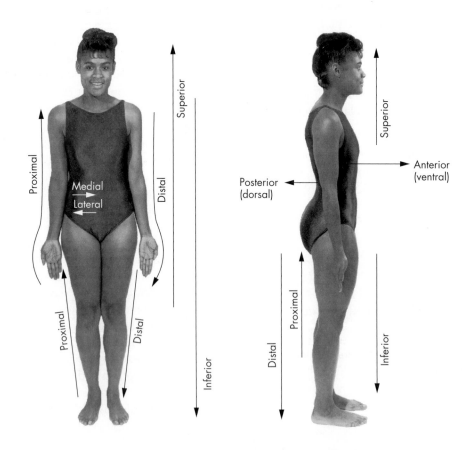

FIG. 1.1 ● Anatomical position and anatomical directions. Anatomical directions refer to the position of one body part in relation to another.

From Van De Graaff KM: *Human anatomy,* ed 6, New York, 2002, McGraw-Hill.

cal directions. With geographic directions we may use *west* to indicate the west end of a street or the western United States. The same is true when using anatomical directions. We may use *superior* to indicate the end of a bone in our lower leg closest to the knee, or we may be speaking about the top of skull. It all depends on the context at the time. Just as we combine *south* and *east* to get *southeast* for the purpose of indicating somewhere in between each direction, we may combine *anterior* and *lateral* to get *anterolateral* for the purpose of describing the general direction or location of in the front and to the outside. Figs. 1.2 and 1.3 provide further examples.

Anterior
In front or in the front part

Anteroinferior
In front and below

Anterolateral
In front and to the outside

Anteromedial
In front and toward the inner side or midline

Anteroposterior
Relating to both front and rear

Anterosuperior
In front and above

Bilateral
Relating to the right and left sides of the body or of a body structure such as the right and left extremities

Caudal
Below in relation to another structure; inferior

Cephalic
Above in relation to another structure; higher, superior

Contralateral
Pertaining or relating to the opposite side

Deep
Beneath or below the surface; used to describe relative depth or location of muscles or tissue

Distal
Situated away from the center or midline of the body, or away from the point of origin

Dorsal
Relating to the back, being or located near, on, or toward the back, posterior part, or upper surface of

Inferior
(infra) Below in relation to another structure; caudal

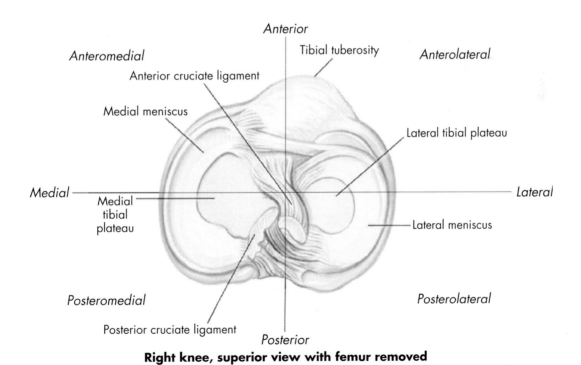

Right knee, superior view with femur removed

FIG. 1.2 ● Anatomical directional terminology.

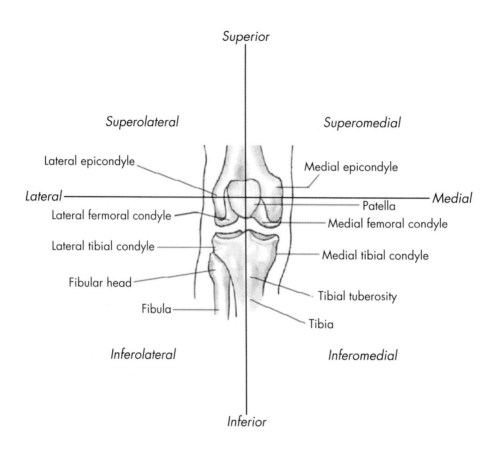

FIG. 1.3 ● Anatomical directional terminology.

Inferolateral

Below and to the outside

Inferomedial

Below and toward the midline or inside

Ipsilateral

On the same side

Lateral

On or to the side; outside, farther from the median or midsagittal plane

Medial

Relating to the middle or center; nearer to the median or midsagittal plane

Median

Relating to, located in, or extending toward the middle, situated in the middle, mesial

Palmar

Relating to the palm or volar aspect of the hand

Plantar

Relating to the sole or undersurface of the foot

Posterior

Behind, in back, or in the rear

Posteroinferior

Behind and below; in back and below

Posterolateral

Behind and to one side, specifically to the outside

Posteromedial

Behind and to the inner side

Posterosuperior

Behind and at the upper part

Prone

Face downward position of the body; stomach lying

Proximal

Nearest the trunk or the point of origin

Superficial

Near the surface; used to describe relative depth or location of muscles or tissue

Superior

(supra) Above in relation to another structure; higher, cephalic

Superolateral

Above and to the outside

Superomedial

Above and toward the midline or inside

Supine

Lying on the back; face upward position of the body

Ventral

Relating to the belly or abdomen, on or toward the front, anterior part of

Volar

Relating to palm of the hand or sole of the foot

Body regions

As mentioned later under the skeletal system, the body can be divided into axial and appendicular regions. Each of these regions may be further divided into different regions such as the cephalic, cervical, trunk, upper limbs, and lower limbs. Within each of these regions there are many more subregions and specific regions. Table 1.1 details a breakdown of these regions and their common names, which are illustrated in Fig. 1.4.

TABLE 1.1 • **Body parts and regions**

	Region name	Common name	Subregion	Specific region name	Common name for specific region
Axial	Cephalic	Head	Cranial (skull)	Frontal	Forehead
				Occipital	Base of skull
			Facial (face)	Orbital	Eye
				Otic	Ear
				Nasal	Nose
				Buccal	Cheek
				Oral	Mouth
				Mental	Chin
	Cervical	Neck		Nuchal	Back of neck
				Throat	Anterior neck
	Trunk	Thoracic	Thorax	Clavicular	Collar bone
				Pectoral	Chest
				Sternal	Breastbone
				Costal	Ribs
				Mammary	Breast
		Dorsal	Back	Scapular	Shoulder blade
				Vertebral	Spinal column
				Lumbar	Low back or loin
		Abdominal	Abdomen	Celiac	Abdomen
				Umbilical	Naval
		Pelvic	Pelvis	Inguinal	Groin
				Pubic	Genital
				Coxal	Hip
				Sacral	Between hips
				Gluteal	Buttocks
				Perineal	Perineum

TABLE 1.1 (continued) • **Body parts and regions**

	Region name	Common name	Subregion	Specific region name	Common name for specific region
Appendicular	Upper limb	Shoulder		Acromial	Point of shoulder
				Omus	Deltoid
				Axillary	Arm pit
		Manual		Brachial	Arm
				Olecranon	Point of elbow
				Cubital	Elbow
				Antecubital	Front of elbow
				Antebrachial	Forearm
				Carpal	Wrist
				Palmar	Palm
				Dorsal	Back of hand
				Digital	Fingers
	Lower limb	Pedal	Foot	Femoral	Thigh
				Patella	Kneecap
				Popliteal	Back of knee
				Sural	Calf
				Crural	Leg
				Talus	Ankle
				Calcaneal	Heel
				Dorsum	Top of foot
				Tarsal	Instep
				Plantar	Sole
				Digital	Toes

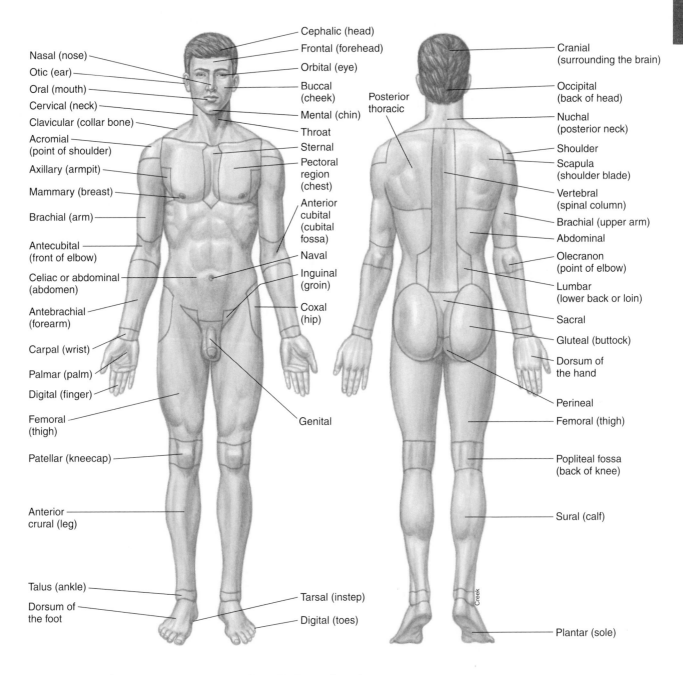

Nasal (nose)
Otic (ear)
Oral (mouth)
Cervical (neck)
Clavicular (collar bone)
Acromial
(point of shoulder)
Axillary (armpit)
Mammary (breast)
Brachial (arm)
Antecubital
(front of elbow)
Celiac or abdominal
(abdomen)
Antebrachial
(forearm)
Carpal (wrist)
Palmar (palm)
Digital (finger)
Femoral
(thigh)
Patellar (kneecap)
Anterior
crural (leg)
Talus (ankle)
Dorsum of
the foot

Cephalic (head)
Frontal (forehead)
Orbital (eye)
Buccal
(cheek)
Mental (chin)
Throat
Sternal
Pectoral
region
(chest)
Anterior
cubital
(cubital
fossa)
Naval
Inguinal
(groin)
Coxal
(hip)
Genital
Tarsal (instep)
Digital (toes)

Posterior
thoracic

Cranial
(surrounding the brain)
Occipital
(back of head)
Nuchal
(posterior neck)
Shoulder
Scapula
(shoulder blade)
Vertebral
(spinal column)
Brachial (upper arm)
Abdominal
Olecranon
(point of elbow)
Lumbar
(lower back or loin)
Sacral
Gluteal (buttock)
Dorsum of
the hand
Perineal
Femoral (thigh)
Popliteal fossa
(back of knee)
Sural (calf)
Plantar (sole)

Creek

FIG. 1.4 ● Body regions. **A,** Anterior view; **B,** Posterior view.

Planes of motion

When studying the various joints of the body and analyzing their movements, it is helpful to characterize them according to specific planes of motion (Fig. 1.5). A plane of motion may be defined as an imaginary two-dimensional surface through which a limb or body segment is moved.

There are three specific, or **cardinal**, planes of motion in which the various joint movements can

be classified. The specific planes that divide the body exactly into two halves are often referred to as cardinal planes. The cardinal planes are the sagittal, frontal, and transverse planes. There are an infinite number of planes within each half that are parallel to the cardinal planes. This is best understood in the following examples of movements in the sagittal plane. Sit-ups involve the spine and, as a result, are performed in the cardinal sagittal plane, which is also known as the **midsagittal** plane. Biceps curls and knee extensions are performed in **parasagittal** planes, which are parallel to the midsagittal plane. Even though these examples are not in the cardinal plane, they are thought of as movements in the sagittal plane.

Although each specific joint movement can be classified as being in one of the three planes of motion, our movements are usually not totally in one specific plane but occur as a combination of motions from more than one plane. These movements from the combined planes may be described as occurring in diagonal, or oblique, planes of motion.

Sagittal, anteroposterior, or AP plane

The sagittal, anteroposterior, or AP plane bisects the body from front to back, dividing it into right and left symmetrical halves. Generally, flexion and extension movements such as biceps curls, knee extensions, and sit-ups occur in this plane.

Frontal, lateral, or coronal plane

The frontal plane, also known as the lateral or coronal plane, bisects the body laterally from side to side, dividing it into front and back halves. Abduction and adduction movements such as jumping jacks and spinal lateral flexion occur in this plane.

Transverse or horizontal plane

The transverse plane divides the body horizontally into superior and inferior halves. Generally, rotational movements such as forearm pronation and supination and spinal rotation occur in this plane.

Diagonal or oblique plane   FIG. 1.6

The diagonal or oblique plane is a combination of more than one plane. In reality, most of our movements in sporting activities fall somewhere less than parallel or perpendicular to the previously described planes and occur in a diagonal plane.

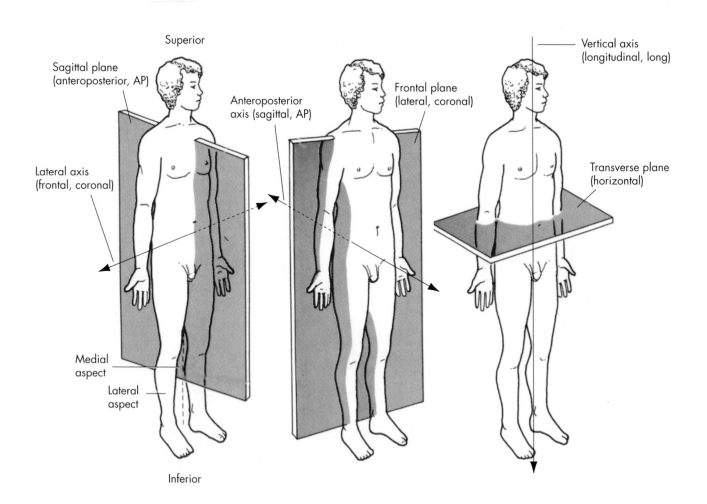

FIG. 1.5 ● Planes of motion and axes of rotation.

Modified from Booher JM, Thibodeau GA: *Athletic injury assessment*, ed 4, New York, 2000, McGraw-Hill.

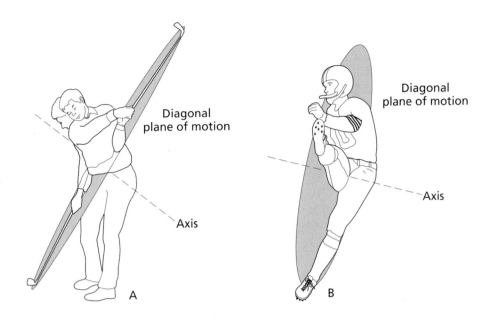

FIG. 1.6 ● Diagonal plane and axis of rotation.

TABLE 1.2 • Planes of motion and their axes of rotation

Plane	Description of plane	Axis of rotation	Description of axis	Common movements
Sagittal (anteroposterior or AP)	Divides the body into right and left halves	Frontal (lateral or coronal) (medial-lateral)	Runs medial/lateral	Flexion, extension
Frontal (lateral or coronal)	Divides the body into anterior and posterior halves	Sagittal (anteroposterior or AP)	Runs anterior/posterior	Abduction, adduction
Transverse (horizontal)	Divides the body into superior and inferior halves	Vertical (longitudinal or long)	Runs superior/inferior	Internal rotation, external rotation

Axes of rotation

As movement occurs in a given plane, the joint moves or turns about an axis that has a 90-degree relationship to that plane. The axes are named in relation to their orientation (Fig. 1.5). Table 1.2 lists the planes of motion with their axes of rotation.

Frontal, lateral, or coronal axis

If the sagittal plane runs from anterior to posterior, then its axis must run from side to side. Since this axis has the same directional orientation as the frontal plane of motion, it is named similarly. As the elbow flexes and extends in the sagittal plane when performing a biceps curl, the forearm is actually rotating about a frontal axis that runs laterally through the elbow joint. The frontal axis may also be referred to as the bilateral axis.

Sagittal or anteroposterior axis

Movement occurring in the frontal plane rotates about a sagittal axis. This sagittal axis has the same directional orientation as the sagittal plane

of motion and runs from front to back at a right angle to the frontal plane of motion. As the hip abducts and adducts during jumping jacks, the femur rotates about an axis that runs front to back through the hip joint.

Vertical or longitudinal axis

The vertical axis, also known as the longitudinal or long axis, runs straight down through the top of the head and is at a right angle to the transverse plane of motion. As the head rotates or turns from left to right when indicating disapproval, the skull and cervical vertebrae are rotating around an axis that runs down through the spinal column.

Diagonal or oblique axis   FIG. 1.6

The diagonal axis, also known as the oblique axis, runs at a right angle to the diagonal plane. As the glenohumeral joint moves from diagonal abduction to diagonal adduction in overhand throwing, its axis runs perpendicular to the plane through the humeral head.

Skeletal systems

Fig. 1.7 shows anterior and posterior views of the skeletal system. Two hundred six bones make up the skeletal system, which provides support and protection for other systems of the body and provides for attachments of the muscles to the bones by which movement is produced. Additional skeletal functions are mineral storage and hemopoiesis, which involves blood cell formation in the red bone marrow. The skeleton may be divided into the appendicular and the axial skeleton. The appendicular skeleton is composed of the appendages, or the upper and lower extremities, and the shoulder and pelvic girdles, while the axial skeleton consists of the skull, vertebral column, ribs, and sternum. Most students who take this course will have had a course in human anatomy, but a brief review is desirable before beginning the study of kinesiology. Other chapters provide additional information and more detailed illustrations of specific bones.

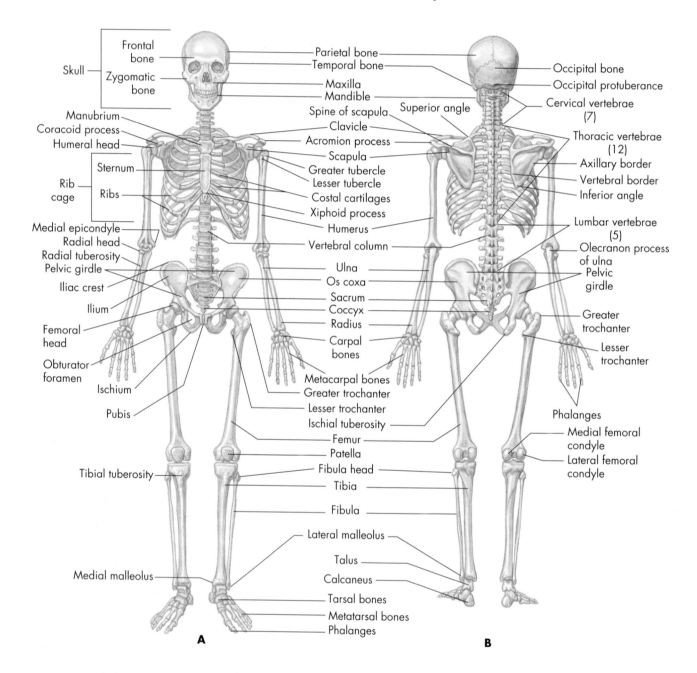

FIG. 1.7 ● Skeleton. **A,** Anterior view; **B,** Posterior view.

Modified from Van De Graaff KM: *Human anatomy,* ed 6, New York, 2002, McGraw-Hill.

Osteology

The adult skeleton, consisting of approximately 206 bones, may be divided into the axial skeleton and the appendicular skeleton. The axial skeleton contains 80 bones, which include the skull, spinal column, sternum, and ribs. The appendicular skeleton contains 126 bones, which include all of the bones of the upper and lower extremities. The pelvis is sometimes classified as being part of the axial skeleton due to its importance in linking the axial skeleton with the lower extremities of the appendicular skeleton. The exact number of bones as well as their specific features occasionally vary from person to person.

Skeletal functions

The skeleton has five major functions:

1. protection of vital soft tissues such as the heart, lungs, and brain
2. support to maintain posture
3. movement by serving as points of attachment for muscles and acting as levers
4. storage for minerals such as calcium and phosphorus
5. hemopoiesis, which is the process of blood formation that occurs in the red bone marrow located in the vertebral bodies, femur, humerus, ribs, and sternum

Types of bones

Bones vary greatly in shape and size but can be categorized in five major categories (Fig. 1.8).

Long bones

Composed of a long cylindrical shaft with relatively wide, protruding ends; serve as levers. The shaft contains the medullary canal. Examples include phalanges, metatarsals, metacarpals, tibia, fibula, femur, radius, ulna, and humerus.

Short bones

Small, cube-shaped, solid bones that usually have a proportionally large articular surface in order to articulate with more than one bone. Short bones provide some shock absorption and include the carpals and tarsals.

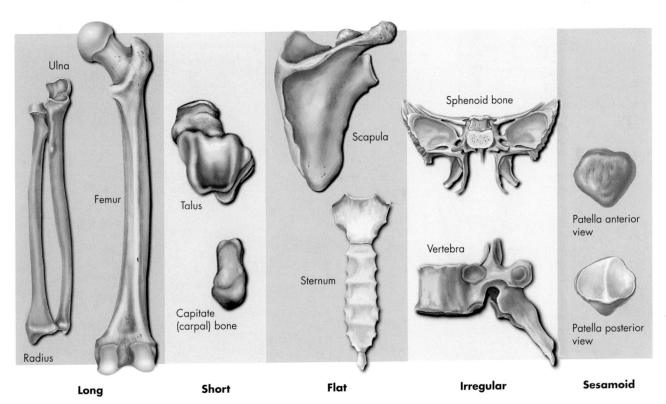

FIG. 1.8 ● Classification of bones by shape.

Modified from Booher JM, Thibodeau GA: *Athletic injury assessment,* ed 4, New York, 2000, McGraw-Hill; Shier D, Butler J, Lewis R: *Hole's human anatomy & physiology,* ed 9, New York, 2002, McGraw-Hill; and Seeley RR, Stephens TD, Tate P: *Anatomy & physiology,* ed 7, New York, 2006, McGraw-Hill.

Flat bones

Usually having a curved surface and varying from thick (where tendons attach) to very thin. Flat bones generally provide protection and include the ilium, ribs, sternum, clavicle, and scapula.

Irregular bones

Irregular shaped bones serve a variety of purposes and include the bones throughout the entire spine and the ischium, pubis, and maxilla.

Sesamoid bones

Small bones embedded within the tendon of a musculotendinous unit that provide protection as well as improve the mechanical advantage of musculotendinous units. In addition to the patella, there are small sesamoid bones within the flexor tendons of the great toe and the thumb.

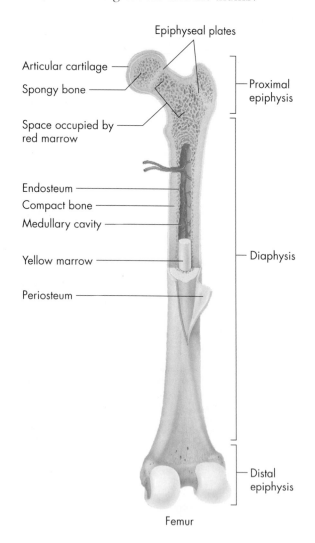

FIG. 1.9 ● Major parts of a long bone.

From Shier D, Butler J, Lewis R: *Hole's human anatomy & physiology,* ed 9, New York, 2002, McGraw-Hill.

Typical bony features

Long bones possess features that are typical of bones in general as illustrated in Fig. 1.9. Long bones have a shaft or **diaphysis**, which is the long cylindrical portion of the bone. The diaphysis wall, formed from hard, dense compact bone, is the **cortex**. The outer surface of the diaphysis is covered by a dense, fibrous membrane known as the **periosteum**. A similar fibrous membrane known as the **endosteum** covers the inside of the cortex. Between the walls of the diaphysis lies the **medullary** or marrow cavity, which contains yellow or fatty marrow. At each end of a long bone is the **epiphysis**, which is usually enlarged and specifically shaped to join with the epiphysis of an adjacent bone at a joint. The epiphysis is formed from spongy or **cancellous** or **trabecular** bone. During bony growth the diaphysis and the epiphysis are separated by a thin plate of cartilage known as the **epiphyseal plate**, commonly referred to as a growth plate (Fig. 1.10). As skeletal maturity is reached, on a timetable that varies from bone to bone as detailed in Table 1.3, the plates are replaced by bone and are closed.

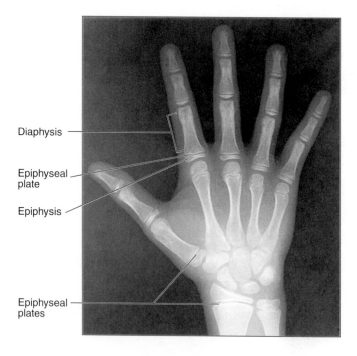

FIG. 1.10 ● The presence of epiphyseal plates, as seen in a radiograph of a child's hand, indicates that the bones are still growing in length. Classification of bones by shape.

Modified from Van De Graaff KM: *Human anatomy,* ed 6, New York, 2002, McGraw-Hill.

TABLE 1.3 • Epiphyseal closure timetables

Approximate age	Bones
7–8	Inferior rami of pubis and ischium (almost complete)
15–17	Scapula, lateral epicondyle of humerus, olecranon process of ulna
18–19	Medial epicondyle of humerus, head and shaft of radius
About 20	Humeral head, distal ends of radius and ulna, distal ends of femur and fibula, proximal end tibia
20–25	Acetabulum in pelvis
25	Vertebrae and sacrum, clavicle, proximal end of fibula, sternum and ribs

Adapted from Goss CM: *Gray's anatomy of the human body,* ed 29, Philadelphia, 1973, Lea & Febiger.

To facilitate smooth, easy movement at joints, the epiphysis is covered by articular or **hyaline** cartilage, which provides a cushioning effect and reduces friction.

Bone development and growth

Most of the skeletal bones of a concern to us in structural kinesiology are **endochondral bones**, which develop from hyaline cartilage. As we develop from an embryo, these hyaline cartilage masses grow rapidly into structures shaped similar to the bones that they will eventually become. This growth continues, and the cartilage gradually undergoes significant change to develop into long bone as detailed in Fig. 1.11.

Bones continue to grow longitudinally as long as the epiphyseal plates are open. These plates begin closing around adolescence and disappear. Most close by age 18, but some may be present until 25. Growth in diameter continues throughout life. This is done by an internal layer of periosteum building new concentric layers on old layers. Simultaneously, bone around the sides of the medullary cavity is resorbed so that the diameter is continually increased. New bone is formed by specialized cells known as **osteoblasts**, whereas the cells that resorb old bone are **osteoclasts**. This bone remodeling, as depicted in Fig. 1.12, is necessary for continued bone growth, changes in bone shape, adjustment of bone to stress, and bone repair.

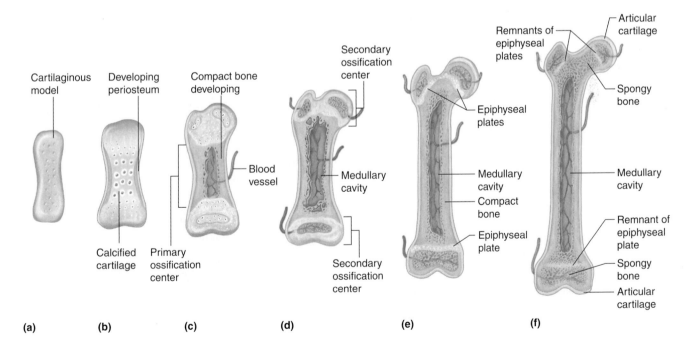

(a)   (b)   (c)   (d)   (e)   (f)

FIG. 1.11 ● Major stages **a–f** in the development of an endochondral bone (relative bone sizes are not to scale).

From Shier D, Butler J, Lewis R: *Hole's essentials of human anatomy and physiology,* ed 9, New York, 2006, McGraw-Hill.

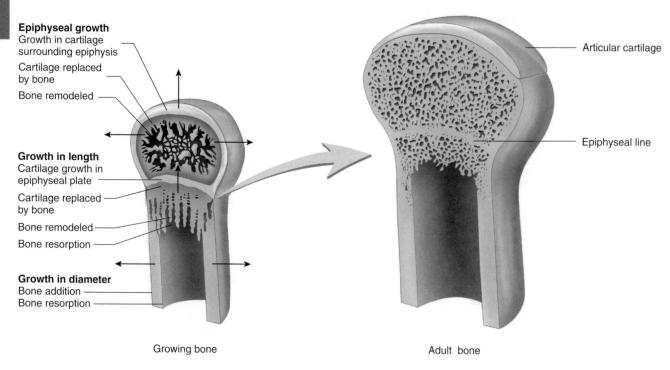

Epiphyseal growth
Growth in cartilage
surrounding epiphysis

Cartilage replaced
by bone

Bone remodeled

Growth in length
Cartilage growth in
epiphyseal plate

Cartilage replaced
by bone

Bone remodeled

Bone resorption

Growth in diameter
Bone addition
Bone resorption

Articular cartilage

Epiphyseal line

Growing bone

Adult bone

FIG. 1.12 ● Remodeling of a long bone.

From Seeley RR, Stephens TD, Tate P: *Anatomy & physiology,* ed 7, New York, 2006, McGraw-Hill.

Bone properties

Calcium carbonate, calcium phosphate, collagen, and water are the basis of bone composition. About 60% to 70% of bone weight is made up of calcium carbonate and calcium phosphate, with water making up approximately 25% to 30% of bone weight. Collagen provides some flexibility and strength in resisting tension. Aging causes progressive loss of collagen and increases bone brittleness, resulting in increased likelihood of fractures.

Most outer bone is cortical; cancellous is underneath. Cortical bone is harder and more compact, with only about 5% to 30% of its volume being porous, with nonmineralized tissue. In contrast, cancellous bone is spongy, with around 30% to 90% of its volume being porous. Cortical bone is stiffer; it can withstand greater stress, but less strain, than cancellous bone. Due to its sponginess, cancellous bone can undergo greater strain before fracturing.

Bone size and shape are influenced by the direction and magnitude of forces that are habitually applied to them. Bones reshape themselves based on the stresses placed upon them, and their mass increases over time with increased stress.

Bone markings

Bones have specific markings that exist to enhance their functional relationship with joints, muscles, tendons, nerves, and blood vessels. Many of these markings serve as important bony landmarks in determining muscle location and attachments and joint function. Essentially, all bone markings may be divided into

1. processes (including elevations and projections), which either form joints or serve as a point of attachment for muscles, tendons, or ligaments, and
2. cavities (depressions), which include openings and grooves that contain tendons, vessels, nerves, and spaces for other structures.

Detailed descriptions and examples of many bony markings are provided in Table 1.4.

TABLE 1.4 • **Bone markings**

	Marking	Description	Examples	Page
Processes that form joints	Condyle	Large, rounded projection that usually articulates with another bone	Medial or lateral condyle of femur	261
	Facet	Small flat or nearly flat surface	Articular facet of vertebra	315
	Head	Prominent, rounded projection of the proximal end of a bone, usually articulating	Head of femur, Head of humerus	217, 218, 108
Processes to which muscles, tendon, or ligaments attach	Angle	Bend or protruding angular projection	Superior and inferior angle of scapula	87
	Border or margin	Edge or boundary line of a bone	Lateral and medial border of scapula	87
	Crest	Prominent, narrow, ridgelike projection	Iliac crest of pelvis	217, 218
	Epicondyle	Projection located above a condyle	Medial or lateral epicondyle of humerus	136
	Line	Ridge of bone less prominent than a crest	Linea aspera of femur	218
	Process	Any prominent projection	Acromion process of scapula, Olecranon process of humerus	87, 108, 136
	Ramus	Part of an irregularly shaped bone that is thicker than a process and forms an angle with the main body	Superior and inferior ramus of pubis	217
	Spine (spinous process)	Sharp, slender projection	Spinous process of vertebra, Spine of scapula	314, 315, 87
	Suture	Line of union between bones	Sagittal suture between parietal bones of skull	17
	Trochanter	Very large projection	Greater or lesser trochanter of femur	217, 218
	Tubercle	Small rounded projection	Greater and lesser tubercles of humerus	108
	Tuberosity	Large rounded or roughened projection	Radial tuberosity, tibial tuberosity	136, 261

TABLE 1.4 (continued) • Bone markings

	Marking	Description	Examples	Page
Cavities (depressions)	Facet	Flatted or shallow articulating surface	Intervertebral facets in cervical, thoracic, and lumbar spine	315
	Foramen	Rounded hole or opening in bone	Obturator foramen in pelvis	217, 218
	Fossa	Hollow, depression, or flattened surface	Supraspinatus fossa, Iliac fossa	87, 217
	Fovea	Very small pit or depression	Fovea capitis of femur	
	Meatus	Tubelike passage within a bone	External auditory meatus of temporal bone	325
	Notch	Depression in the margin of a bone	Trochlear and radial notch of the ulna	136
	Sinus	Cavity or hollow space within a bone	Frontal sinus	
	Sulcus (groove)	Furrow or groovelike depression on a bone	Intertubercular (bicipital) groove of humerus	108

Types of joints

The articulation of two or more bones allows various types of movement. The extent and type of movement determine the name applied to the joint. Bone structure limits the kind and amount of movement in each joint. Some joints or **arthrosis** have no movement, others are only very slightly moveable, and others are freely movable with a variety of movement ranges. The type and range of movements are similar in all humans; but the freedom, range, and vigor of movements are limited by ligaments and muscles.

Articulations may be classified according to the structure or function. Classification by structure places joints into one of three categories: fibrous, cartilaginous, or synovial. Functional classification also results in three categories: synarthrosis (synarthrodial), amphiarthrosis (amphiarthrodial), and diarthrosis (diarthrodial). There are subcategories in each classification. Due to the strong relationship between structure and function, there is significant overlap between the classification systems. That is, there is more similarity than difference between the two members in each of the following pairs: fibrous and synarthrodial joints, cartilaginous and amphiarthrodial joints, and synovial and diarthrodial joints. However, not all joints fit neatly into both systems. Table 1.5 provides a detailed listing of all joint types according to both classification systems. Since this text is concerned primarily with movement, the more functional

TABLE 1.5 • **Joint classification by structure and function**

		Structural classification		
		Fibrous	Cartilagenous	Synovial
Functional classification	Synarthrodial	Gomphosis Suture	——	——
	Amphiarthrodial	Syndesmosis	Symphysis Synchondrosis	——
	Diarthrodial	——	——	Arthrodial Condyloidal Enarthrodial Ginglymus Sellar Trochoidal

system (synarthrodial, amphiarthrodial, and diarthrodial joints) will be used throughout, following a brief explanation of structural classification.

Fibrous joints are joined together by connective tissue fibers and are generally immovable. Subcategories are sutures and gomphosis, which are immovable, and syndesmosis, which allow a slight amount of movement. Cartilaginous joints are joined together by hyaline cartilage or fibrocartilage, which allows very slight movement. Subcategories include synchondrosis and symphysis. Synovial joints are freely movable and generally are diarthrodial. Their structure and subcategories are discussed in detail under diarthrodial joints.

The articulations are grouped into three classes based primarily on the amount of movement possible, with consideration given to their structure.

Synarthrodial (immovable) joints FIG. 1.13

Structurally, these articulations are divided into two types:

Suture

Found in the sutures of the cranial bones. The sutures of the skull are truly immovable beyond infancy.

Gomphosis

Found in the sockets of the teeth. The socket of a tooth is often referred to as a gomphosis (type of a joint in which a conical peg fits into a socket). Normally, there should be an absolutely minimal amount of movement of the teeth in the mandible or maxilla.

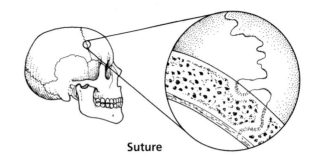

Suture

Gomphosis

FIG. 1.13 ● Synarthrodial joints.

Modified from Booher JM, Thibodeau GA: *Athletic injury assessment,* ed 4, New York, 2000, McGraw-Hill.

Amphiarthrodial (slightly movable) joints FIG. 1.14

Structurally, these articulations are divided into three types:

Syndesmosis

Type of joint held together by strong ligamentous structures that allow minimal movement between

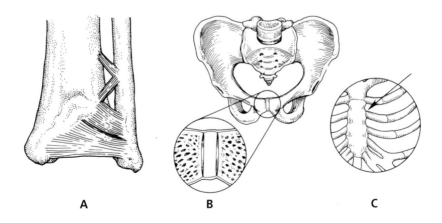

A   B   C

FIG. 1.14 ● Amphiarthrodial joints. **A,** Syndesmosis joint; **B,** Symphysis joint; **C,** Synchondrosis joint.

Modified from Booher JM, Thibodeau GA: *Athletic injury assessment,* ed 4, New York, 2000, McGraw-Hill.

the bones. Examples are the coracoclavicular joint and the inferior tibiofibular joint.

Symphysis

Type of joint separated by a fibrocartilage pad that allows very slight movement between the bones. Examples are the symphysis pubis and the intervertebral disks.

Synchondrosis

Type of joint separated by hyaline cartilage that allows very slight movement between the bones. Examples are the costochondral joints of the ribs with the sternum.

Diarthrodial (freely movable) joints   FIG. 1.15

Diarthrodial joints, also known as synovial joints, are freely movable. A sleevelike covering of ligamentous tissue known as the **joint capsule** surrounds the bony ends forming the joints. This ligamentous capsule is lined with a thin vascular synovial capsule that secretes synovial fluid to lu-

bricate the area inside the joint capsule, known as the **joint cavity**. In certain areas the capsule is thickened to form tough, nonelastic ligaments that provide additional support against abnormal movement or joint opening. These ligaments vary in location, size, and strength depending upon the particular joint.

In many cases, additional ligaments, not continuous with the joint capsule, provide further support. In some cases, these additional ligaments may be contained entirely within the joint capsule, or intraticularly such as the anterior cruciate ligament in the knee, or extraticularly such as the fibula collateral ligament of the knee, which is outside the joint capsule. The articular surfaces on the ends of the bones inside the joint cavity are covered with layers of articular or **hyaline cartilage**. This resilient cartilage absorbs shock to protect the bone it covers. When the joint surfaces are unloaded or distracted, this articular cartilage slowly absorbs a slight amount of the joint synovial fluid, only to slowly secrete it during subsequent weight bearing

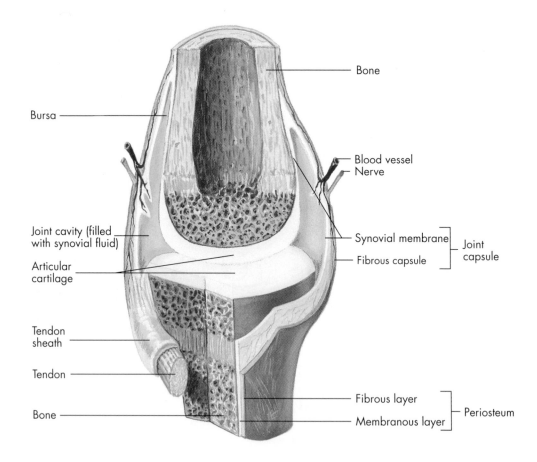

FIG. 1.15 ● Structure of a diarthrodial synovial joint.

From Seeley RR, Stephens TD, Tate P: *Anatomy & physiology,* ed 6, New York, 2000, McGraw-Hill.

and compression. Additionally, some diarthrodial joints have a fibrocartilage disk between their articular surfaces. Structurally, this type of articulation can be divided into six groups, as shown in Fig. 1.16.

Diarthrodial joints have motion possible in one or more planes. Those joints having motion possible in one plane are said to have one degree of freedom of motion, whereas joints having motion in two or three planes of motion are described as having two and three degrees of freedom of motion, respectively. Refer to Table 1.6 for a comparison of diarthrodial joint features by subcategory.

Arthrodial (gliding, plane) joint

This joint type is characterized by two flat, or plane, bony surfaces that butt against each other. This type of joint permits limited gliding movement. Examples are the carpal bones of the wrist and the tarsometatarsal joints of the foot.

Condyloidal (ellipsoid, ovoid, biaxial ball-and-socket) joint

This is a type of joint in which the bones permit movement in two planes without rotation. Examples are the wrist between the radius and the proximal row of the carpal bones or the second, third, fourth, and fifth metacarpophalangeal joints.

Enarthrodial (spheroidal, multiaxial ball-and-socket) joint

This is a type of joint that permits movement in all planes. Examples are the shoulder (glenohumeral) and hip joints.

Ginglymus (hinge) joint

This is a type of joint that permits a wide range of movement in only one plane. Examples are the elbow, ankle, and knee joints.

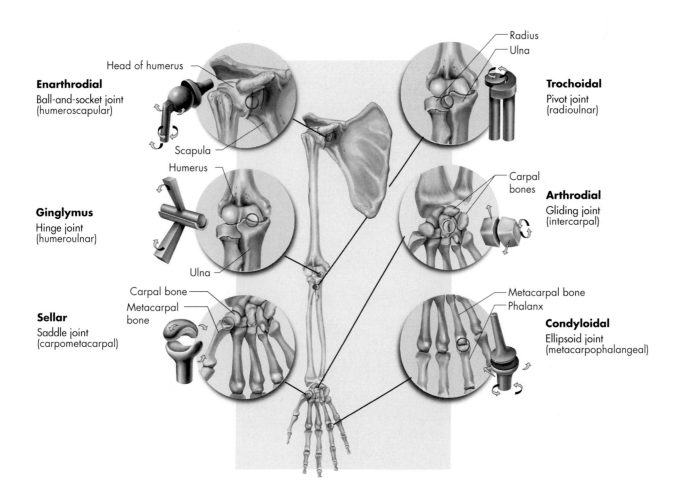

Enarthrodial
Ball-and-socket joint
(humeroscapular)

Head of humerus

Scapula

Humerus

Ginglymus
Hinge joint
(humeroulnar)

Ulna

Sellar
Saddle joint
(carpometacarpal)

Carpal bone
Metacarpal bone

Radius
Ulna

Trochoidal
Pivot joint
(radioulnar)

Carpal bones

Arthrodial
Gliding joint
(intercarpal)

Metacarpal bone
Phalanx

Condyloidal
Ellipsoid joint
(metacarpophalangeal)

FIG. 1.16 ● Types of diarthrodial or synovial joints.

From Booher JM, Thibodeau GA: *Athletic injury assessment,* ed 4, New York, 2000, McGraw-Hill.

Sellar (saddle) joint

This type of reciprocal reception is found only in the thumb at the carpometacarpal joint and permits ball-and-socket movement, with the exception of slight rotation.

Trochoidal (pivot, screw) joint

This is a type of joint with a rotational movement around a long axis. An example is the rotation of the radius at the radioulnar joint.

TABLE 1.6 • **Diarthrodial joint classification**

Classification name	Number of axes	Degrees of freedom	Typical movements	Joint examples	Plane for examples	Axis for examples
Ginglymus (hinge)	Uniaxial	One	Flexion, extension	Elbow joint (humero-ulnar)	Sagittal	Frontal
Trochoid (pivot, screw)			Internal rotation, external rotation	Proximal and distal radioulnar joint	Transverse	Vertical
Condyloid (ellipsoid, ball-and-socket, ovoid)	Biaxial	Two	Flexion, extension, Abduction, adduction	2nd–5th metacarpophalangeal joints	Sagittal Frontal	Frontal Sagittal
Arthrodial (gliding, plane)	Multiaxial	Three	Flexion, extension, Abduction, adduction, Internal rotation, external rotation	Transverse tarsal joint	Variable Frontal	Variable Sagittal
Enarthrodial (ball-and-socket, spheroidal)				Glenohumeral joint	Sagittal Frontal Transverse	Frontal Sagittal Vertical
Sellar (saddle)				1st carpometacarpal joint	Sagittal Frontal Transverse	Frontal Sagittal Vertical

Movements in joints

In many joints, several different movements are possible. Some joints permit only flexion and extension; others permit a wide range of movements, depending largely on the joint structure. We refer to the area through which a joint may normally be freely and painlessly moved as the **range of motion**. The specific amount of movement possible in a joint or range of motion may be measured by using an instrument known as a **goniometer** to compare the change in joint angles.

The goniometer axis, or hinge point, is placed even with the axis of rotation at the joint line. As the joint is moved, both arms of the goniometer are held in place either along or parallel to the long axis of the bones on either side of the joint. The joint angle can then be read from the goniometer, as shown in Fig. 1.17. As an example, we could measure the angle between the femur and tibia in full knee extension (which would usually be zero), and then ask the person to flex the knee as far as possible. If we measured the angle again upon reaching full knee flexion, we would find a goniometer reading of around 140 degrees.

Depending upon the size of the joint and its movement potential, different goniometers may be more or less appropriate. Fig. 1.18 depicts a variety of goniometers that may be utilized to determine the range of motion for a particular joint.

Please note that the normal range of motion for a particular joint varies to some degree from person to person. Appendixes 1 and 2 provide the average normal ranges of motion for all joints.

When using movement terminology, it is important to understand that the terms are used to describe the actual change in position of the bones relative to each other. That is, the angles between the bones change, whereas the movement occurs between the articular surfaces of the joint. We may say, in describing knee movement, "flex the knee"; this movement results in the leg moving closer to the thigh. Some describe this as leg flexion occurring at the knee joint and may say "flex the leg," meaning flex the knee. Additionally, movement terms are utilized to describe movement occurring throughout the full range of motion or through a very small range. Using the knee flexion example again, we may flex the knee through the full range by beginning in full knee extension (zero degrees of knee flexion) and flex it fully, so that the heel comes in contact with the buttocks; this would be approximately 140 degrees of flexion. We may also begin with the knee in 90 degrees of flexion and then flex it 30 degrees more; this movement results in a knee flexion angle of 120 degrees, even though the knee only flexed 30 degrees. In both examples, the knee is in different degrees of flexion. We may also begin with the knee in 90 degrees of flexion and extend it 40 degrees, which would result in a flexion angle of 50 degrees. Even though we extended the knee, it is still flexed, only less so than before.

Some movement terms may be used to describe motion at several joints throughout the body, whereas other terms are relatively specific to a joint or group of joints (Fig. 1.19). Rather than list the terms alphabetically, we have chosen to group them according to the body area and pair them with opposite terms where applicable. Additionally, prefixes such as **hyper-** or **hypo-** may be combined with these terms to emphasize motion beyond or below normal, respectively. Of these combined terms, **hyperextension** is the most commonly used.

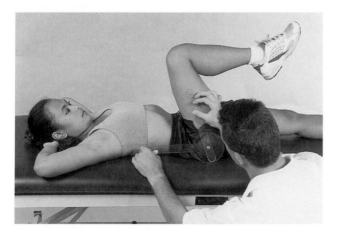

FIG. 1.17 ● Goniometric measurement of hip joint flexion.

From Prentice WE: *Arnheim's principles of athletic training,* ed 11, New York, 2003, McGraw-Hill.

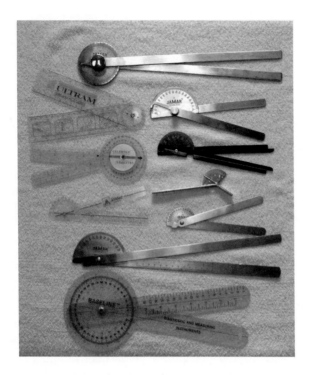

FIG. 1.18 ● Various goniometers used for measuring joint range of motion.

A   B   C

FIG. 1.19 ● Joint movements. **A**, illustrates examples of sagittal plane movements; **extension** of left toes, ankle (plantar flexion), knee, hip, shoulder, elbow, wrist, fingers, and cervical spine; **flexion** of right toes, ankle (dorsiflexion), knee, hip, shoulder, elbow, wrist, and fingers. **B**, illustrates examples of lateral plane movements; **abduction** of left transverse tarsal/subtalar joints (eversion), shoulder, wrist, fingers, and shoulder girdle upward rotation, cervical spine (lateral flexion to left), and right hip; **adduction** of right transverse tarsal/subtarsal joints (inversion), shoulder, wrist, fingers, and shoulder girdle downward rotation. **C**, illustrates examples of transverse plane movements; **internal rotation** of right hip, left shoulder, radioulnar joints (pronation); **external rotation** of left knee, hip, right shoulder, radioulnar joints (supination), and cervical spine (right rotation).

Terms describing general movements

Abduction

Lateral movement away from the midline of the trunk in the frontal plane. An example is raising the arms or legs to the side horizontally.

Adduction

Movement medially toward the midline of the trunk in the frontal plane. An example is lowering the arm to the side or the thigh back to the anatomical position.

Flexion

Bending movement that results in a decrease of the angle in a joint by bringing bones together, usually in the sagittal plane. An example is the elbow joint when the hand is drawn to the shoulder.

Extension

Straightening movement that results in an increase of the angle in a joint by moving bones apart, usually in the sagittal plane. Using the elbow, an example is when the hand moves away from the shoulder.

Circumduction

Circular movement of a limb that delineates an arc or describes a cone. It is a combination of flexion, extension, abduction, and adduction. Sometimes referred to as circumflexion. An example occurs when the shoulder joint and the hip joint move in a circular fashion around a fixed point, either clockwise or counterclockwise.

Diagonal abduction

Movement by a limb through a diagonal plane away from the midline of the body such as in the hip or glenohumeral joint.

Diagonal adduction

Movement by a limb through a diagonal plane toward and across the midline of the body such as in the hip or glenohumeral joint.

External rotation

Rotary movement around the longitudinal axis of a bone away from the midline of the body. Occurs in the transverse plane and is also known as rotation laterally, outward rotation, and lateral rotation.

Internal rotation

Rotary movement around the longitudinal axis of a bone toward the midline of the body. Occurs in the transverse plane and is also known as rotation medially, inward rotation, and medial rotation.

Terms describing ankle and foot movements

Eversion

Turning the sole of the foot outward or laterally in the frontal plane; abduction. An example is standing with the weight on the inner edge of the foot.

Inversion

Turning the sole of the foot inward or medially in the frontal plane; adduction. An example is standing with the weight on the outer edge of the foot.

Dorsal flexion (dorsiflexion)

Flexion movement of the ankle that results in the top of the foot moving toward the anterior tibia bone in the sagittal plane.

Plantar flexion

Extension movement of the ankle that results in the foot and/or toes moving away from the body in the sagittal plane.

Pronation

A combination of ankle dorsiflexion, subtalar eversion, and forefoot abduction (toe-out).

Supination

A combination of ankle plantar flexion, subtalar inversion, and forefoot adduction (toe-in).

Terms describing radioulnar joint movements

Pronation

Internally rotating the radius in the transverse plane so that it lies diagonally across the ulna, resulting in the palm-down position of the forearm.

Supination

Externally rotating the radius in the transverse plane so that it lies parallel to the ulna, resulting in the palm-up position of the forearm.

Terms describing shoulder girdle movements

Depression

Inferior movement of the shoulder girdle in the frontal plane. An example is returning to the normal position from a shoulder shrug.

Elevation

Superior movement of the shoulder girdle in the frontal plane. An example is shrugging the shoulders.

Protraction (abduction)

Forward movement of the shoulder girdle in the horizontal plane away from the spine. Abduction of the scapula.

Retraction (adduction)

Backward movement of the shoulder girdle in the horizontal plane toward the spine. Adduction of the scapula.

Rotation downward

Rotary movement of the scapula in the frontal plane with the inferior angle of the scapula moving medially and downward. Occurs primarily in the return from upward rotation. The inferior angle may actually move upward slightly as the scapula continues in extreme downward rotation.

Rotation upward

Rotary movement of the scapula in the frontal plane with the inferior angle of the scapula moving laterally and upward.

Terms describing shoulder joint movements

Horizontal abduction

Movement of the humerus in the horizontal plane away from the midline of the body. Also known as horizontal extension or transverse abduction.

Horizontal adduction

Movement of the humerus in the horizontal plane toward the midline of the body. Also known as horizontal flexion or transverse adduction.

Terms describing spine movements

Lateral flexion (side bending)

Movement of the head and/or trunk in the frontal plane laterally away from the midline. Abduction of the spine.

Reduction

Return of the spinal column in the frontal plane to the anatomic position from lateral flexion. Adduction of the spine.

Terms describing wrist and hand movements

Dorsal flexion (dorsiflexion)

Extension movement of the wrist in the sagittal plane with the dorsal or posterior side of the hand moving toward the posterior side of the lateral forearm.

Palmar flexion

Flexion movement of the wrist in the sagittal plane with the volar or anterior side of the hand moving toward the anterior side of the forearm.

Radial flexion (radial deviation)

Abduction movement at the wrist in the frontal plane of the thumb side of the hand toward the lateral forearm.

Ulnar flexion (ulnar deviation)

Adduction movement at the wrist in the frontal plane of the little finger side of the hand toward the medial forearm.

Opposition of the thumb

Diagonal movement of the thumb across the palmar surface of the hand to make contact with the fingers.

Reposition of the thumb

Diagonal movement of the thumb as it returns to the anatomical position from opposition with the hand and/or fingers.

These movements are considered in detail in the chapters that follow as they apply to the individual joints.

Combinations of movements can occur. Flexion or extension can occur with abduction, adduction, or rotation.

Throughout this text a series of movement icons will be utilized to represent different joint movements. These icons will be displayed in the page margins to indicate the joint actions of the muscles displayed on that page. As further explained in Chapter 2, the actions displayed represent the movements that occur when the muscle contracts concentrically. Table 1.7 provides a complete listing of the icons. Refer back to them as needed when reading Chapters 4, 5, 6, 7, 9, 10, 11, and 12.

TABLE 1.7 • **Movement icons representing joints actions**

Shoulder girdle					
Scapula elevation	Scapula depression	Scapula abduction	Scapula adduction	Scapula upward rotation	Scapula downward rotation

Glenohumeral						
Shoulder flexion	Shoulder extension	Shoulder abduction	Shoulder adduction	Shoulder external rotation	Shoulder internal rotation	Shoulder horizontal abduction	Shoulder horizontal adduction

Elbow		Radioulnar joints	
Elbow flexion	Elbow extension	Radioulnar supination	Radioulnar pronation

Wrist			
Wrist extension	Wrist flexion	Wrist abduction	Wrist adduction

Thumb carpometacarpal joint			Thumb metacarpophalangeal joint		Thumb interphalangeal joint	
Thumb CMC flexion	Thumb CMC extension	Thumb CMC abduction	Thumb MCP flexion	Thumb MCP extension	Thumb IP flexion	Thumb IP extension

TABLE 1.7 (continued) • Movement icons representing joints actions

2nd, 3rd, 4th, and 5th MCP, PIP, and DIP joints		2nd, 3rd, 4th, and 5th MCP and PIP joints	2nd, 3rd, 4th, and 5th metacarpophalangeal joints		2nd, 3rd, 4th, and 5th PIP joints	2nd, 3rd, 4th, and 5th DIP joints
2nd–5th MCP, PIP, and DIP flexion	2nd–5th MCP, PIP, and DIP extension	2nd–5th MCP and PIP flexion	2nd–5th MCP flexion	2nd–5th MCP extension	2nd–5th PIP flexion	2nd–5th DIP flexion

Hip					
Hip flexion	Hip extension	Hip abduction	Hip adduction	Hip external rotation	Hip internal rotation

Knee			
Knee flexion	Knee extension	Knee external rotation	Knee internal rotation

Ankle		Transverse tarsal and subtalar joints	
Ankle plantar flexion	Ankle dorsal flexion	Transverse tarsal and subtalar inversion	Transverse tarsal and subtalar eversion

Great toe metatarsophalangeal and interphalangeal joints		2nd–5th metatarsophalangeal, proximal interphalangeal, and distal interphalangeal joints	
Great toe MTP and IP flexion	Great toe MTP and IP extension	2nd–5th MTP, PIP and DIP flexion	2nd–5th MTP, PIP, and DIP extension

Cervical spine			
Cervical flexion	Cervical extension	Cervical lateral flexionn	Cervical rotation unilaterally

TABLE 1.7 (continued) • Movement icons representing joints actions

Lumbar spine			
Lumbar flexion	Lumbar extension	Lumbar lateral flexion	Lumbar rotation unilaterally

Physiological movements versus accessory motions

Movements such as flexion, extension, abduction, adduction, and rotation occur by the bones moving through planes of motion about an axis of rotation at the joint. These movements may be referred to as physiological movements. The motion of the bones relative to the three cardinal planes resulting from these physiological movements is referred to **osteokinematic** motion. In order for these osteokinematic motions to occur, there must be movement between the actual articular surfaces of the joint. This motion between the articular surfaces is known as **arthrokinematics**, and includes three specific types of **accessory motion**. These accessory motions, named specifically to describe the actual change in relationship between the articular surface of one bone relative to another, are **spin**, **roll**, and **glide** (Fig. 1.20).

Roll is sometimes referred to as rock or rocking, whereas glide is sometimes referred to as slide or translation. If accessory motion is prevented from occurring, then physiological motion cannot occur to any substantial degree other than by joint compression or distraction. Because most diarthrodial joints in the body are composed of a concave surface articulating with a convex surface, roll and glide must occur together to some degree. For example, as illustrated in Fig. 1.21, as a person stands from a squatting position, in order for the knee to extend, the femur must roll forward and simultaneously slide backward on the tibia. If not for the slide, the femur would roll off the front of the tibia, and if not for the roll, the femur would slide off the back of the tibia.

Spin may occur in isolation or in combination with roll and glide, depending upon the joint structure. To some degree, spin occurs at the knee as it flexes and extends. In the squatting to standing example, the femur spins medially or internally rotates as the knee reaches full extension. Table 1.8 provides examples of accessory motion.

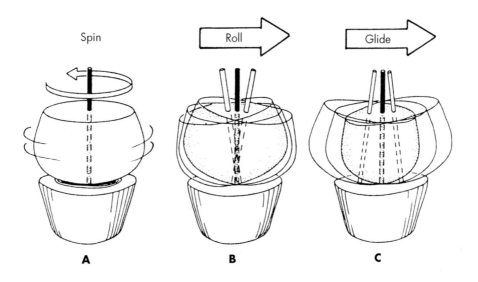

A **B** **C**

FIG. 1.20 ● Joint arthrokinematics. **A,** Spin; **B,** Roll; **C,** Glide.

From Prentice WE: *Rehabilitation techniques for sports medicine and athletic training*, ed 4, New York, 2004, WCB/McGraw-Hill.

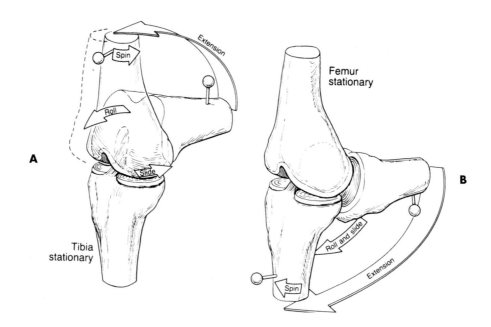

FIG. 1.21 ● Knee joint arthrokinematics. **A,** Standing from squatting; **B,** Flexing from non–weight-bearing position.

Modified Prentice WE: *Rehabilitation techniques for sports medicine and athletic training,* ed 4, New York, 2004, WCB/McGraw-Hill.

TABLE 1.8 • Accessory motion

Accessory motion	Anatomical joint example	Analogy	
Roll (rocking)	Knee extension occurring from femoral condyles rolling forward on tibia as a person stands from squatted position	Tire rolling across a road surface, as in normal driving with good traction	*Combination of roll and glide:* Tire spinning on slick ice (i.e., poor traction) but still resulting in movement across the road surface
Glide (slide or translation)	Knee extension occurring from femoral condyles sliding backward on tibia as a person stands from squatted position	Tire skidding across a slick surface with the brakes locked	
Spin	Radioulnar pronation/supination occurring from spinning of radial head against humeral capitulum	Point of a toy top spinning around in one spot on the floor	

Roll (rock)

A series of points on one articular surface contacts with a series of points on another articular surface.

Glide (slide, translation)

A specific point on one articulating surface comes in contact with a series of points on another surface.

Spin

A single point on one articular surface rotates about a single point on another articular surface. Motion occurs around some stationary longitudinal mechanical axis in either a clockwise or a counterclockwise direction.

Web sites

Anatomy & Physiology Tutorials

www.gwc.maricopa.edu/class/bio201/index.htm

BBC Science & Nature

www.bbc.co.uk/science/humanbody/body/factfiles/skeleton_anatomy.shtml

Describes each bone and allows viewing of each from different angles.

BBC Science & Nature

www.bbc.co.uk/science/humanbody/body/interactives/3djigsaw_02/index.shtml?skeleton

Allows interactive placement of bone and joint structures.

BBC Science & Nature

www.bbc.co.uk/science/humanbody/factfiles/joints/ball_and_socket_joint.shtml

Describes each type of joint and allows viewing of how the joint moves within the body.

University of Michigan Learning Resource Center, Hypermuscle: Muscles in Action

www.med.umich.edu/lrc/Hypermuscle/Hyper.html#flex

Describes each motion and allows viewing of the motion performed.

Articulations

http://basic-anatomy.net

A thorough discussion of the articulations.

Foss Human Body

http://sv.berkeley.edu/showcase/pages/bones.html

An interactive site that allows assembly of the skeleton.

Functions of the Skeletal System

http://training.seer.cancer.gov/module_anatomy/unit3_1_bone_functions.html

Several pages with information on bone tissue, bone development and growth, and the joints.

Wireframe Skeleton

www.2flashgames.com/f/f-220.htm

Move around the skeleton's limbs, arms, legs, body, and make it do funny things.

Skeletal System

www.bio.psu.edu/faculty/strauss/anatomy/skel/skeletal.htm

Pictures of dissected bones and their anatomical landmarks.

Articulations

www.douglas.bc.ca/biology/project/articulations

Details all of the joint types with pictures and review questions.

eSkeletons Project

www.eskeletons.org

An interactive site with a bone viewer showing the morphology, origins, insertions, and articulations of each bone.

ExRx Articulations

www.exrx.net/Lists/Articulations.html

Detailed common exercises demonstrating movements of each joint and listing the muscles involved.

Skeleton Shakedown

www.harcourtschool.com/activity/skel/skel.html

Help put a disarticulated skeleton back together.

Human Anatomy Online

www.innerbody.com/image/skelfov.html

Interactive skeleton labeling.

KLB Science Department Interactives

www.kibschool.org.uk/interactive/science/skeleton.htm

Skeleton labeling exercises.

Introductory Anatomy: Joints

www.leeds.ac.uk/chb/lectures/anatomy4.html

Notes on joint articulations.

The Interactive Skeleton

www.pdh-odp.co.uk/skeleton.htm

Point and click to detailed skeletal illustrations.

Radiographic Anatomy of the Skeleton

www.rad.washington.edu/radanat

X-rays with and without labels of bony landmarks.

Radiographic Anatomy of the Skeleton

www.szote.u-szeged.hu/Radiology/Anatomy/skeleton.htm

X-rays with and without labels of bony landmarks.

Virtual Skeleton

www.uwyo.edu/RealLearning/4210qtvr.html

A 3-dimensional human osteology with Quicktime movies of each bone.

Skeleton: The Joints

www.zoology.ubc.ca/~biomania/tutorial/bonejt/outline.htm

Point and click to detailed joint illustrations.

Forensic Anthropology

www-personal.une.edu.au/~pbrown3/skeleton.pdf

WORSHEET EXERCISES

As an aid to learning, for in-class or out-of-class assignments, or for testing, tear-out worksheets are found at the end of the text (see pp. 372 and 373).

Posterior skeletal worksheet (no. 1)

On the posterior skeletal worksheet, list the names of the bones and all of the prominent features of each bone.

Anterior skeletal worksheet (no. 2)

On the anterior skeletal worksheet, list the names of the bones and all of the prominent features of each bone.

LABORATORY AND REVIEW EXERCISES

1. Choose several different locations at random on your body and specifically describe the locations, using the correct anatomical directional terminology.
2. Complete the blanks in the following paragraphs using each word from the list below only once except for the ones marked with two arterisks **, which are used twice. The number of blanks indicates the number of letters of the word for each blank.

a.	anterior**	s.	medial
b.	anteroinferior	t.	palmar
c.	anterolateral	u.	plantar
d.	anteromedial	v.	posterior**
e.	anteroposterior	w.	posteroinferior
f.	anterosuperior	x.	posterolateral
g.	bilateral	y.	posteromedial
h.	caudal	z.	posterosuperior
i.	cephalic	aa.	prone
j.	contralateral	bb.	proximal
k.	deep	cc.	superficial
l.	distal	dd.	superior
m.	dorsal	ee.	superolateral**
n.	inferior	ff.	superomedial
o.	inferolateral	gg.	supine
p.	inferomedial	hh.	ventral
q.	ipsilateral	ii.	volar
r.	lateral		

When Jacob greeted Stephanie at the beach, he reached out with the _____ surface of his hand to grasp the _____ surface of her hand for a handshake. As the _____ aspect of their bodies faced each other, Jacob noticed that the hair located on the most _____ part of Stephanie's head appeared to be a different color than he remembered. He then asked her to turn around so that he could see it from a _____ view. As she did so, it became obvious to him that she had blonde streaks running from her _____ region in an _____ direction all the way down to her _____ region.

Stephanie then asked Jacob if the sunburn on the _____ portions of his shoulders was due to the exposure that his tank-top shirt provided. He replied yes but that it was only a _____ burn and did not go too ____. He then said, "I wished I would have had my shirt off so that I would have gotten some more sun on the _____ portion of my shoulders up to my neck."

Stephanie said that she recently got sunburned on her back while lying _____ at the beach. She then flipped her hair around the _____ side of her neck toward the _____ portion of her trunk to expose her _____ region. Jacob remarked, "Wow, instead of the bikini you have on today with straps over your shoulders running from your _____ chest to your _____ shoulders, you must have been wearing one with crossing straps as I see you have tan lines running in an _____ direction to your _____ low back from the _____ aspect of your _____ shoulders. You should have spent more time lying _____." She replied "Well, I did lie partially on my back and my right side for a while. See where the _____ portion of my right thigh and the _____ portion of my left thigh are tanned just right, but unfortunately in that position the _____ right thigh and _____ thigh received relatively little exposure." Jacob commented, "Yep, when you lie on one side most of the time, you get all the sun on the _____ side and none on the _____ side. It looks like you must have had a towel covering your feet and ankles since your _____ lower extremities are not nearly as tan as your _____ lower extremities." Stephanie replied, "You are correct. I kept the bottom of my lower legs covered almost all of the time while lying on both sides so that the sensitive skin on my _____ __ and _____ shins would not burn. But, I did get a good _____ tan on my _____ trunk, except for the _____ _____ aspect of my right elbow I was resting on." As Jacob slipped his sandals on to protect the _____ aspect of his feet from the hot sand, he said, "Well, nice to see you. I have to go by the doctor's office and get an _____ _____ chest X-ray to make sure my pneumonia has cleared up."

3. The specific body area joint movement terms arise from the basic motions in the three specific planes: flexion/extension in the sagittal plane, abduction/adduction in the frontal plane, and rotation in the transverse plane. With this in mind, complete the joint movement terminology chart below.

Joint movement terminology chart

For each specific motion in the left column, provide the basic motion that it represents in the right column by using *flexion, extension, abduction, adduction,* or *rotation* (*external* or *internal*).

Specific motion	Basic motion
Eversion	
Inversion	
Dorsal flexion	
Plantar flexion	
Pronation	
Supination	
Lateral flexion	
Radial flexion	
Ulnar flexion	

4. Determine which joints have movements possible in each of the following planes:
   a. Sagittal
   b. Frontal
   c. Transverse
5. Determine the planes in which the following activities occur. Also, use a pencil to visualize the axis for each of the following activities.
   a. Walking up stairs
   b. Turning a knob to open a door
   c. Nodding the head to agree
   d. Shaking the head to disagree
   e. Shuffling the body from side to side
   f. Looking over your shoulder to see behind you
6. What are the five functions of the skeleton?
7. List the bones of the upper extremity.
8. List the bones of the lower extremity.
9. List the bones of the shoulder girdle.
10. List the bones of the pelvic girdle.
11. Describe and explain the differences and similarities between the radius and ulna.
12. Describe and explain the differences and similarities between the humerus and femur.

13. Utilizing Fig. 1.7 and other resources, place an
X in the appropriate column of the bone typing chart to indicated its classification.

Bone typing chart

Bone	Long	Short	Flat	Irregular	Sesamoid
Frontal					
Zygomatic					
Parietal					
Temporal					
Occipital					
Maxilla					
Mandible					
Cervical vertebrae					
Clavicle					
Scapula					
Humerus					
Ulna					
Radius					
Carpal bones					
Metacarpals					
Phalanges					
Ribs					
Sternum					
Lumbar vertebrae					
Ilium					
Ischium					
Pubis					
Femur					
Patella					
Fabella					
Tibia					
Fibula					
Talus					
Calacaneus					
Navicular					
Cuneiforms					
Metatarsals					

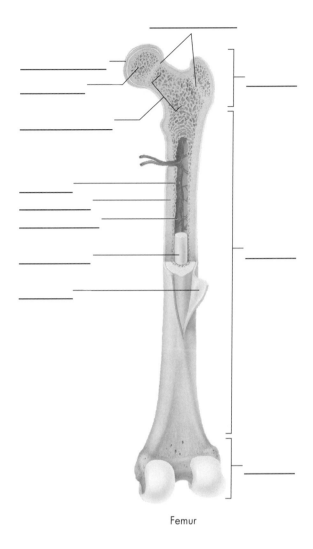

Femur

14. Label the parts of a long bone.
15. Using body landmarks, how would you suggest determining the length of each lower extremity for comparison to decide if someone had a true total leg length discrepancy?

16. Explain why the fibula is more susceptible to fractures than the tibia.
17. Locate the various types of joints on a human skeleton and palpate their movements on a living subject.
18. Individually practice the various joint movements, on yourself or with another subject.
19. Stand facing a closed door. Reach out and grasp the knob. Turn it and open the door widely toward you. Determine all of the joints involved in this activity and list the movements for each joint.
20. Utilize a goniometer to measure the joint ranges of motion for several students in your class for each of the following movements. Compare your results with the average ranges provided in Appendixes 1 and 2.
    a. External and internal rotation of the shoulder with the shoulder in 90 degrees of abduction while supine
    b. Elbow flexion in the supine position
    c. Wrist extension with the forearm in neutral and the elbow in 90 degrees of flexion
    d. Hip external and internal rotation in the sitting position with the hip and knee each in 90 degrees of flexion
    e. Knee flexion in the prone position
    f. Ankle dorsiflexion with the knee in 90 degrees of flexion versus knee in full extension
21. Complete the joint type, movement, and plane of motion chart (p. 34) by
    a. Filling in the type of diarthrodial joint
    b. Listing the movements of the joint under the plane of motion in which they occur
22. Using proper terminology, complete the joint position chart below by listing the name of each joint involved and its position upon completion of the multiple joint movement.

Joint position chart

Multiple joint movement	Joints and respective position of each
Reach straight over the superior aspect of your head to touch the contralateral ear	
Place the toe of one foot against the posterior aspect of the contralateral calf	
Reach behind the back and use your thumb to touch a spinous process	
Pull the knee as far as possible to the ipsilateral shoulder	
Place the plantar aspect of both feet against each other	

Joint type, movement, and plane of motion chart

Joint		Plane of motion			
		Type	Sagittal	Lateral	Transverse
Scapulothoracic joint	Sternoclavicular				
	Acromioclavicular				
Glenohumeral joint					
Elbow					
Radioulnar joint					
Wrist					
1st carpometacarpal joint					
1st metacarpophalangeal joint					
Thumb interphalangeal joint					
2nd, 3rd, 4th, and 5th metacarpophalangeal joints					
2nd, 3rd, 4th, and 5th proximal interphalangeal joints					
2nd, 3rd, 4th, and 5th distal interphalangeal joints					
Cervical spine C1–C2					
Cervical spine C2–C7					
Lumbar spine					
Hip					
Knee (tibiofemoral joint)					
Knee (patellofemoral joint)					
Ankle					
Transverse tarsal and subtalar joints					
Metatarsophalangeal joints					
Great toe interphalangeal					
2nd, 3rd, 4th, and 5th proximal interphalangeal joints					
2nd, 3rd, 4th, and 5th distal interphalangeal joints					

23. List two sport skills that involve movements that are best viewed from the side. List the primary movements that occur in the ankle, knee, hip, spine, glenohumeral joint, elbow, and wrist. Which plane are these movements occurring in primarily? What axis of rotation is involved primarily?

24. List two sport skills that involve movements that are best viewed from the front or rear. List the primary movements that occur in the transverse tarsal/subtalar joint, hip, spine, glenohumeral joint, and wrist. Which plane are these movements occurring in primarily? What axis of rotation is involved primarily?

25. List the similarities between the ankle/foot/toes and the wrist/hand/fingers regarding the bones, joint structures, and movements. What are the differences?

26. Compare and contrast the glenohumeral and acetabulofemoral joints. Which one is more susceptible to dislocations and why?
27. Why is the anatomical position so important in understanding anatomy and joint movements?
28. Discuss the following joints among your classmates and place in order from the one with the least total range of motion to the most. Be prepared to defend your answer.
    a. Ankle   d. Hip
    b. Elbow   e. Knee
    c. Glenohumeral   f. Wrist
29. Is there more inversion or eversion possible in the transverse tarsal and subtalar joint? Explain why this occurs based on anatomy.
30. Is there more abduction or adduction possible in the wrist joint? Explain why this occurs based on anatomy.
31. For each of the joint motions listed in the plane of motion and axis of rotation chart, list the plane of motion in which the motion occurs and its axis of rotation.

Plane of motion and axis of rotation chart

Motion	Plane of motion	Axis of rotation
Cervical rotation		
Shoulder girdle elevation		
Glenohumeral horizontal adduction		
Elbow flexion		
Radioulnar pronation		
Wrist radial deviation		
Metacarpophalangeal abduction		
Lumbar lateral flexion		
Hip internal rotation		
Knee extension		
Ankle inversion		
Great toe extension		

References

Anthony C, Thibodeau G: *Textbook of anatomy and physiology,* ed 10, St. Louis, 1979, Mosby.

Booher JM, Thibodeau GA: *Athletic injury assessment,* ed 4, New York, 2000, McGraw-Hill.

Goss CM: *Gray's anatomy of the human body,* ed 29, Philadelphia, 1973, Lea & Febiger.

Hamilton N, Luttgens K: *Kinesiology: scientific basis of human motion,* ed 10, New York, 2002, McGraw-Hill.

Lindsay DT: *Functional human anatomy,* St. Louis, 1996, Mosby.

Logan GA, McKinney WC: *Anatomic kinesiology,* ed 3, Dubuque, IA, 1982, Brown.

National Strength and Conditioning Association; Baechle TR, Earle RW: *Essentials of strength training and conditioning,* ed 2, Champaign, IL, 2000, Human Kinetics.

Neumann, DA: *Kinesiology of the musculoskeletal system: foundations for physical rehabilitation,* St. Louis, 2002, Mosby.

Northrip JW, Logan GA, McKinney WC: *Analysis of sport motion: anatomic and biomechanic perspectives,* ed 3, Dubuque, IA, 1983, Brown.

Prentice WE: *Arnheim's principles of athletic training,* ed 11, New York, 2003, McGraw-Hill.

Prentice WE: *Rehabilitation techniques in sports medicine,* ed 4, New York, 2004, McGraw-Hill.

Seeley RR, Stephens TD, Tate P: *Anatomy & physiology,* ed 7, New York, 2006, McGraw-Hill.

Shier D, Butler J, Lewis R: *Hole's essentials of human anatomy and physiology,* ed 9, New York, 2006, McGraw-Hill.

Stedman TL: *Stedman's medical dictionary,* ed 27, Baltimore, 2000, Lippincott Williams & Wilkins.

Steindler A: *Kinesiology of the human body,* Springfield, IL, 1970, Thomas.

Van De Graaff KM: *Human anatomy,* ed 6, New York, 2002, McGraw-Hill.

Van De Graaff KM, Fox SI, LaFleur KM: *Synopsis of human anatomy & physiology,* Dubuque, IA, 1997, Brown.

Neuromuscular Fundamentals

Objectives

- To review the basic anatomy and function of the muscular and nervous systems

- To review and understand the basic terminology used to describe muscular locations, arrangements, characteristics, and roles, as well as neuromuscular functions

- To learn and understand the different types of muscle contraction and the factors involved in each

- To learn and understand basic neuromuscular concepts in relation to how muscles function in joint movement and work together to achieve motion

- To develop a basic understanding of the neural control mechanisms for movement

---

Online Learning Center Resources

Visit *Manual of Structural Kinesiology's* Online Learning Center at **www.mhhe.com/floyd16e** for additional information and study material for this chapter, including:

- *Self-grading quizzes*
- *Anatomy flashcards*
- *Animations*

Skeletal muscles are responsible for movement of the body and all of its joints. Muscle contraction produces the force that causes joint movement in the human body. In addition to the function of movement, muscles also provide protection, contribute to posture and support, and produce a major portion of total body heat. There are over 600 skeletal muscles, which constitute approximately 40% to 50% of body weight. Of these, there are 215 pairs of skeletal muscles. These pairs of muscles usually work in cooperation with each other to perform opposite actions at the joints they cross. In most cases, muscles work in groups rather than independently to achieve a given joint motion. This is known as **aggregate muscle action**.

Muscle nomenclature

In attempting to learn the skeletal muscles, it is helpful to have an understanding of how they are named. Muscles are usually named because of one or more distinctive characteristics, such as their visual appearance, anatomical location, or function. Examples of skeletal muscle naming are as follows:

Shape—deltoid, rhomboid
Size—gluteus maximus, teres minor
Number of divisions—triceps brachii
Direction of its fibers—external oblique
Location—rectus femoris, palmaris longus
Points of attachment—coracobrachialis, extensor hallucis longus, flexor digitorum longus

Action—erector spinae, supinator, extensor digiti minimi
Action and shape—pronator quadratus
Action and size—adductor magnus
Shape and location—serratus anterior
Location and attachment—brachioradialis
Location and number of divisions—biceps femoris

In discussions regarding the muscles, they are often grouped together for brevity of conversation and clearer understanding. The naming of muscle groups follows a similar pattern. Next are some muscle groups according to different naming rationales.

Shape—hamstrings
Number of divisions—quadriceps, triceps surae

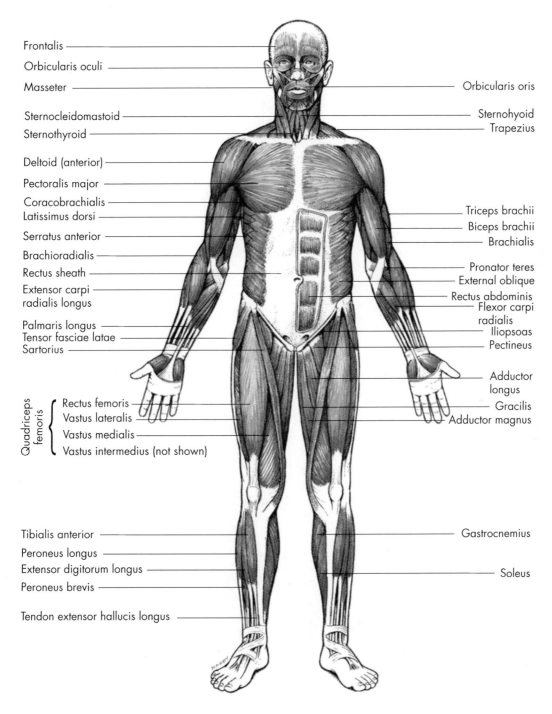

Frontalis
Orbicularis oculi
Masseter
Orbicularis oris
Sternocleidomastoid
Sternothyroid
Sternohyoid
Trapezius
Deltoid (anterior)
Pectoralis major
Coracobrachialis
Latissimus dorsi
Triceps brachii
Biceps brachii
Brachialis
Serratus anterior
Brachioradialis
Rectus sheath
Pronator teres
External oblique
Extensor carpi radialis longus
Rectus abdominis
Flexor carpi radialis
Palmaris longus
Tensor fasciae latae
Sartorius
Iliopsoas
Pectineus
Adductor longus
Quadriceps femoris {
Rectus femoris
Vastus lateralis
Vastus medialis
Vastus intermedius (not shown)
Gracilis
Adductor magnus
Tibialis anterior
Peroneus longus
Extensor digitorum longus
Peroneus brevis
Gastrocnemius
Soleus
Tendon extensor hallucis longus

FIG. 2.1 ● Superficial muscles of the human body, anterior view from anatomical position.

From Thibodeau GA: *Anatomy and physiology,* St. Louis, 1987, Mosby.

Location—peroneals, abdominal, shoulder girdle
Action—hip flexors, rotator cuff

Figs. 2.1 and 2.2 depict the muscular system from a superficial point of view. Not shown in these figures are the deep muscles.

Muscles shown in these figures, and many other muscles, will be studied in more detail as each joint of the body is considered in other chapters of this book.

Shape of muscles and fiber arrangement

Various muscles have different shapes, and their fibers may be arranged differently in relation to each other and to the tendons to which they join. The shape and fiber arrangement play a role in the muscle's ability to exert force and the range through which it can effectively exert force onto the bones to which it is attached. A factor in the

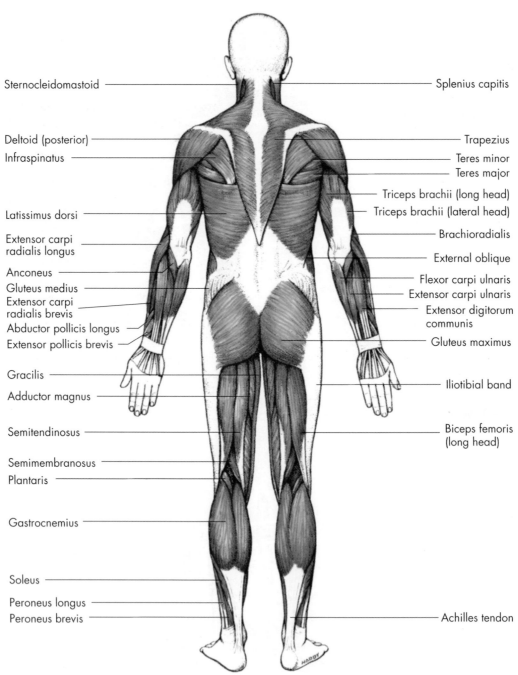

Sternocleidomastoid

Deltoid (posterior)
Infraspinatus

Latissimus dorsi

Extensor carpi
radialis longus

Anconeus
Gluteus medius
Extensor carpi
radialis brevis
Abductor pollicis longus
Extensor pollicis brevis

Gracilis

Adductor magnus

Semitendinosus

Semimembranosus

Plantaris

Gastrocnemius

Soleus

Peroneus longus
Peroneus brevis

Splenius capitis

Trapezius
Teres minor
Teres major
Triceps brachii (long head)
Triceps brachii (lateral head)
Brachioradialis
External oblique
Flexor carpi ulnaris
Extensor carpi ulnaris
Extensor digitorum
communis
Gluteus maximus

Iliotibial band

Biceps femoris
(long head)

Achilles tendon

FIG. 2.2 ● Superficial muscles of the human body, posterior view.

From Thibodeau GA: *Anatomy and physiology,* St. Louis, 1987, Mosby.

ability of a muscle to exert force is its cross-section diameter. Keeping all other factors constant, a muscle with a greater cross-section diameter will be able to exert a greater force. A factor in the ability of a muscle to move a joint through a large range of motion is its ability to shorten. Generally, longer muscles can shorten through a greater range and therefore are more effective in moving joints through large ranges of motion.

Essentially, all skeletal muscles may be grouped into two major types of fiber arrangements: parallel and pennate. Each may be subdivided further according to shape.

Parallel muscles have their fibers arranged parallel to the length of the muscle. Generally, parallel muscles will produce a greater range of movement than similar-sized muscles with a pennate arrangement. Parallel muscles are categorized into the following shapes:

Flat muscles are usually thin and broad, originating from broad, fibrous, sheetlike aponeuroses that allow them to spread their forces over a broad area. Examples include the rectus abdominus and external oblique.

Fusiform muscles are spindle-shaped with a central belly that tapers to tendons on each end; this allows them to focus their power onto small, bony targets. Examples are the brachialis and the brachioradialis.

Strap muscles are more uniform in diameter with essentially all of their fibers arranged in a long parallel manner. This also enables a focusing of power onto small, bony targets. The sartorius is an example.

Radiate muscles are also described sometimes as being triangular, fan-shaped, or convergent. They have the combined arrangement of flat and fusiform muscles, in that they originate on broad aponeuroses and converge onto a tendon. Examples are the pectoralis major or trapezius.

Sphincter or circular muscles are technically endless strap muscles that surround openings and function to close them upon contraction. An example is the orbicularis oris, surrounding the mouth.

Pennate muscles have shorter fibers that are arranged obliquely to their tendons in a structure similar to that of a feather. This arrangement increases the cross-sectional area of the muscle, thereby increasing its force production capability. Pennate muscles are categorized based upon the exact arrangement between the fibers and the tendon.

Unipennate muscle fibers run obliquely from a tendon on one side only. Examples are seen in the biceps femoris, extensor digitorum longus, and tibialis posterior.

Bipennate muscle fibers run obliquely on both sides from a central tendon, as in the rectus femoris and flexor hallucis longus.

Multipennate muscles have several tendons with fibers running diagonally between them, as in the deltoid.

Bipennate and unipennate produce strongest contraction. Review Table 2.1 and Fig. 2.3 regarding muscle shapes and fiber arrangements.

TABLE 2.1 • Muscle shape and fiber arrangement

Fiber arrangement	Advantage	Shape	Characteristics/description	Example
Parallel (fibers arranged parallel to the length of the muscle)	Produces greater range of movement than similar-sized pennate muscles	Flat	Usually thin and broad, originating from broad, fibrous, sheetlike aponeuroses that allow them to spread their forces over a broad area	Rectus abdominus, external oblique
		Fusiform	Spindle-shaped with central belly that tapers to tendons on each end; can focus their power onto small, bony targets	Brachialis, brachioradialis
		Strap	More uniform in diameter with essentially all of their fibers arranged in a long parallel manner; can focus their power onto small, bony targets	Sartorius
		Radiate (triangular, fan-shaped, or convergent)	Combined arrangement of flat and fusiform muscles, originate on broad aponeuroses and converge onto a tendon	Pectoralis major, trapezius
		Sphincter (circular)	Technically endless strap muscles, surround openings and function to close them upon contraction	Orbicularis oris, orbicularis oculi
Pennate (shorter fibers, arranged obliquely to their tendons)	Produces greater force than similar-sized parallel muscles due to increased cross-sectional area	Unipennate	Run obliquely from a tendon on one side only	Biceps femoris, extensor digitorum longus, tibialis posterior
		Bipennate	Run obliquely on both sides from a central tendon	Rectus femoris, flexor hallucis longus
		Multipennate	Several tendons with fibers running diagonally between them	Deltoid

Type and Description	Appearance
Parallel—straplike; long excursion (contract over a great distance); good endurance; not especially strong; e.g., sartorius and rectus abdominis muscles	
Radiate—fan-shaped; force of contraction focused onto a single point of attachment; stronger than parallel type; e.g, deltoid and pectoralis major	
Sphincteral—fibers concentrically arranged around a body opening (orifice); act as a sphincter when contracted; e.g., orbicularis oculi and orbicularis oris	
Pennate—many fibers per unit area; strong muscles; short excursions; highly dexterous; tire quickly; three types: (a) unipennate, (b) bipennate, and (c) multipennate	

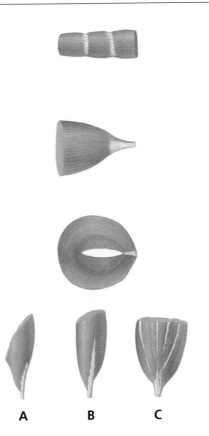

A   B   C

FIG. 2.3 ● Muscle shapes and fiber arrangement.

Modified from Van De Graaff KM: *Human anatomy*, ed 6, Dubuque, IA, 2002, McGraw-Hill.

Muscle tissue properties

Skeletal muscle tissue has four properties related to its ability to produce force and movement about joints. **Irritability** or **excitability** refers to the muscle property of being sensitive or responsive to chemical, electrical, or mechanical stimuli. When an appropriate stimulus is provided, muscle responds by developing tension. **Contractility** refers to the ability of muscle to contract and develop tension or internal force against resistance when stimulated. The ability of muscle tissue to develop tension or contract is unique in that other body tissues do not have this property. **Extensibility** refers to the ability of muscle to be passively stretched beyond its normal resting length. As an example, the triceps brachii displays extensibility when it is stretched beyond its normal resting by the biceps brachii and other elbow flexors contracting to achieve full elbow flexion. **Elasticity** refers to the ability of muscle to return to its original resting length following stretching. In continuing with the elbow example, the triceps brachii displays elasticity by returning to its original resting length when the elbow flexors cease contracting and relax.

Muscle terminology

Locating the muscles, their proximal and distal attachments, and their relationship to the joints they cross is critical to determining the effects that muscles have on the joints. It is also necessary to understand certain terms as body movement is considered.

Intrinsic

Pertaining usually to muscles within or belonging solely to the body part on which they act. The small intrinsic muscles found entirely within the hand are examples.

Extrinsic

Pertaining usually to muscles that arise or originate outside of (proximal to) the body part on
which they act. The forearm muscles that attach
proximally on the distal humerus and insert on
the fingers are examples.

Action

The specific movement of the joint resulting from a
concentric contraction of a muscle that crosses the
joint. An example is the biceps brachii, which has
the action of flexion at the elbow. In most cases a
particular action is caused by a group of muscles
working together. Any of the muscles in the group
can be said to cause the action, even though it is
usually an effort of the entire group. A particular
muscle may cause more than one action either at
the same joint or at a different joint, depending
upon the characteristics of the joints crossed by the
muscle and the exact location of the muscle and its
attachments in relation to the joint(s).

Innervation

The segment of the nervous system defined as being
responsible for providing a stimulus to muscle fibers
within a specific muscle or portion of a muscle. A
particular muscle may be innervated by more than
one nerve, and a particular nerve may innervate
more than one muscle or portion of a muscle.

Amplitude

Range of muscle fiber length between maximal
and minimal lengthening.

Gaster (belly or body)

The central, fleshy portion of the muscle that generally increases in diameter as the muscle contracts. It is the contractile portion of the muscle.

When a particular muscle contracts, it tends to pull
both ends toward the **gaster**, or middle, of the muscle. Consequently, if neither of the bones to which a
muscle is attached were stabilized, then both bones
would move toward each other upon contraction.
The more common case, however, is that one bone
is more stabilized by a variety of factors, and as a
result, the less stabilized bone usually moves toward
the more stabilized bone upon contraction.

Tendon

Fibrous connective tissue, often cordlike in appearance, that connects muscles to bones and other
structures. In some cases, two muscles may share
a common tendon such as in the Achilles tendon
of the gastrocnemius and soleus muscles. In other
cases a muscle may have multiple tendons con-

necting it to one or more bones such as the three
proximal attachments of the triceps brachii.

Origin

From a structural perspective, the proximal attachment of a muscle or the part that attaches
closest to the midline or center of the body is usually considered to be the origin. From a functional
and historical perspective, the least movable part
or attachment of the muscle has generally been
considered to be the origin.

Insertion

Structurally, the distal attachment or the part that
attaches farthest from the midline or center of the
body is considered to be the insertion. Functionally and historically, the most movable part is
generally considered the insertion.

As an example, in the biceps curl exercise, the
biceps brachii muscle in the arm has its origin on
the scapula (least movable bone) and its insertion
on the radius (most movable bone). In some movements this process can be reversed. An example of
this reversal can be seen in the pull-up, where the
radius is relatively stable and the scapula moves
up. Even though in this example the most movable
bone is reversed, the proximal attachment of the biceps brachii is always on the scapula and still considered to be the origin while the insertion is still on
the radius. The biceps brachii would be an extrinsic
muscle of the elbow, whereas the brachialis would
be intrinsic to the elbow. For each muscle studied,
the origin and insertion are indicated.

Types of muscle contraction (action)

When tension is developed in a muscle as a result of a stimulus, it is known as a contraction.
The term *muscle contraction* may be confusing,
because in some types of contractions the muscle
does not shorten in length as the term *contraction*
indicates. As a result, it has become increasingly
common to refer to the various types of muscle
contractions as *muscle actions* instead.

Muscle contractions can be used to *cause, control,* or *prevent* joint movement. To elaborate, muscle contractions can be used to initiate or accelerate
the movement of a body segment, to slow down
or decelerate the movement of a body segment, or
to prevent movement of a body segment by external forces. All muscle contractions or actions can
be classified as being either isometric or isotonic.
An isometric contraction occurs when tension is
developed within the muscle, but the joint angles

remain constant. Isometric contractions may be thought of as **static** contractions, because a significant amount of tension may be developed in the muscle to maintain the joint angle in a relatively static or stable position. Isometric contractions may be used to stabilize a body segment to prevent it from being moved by external forces.

Isotonic contractions involve the muscle developing tension to either cause or control joint movement. They may be thought of as **dynamic** contractions, because the varying degrees of tension in the muscles are causing the joint angles to change. The isotonic type of muscle contraction is classified further as being either concentric or eccentric on the basis of whether shortening or lengthening occurs. Concentric contractions involve the muscle developing tension as it shortens, whereas eccentric contractions involve the muscle lengthening under tension. In Fig. 2.4, *A* and *B* illustrate isotonic and *C* demonstrates isometric contractions.

It is also important to note that movement may occur at any given joint without any muscle contraction whatsoever. This movement is referred to as passive and is solely due to external forces such as those applied by another person, object, or resistance, or the force of gravity in the presence of muscle relaxation.

Concentric contraction

Involves the muscle developing tension as it shortens and occurs when the muscle develops enough force to overcome the applied resistance. Concentric contractions may be thought of as causing movement against gravity or resistance and are described as positive contractions. The force developed by the muscle is greater than that of the resistance. This results in the joint angle being changed in the direction of the applied muscular force and causes the body part to move against gravity or external forces.

Eccentric contraction (muscle action)

Involves the muscle lengthening under tension, and occurs when the muscle gradually lessens in tension to control the descent of the resistance. The weight or resistance may be thought of as overcoming the muscle contraction, but not to the point that the muscle cannot control the descending movement. Eccentric muscle actions control movement with gravity or resistance and are described as negative contractions. The force developed by the muscle is less than that of the resistance; this results in a change in the joint angle in the direction of the resistance or external force and allows the body part to move with gravity or external forces (resistance). Eccentric contractions are used to decelerate the movement of a body segment from a higher speed to a slower speed or stop the movement of a joint already in motion. Since the muscle is lengthening as opposed to shortening, the relatively recent change in terminology from muscle contraction to muscle action is becoming more commonly accepted.

Table 2.2 explains the various types of contraction and resultant joint movements. The terminology utilized in defining and describing these actions is included.

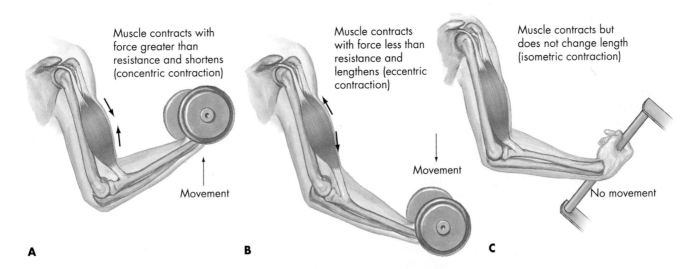

A Muscle contracts with force greater than resistance and shortens (concentric contraction) — Movement

B Muscle contracts with force less than resistance and lengthens (eccentric contraction) — Movement

C Muscle contracts but does not change length (isometric contraction) — No movement

FIG. 2.4 ● Isotonic and isometric contractions. **A,** Concentric contraction; **B,** Eccentric contraction; **C,** Isometric contractions, occurring when a muscle contracts but does not shorten.

Modified from Shier D, Butler J, Lewis R: *Hole's human anatomy & physiology,* ed 9, Dubuque, IA, 2002, McGraw-Hill.

TABLE 2.2 • Muscle contraction and movement matrix

Chapter
2

Definitive and descriptive factors	Type of contraction (muscle action)			Movement without contraction
	Isometric	Isotonic		
		Concentric	Eccentric	
Agonist muscle length	No appreciable change	Shortening ➡⬅	Lengthening ⬅➡	Dictated solely by gravity and/or external forces
Antagonist muscle length	No appreciable change	Lengthening ⬅➡	Shortening ➡⬅	Dictated solely by gravity and/or external forces
Joint angle changes	No appreciable change	In direction of force	In direction of external force (resistance)	Dictated solely by gravity and/or external forces
Direction of body part	Against immovable object or matched external force (resistance)	Against gravity and/or other external force (resistance)	With gravity and/or other external force (resistance)	Consistent with gravity and/or other external forces
Motion	Pressure (force) applied, but no resulting motion	Causes motion	Controls motion	Either no motion or passive motion occurs as a result of gravity and/or other external forces
Description	Static; fixating	Dynamic shortening; positive work	Dynamic lengthening; negative work	Passive; relaxation
Applied muscle force versus resistance	Force = resistance	Force > resistance	Force < resistance	No force, all resistance
Speed relative to gravity or applied resistance including inertial forces	Equal to speed of applied resistance	Faster than the inertia of the resistance	Slower than the speed of gravity or applied inertial forces	Consistent with inertia of applied external forces or the speed of gravity
Acceleration/ deceleration	Zero acceleration	Acceleration ↗	Deceleration ↘	Either zero or acceleration consistent with applied external forces
Descriptive symbol	(=)	(+)	(−)	(0)
Practical application	Prevents external forces from causing movement	Initiates movement or speeds up the rate of movement	Slows down the rate of movement or stops movement, "braking action"	Passive motion by force from gravity and/or other external forces

Various exercises may use any one or all of these contraction types for muscle development. Development of exercise machines has resulted in another type of muscle exercise known as **isokinetics**. Isokinetics is not another type of contraction, as some authorities have mistakenly described; rather, it is a specific technique that may use any or all of the different types of contractions. Isokinetics is a type of dynamic exercise usually using concentric and/or eccentric muscle contractions in which the speed (or velocity) of movement is constant and muscular contraction (ideally, maximum contraction) occurs throughout the movement. Biodex, Cybex, Lido, and other new types of apparatuses are engineered to allow this type of exercise.

Students well educated in kinesiology should be qualified to prescribe exercises and activities for the development of large muscles and muscle groups in the human body. They should be able to read the description of an exercise or observe an exercise and immediately know the most important muscles being used. Descriptive terms of how muscles function in joint movements follow.

Role of muscles

Agonist   FIG. 2.5

Generally described as muscles that, when contracting concentrically, cause joint motion through a specified plane of motion; known as primary or prime movers, or muscles most involved.

Antagonist   FIG. 2.5

Muscles that are usually located on the opposite side of the joint from the agonist and have the opposite concentric action; known as contralateral muscles, they work in cooperation with agonist muscles by relaxing and allowing movement, but when contracting concentrically, they perform the opposite joint motion of the agonist.

Stabilizers

Muscles that surround the joint or body part and contract to fixate or stabilize the area to enable another limb or body segment to exert force and move; known as fixators, they are essential in es-

tablishing a relatively firm base for the more distal joints to work from when carrying out movements.

Synergist

Muscles that assist in the action of agonists but are not necessarily prime movers for the action; known as guiding muscles, they assist in refined movement and rule out undesired motions.

Neutralizers

Muscles that counteract or neutralize the action of other muscles to prevent undesirable movements such as inappropriate muscle substitutions; referred to as neutralizing, they contract to resist specific actions of other muscles.

Tying roles of muscles all together

When a muscle with multiple agonist actions contracts, it attempts to perform all of its actions. Muscles cannot determine which of their actions are appropriate for the task at hand. The resulting actions actually performed depend upon several factors, such as the motor units activated, joint position, muscle length, and the relative contrac-

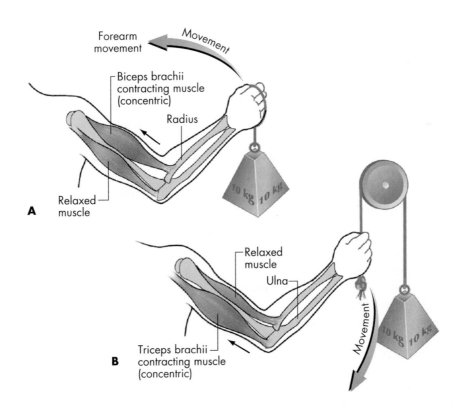

FIG. 2.5 ● Agonist–antagonist relationship. **A,** Biceps is agonist in flexing the elbow while triceps is antagonist; **B,** Triceps is agonist in extending the elbow while biceps is antagonist.

Modified from Shier D, Butler J, Lewis R: *Hole's human anatomy & physiology,* ed 9, Dubuque, IA, 2002, McGraw-Hill.

tion or relaxation of other muscles acting on the joint. In certain instances, two muscles may work in synergy by counteracting their opposing actions to accomplish a common action.

As discussed, agonist muscles are primarily responsible for a given movement, such as that of hip flexion and knee extension during kicking a ball. In this example, the hamstrings are antagonistic and relax to allow the kick to occur. This does not mean all other muscles in the hip area are uninvolved. The preciseness of the kick depends upon the involvement of many other muscles. As the lower extremity swings forward, its route and subsequent angle at the point of contact depend upon a certain amount of relative contraction or relaxation in the hip abductors, adductors, internal rotators, and external rotators. These muscles act in a synergistic fashion to guide the lower extremity in a precise manner. That is, they are not primarily responsible for knee extension and hip flexion, but they do contribute to the accuracy of the total movement. These guiding muscles assist in refining the kick and preventing extraneous motions. Additionally, the muscles in the contralateral hip and pelvic area must be under relative tension to help fixate or stabilize the pelvis on that side in order to provide a relatively stable pelvis for the hip flexors on the involved side to contract against. In kicking the ball, the pectineus and tensor fascia latae are adductors and abductors, respectively, in addition to flexors. The actions of abduction and adduction are neutralized by each other and the common action of the two muscles results in hip flexion.

From a practical point of view, it is not essential that individuals know the exact force exerted by each of the elbow flexors—biceps, brachialis, and brachioradialis—in chinning. It is important to understand that this muscle group is the agonist or primary mover responsible for elbow joint flexion. Similarly, it is important to understand that these muscles contract concentrically when the chin is pulled up to the bar and that they contract eccentrically when the body is lowered slowly. Antagonistic muscles produce actions opposite those of the agonist. For example, the muscles that produce extension of the elbow joint are antagonistic to the muscles that produce flexion of the elbow joint. It is important to understand that specific exercises need to be given for the development of each antagonistic muscle group. The return movement to the hanging position at the elbow joint after chinning is elbow joint extension, but the triceps and anconeus are not being strengthened. A concentric contraction

of the elbow joint flexors occurs, followed by an eccentric contraction of the same muscles.

Reversal of muscle function

A muscle group that is described to perform a given function can contract to control the exact opposite motion. Fig. 2.5, *A* illustrates how the biceps is an agonist by contracting concentrically to flex the elbow. The triceps is an antagonist to elbow flexion, and the pronator teres is considered to be a synergist to the biceps in this example. If the biceps were to slowly lengthen and control elbow extension, it would still be the agonist, but it would be contracting eccentrically. Fig. 2.5, *B* illustrates how the triceps is an agonist by contracting concentrically to extend the elbow. The biceps is an antagonist to elbow extension in this example. If the triceps were to slowly lengthen and control elbow flexion, it would still be the agonist, but it would be eccentrically contracting. In both of these examples, the deltoid, trapezius, and various other shoulder muscles are serving as stabilizers of the shoulder area.

Determination of muscle action

The specific action of a muscle may be determined through a variety of methods. These include reasoning in consideration of anatomical lines of pull, anatomical dissection, palpation, models, electromyography, and electrical stimulation.

With an understanding of a muscle's line of pull relative to a joint, one may determine the muscle's action at the joint. (See lines of pull later in this chapter.) Although not available to all students, cadaver dissection of muscles and joints is an excellent way to further understand muscle action.

For most of the skeletal muscles, **palpation** is a very useful way to determine muscle action. This is done through using the sense of touch to feel or examine a muscle as it is contracted. Palpation is limited to superficial muscles but is helpful in furthering one's understanding of joint mechanics. Models such as long rubber bands may be used to facilitate understanding of lines of pull and simulate muscle lengthening or shortening as joints move through various ranges of motion.

Electromyography (EMG) utilizes either surface electrodes that are placed over the muscle or fine wire/needle electrodes placed into the muscle. As the subject then moves the joint and contracts the muscles, the EMG unit detects the action potentials of muscles and provides an electronic readout of the contraction intensity and duration. EMG is the most accurate way of detecting the presence and extent of muscle activity.

Electrical muscle stimulation is somewhat a reverse approach of electromyography. Instead of electricity being used to detect muscle action, it is used to cause muscle activity. Surface electrodes are placed over a muscle and then the stimulator causes the muscle to contract. The joint's actions may then be observed to see the effect of the muscle's contraction on it.

Lines of pull

Combining the knowledge of a particular joint's functional design and diarthrodial classification with an understanding of the specific location of a musculotendinous unit as it crosses a joint is extremely helpful in understanding its action on the joint. For example, knowing that the rectus femoris has its origin on the anterior inferior iliac spine and its insertion on the tibial tuberosity via the patella, you can then determine that the muscle must have an anterior relationship to the knee and hip. Combining this knowledge with knowing that both joints are capable of sagittal plane movements such as flexion/extension, you can then determine that when the rectus femoris contracts concentrically it should cause the knee to extend and the hip to flex.

Furthermore, knowing that the semitendinosus, semimembranosus, and biceps femoris all originate on the ischial tuberosity and that the semitendinosus and semimembranosus cross the knee posteromedially before inserting on the tibia but that the biceps femoris crosses the knee posterolaterally before inserting on the fibula head, you may determine that all three muscles have posterior relationships to the hip and knee, which would enable them to be hip extensors and knee flexors upon concentric contraction. The specific knowledge related to their distal attachments and the knee's capability to rotate when flexed would allow you to determine that the semitendinosus and semimembranosus would cause internal rotation, whereas the biceps femoris would cause external rotation. A knowledge of the knee's axes of rotation being only frontal and vertical, but not sagittal, enables you to determine that even though the semitendinosus and semimembranosus have a posteromedial line of pull and the biceps femoris has a posterolateral line of pull, they are not capable of causing knee adduction and abduction, respectively.

You also apply this concept in reverse. For example, if the only action of a muscle such as the brachialis is known to be elbow flexion, then you should be able to determine that its line of pull must be anterior to the joint. Additionally, you would know that the origin of the brachialis must be located somewhere on the anterior humerus and the insertion must be somewhere on the anterior ulna.

Consider all of the following factors and their relationships as you study movements of the body to have a more thorough understanding.

1. Exact locations of bony landmarks to which muscles attach proximally and distally and their relationship to joints
2. The planes of motion through which a joint is capable of moving
3. The muscle's relationship or line of pull relative to the joint's axes of rotation
4. As a joint moves through a particular range of motion, the ability of the line of pull of a particular muscle to change and even result in the muscle having a different or opposite action than in the original position
5. The potential effect of other muscles' relative contraction or relaxation on a particular muscle's ability to cause motion
6. The effect of a muscle's relative length on its ability to generate force (See muscle length–tension relationship and active/passive insufficiency.)
7. The effect of the position of other joints on the ability of a biarticular or multiarticular muscle to generate force or allow lengthening (See uniarticular/biarticular/multiarticular muscles.)

Neural control of voluntary movement

When we discuss muscular activity, we should really state it as neuromuscular activity, since muscle cannot be active without nervous innervation. All voluntary movement is a result of both the muscular and nervous systems working together. All muscle contraction occurs as a result of stimulation from the nervous system. Ultimately, every muscle fiber is innervated by a somatic motor neuron, which, when an appropriate stimulus is provided, results in a muscle contraction. Depending upon a variety of factors, this stimulus may be processed in varying degrees at different levels of the **central nervous system** (CNS), which, for the purposes of this discussion, may be divided into five levels of control. This listing, in order from the most general level of control and the most superiorly located to the most specific level of control and the most inferiorly located, includes the cerebral cortex, the basal ganglia, the cerebellum, the brain stem, and the spinal cord.

The **cerebral cortex** is the highest level of control. It provides for the creation of voluntary movement as aggregate muscle action, but not as specific muscle activity. Sensory stimuli from the body also are interpreted here to a degree for the determination of needed responses.

The **basal ganglia** is the next level. It controls the maintenance of postures and equilibrium and learned movements such as driving a car. Sensory integration for balance and rhythmic activities is controlled here.

Third, the **cerebellum** is a major integrator of sensory impulses and provides feedback relative to motion. It controls the timing and intensity of muscle activity to assist in the refinement of movements.

Next, the **brain stem** integrates all central nervous system activity through excitation and inhibition of desired neuromuscular actions and functions in arousal or maintaining a wakeful state.

The **spinal cord** is the common pathway between the CNS and the **peripheral nervous system** (PNS), which contains all the remaining nerves throughout the body. It has the most specific control and integrates various simple and complex spinal reflexes, as well as cortical and basal ganglia activity, with various classifications of spinal reflexes.

Functionally, the PNS can be divided into sensory and motor divisions. The sensory or **afferent** nerves bring impulses from receptors in the skin, joints, muscles, and other peripheral aspects of the body to the CNS, while the motor or **efferent** nerves carry impulses to the outlying regions of the body.

The spinal nerves, illustrated in Fig. 2.6, also provide both motor and sensory function for their respective portions of the body and are named

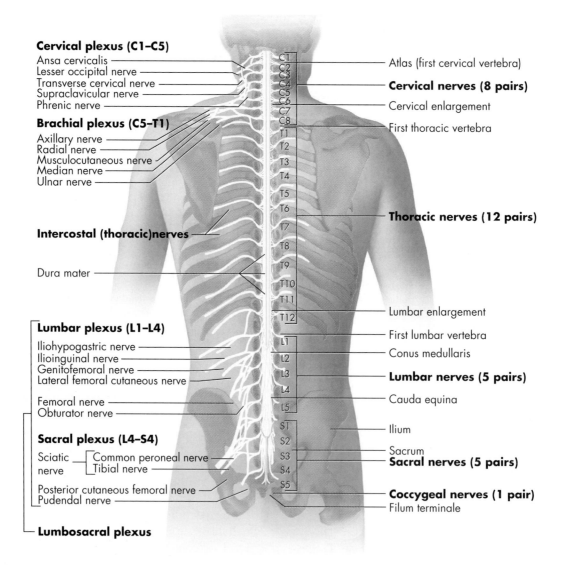

FIG. 2.6 ● Spinal nerve roots and plexuses.

Modified from Booher JM, Thibodeau GA: *Athletic injury assessment,* ed 4, Dubuque, IA, 2000, McGraw-Hill.

for the locations from which they exit the vertebral column. From each side of the spinal column, there are 8 cervical nerves, 12 thoracic nerves, 5 lumbar nerves, 5 sacral, and 1 coccygeal nerve. Cervical nerves 1 through 4 form the cervical plexus, which is generally responsible for sensation from the upper part of the shoulders to the back of the head and front of the neck. The cervical plexus supplies motor innervation to several muscles of the neck. Cervical nerves 5 through 8, along with thoracic nerve 1, form the brachial plexus, which supplies motor and sensory function to the upper extremity and most of the scapula. Thoracic nerves 2 through 12 run directly to specific anatomical locations in the thorax. All of the lumbar, sacral, and coccygeal nerves form the lumbosacral plexus, which supplies sensation and motor function to the lower trunk and the entire lower extremity and perineum.

One aspect of the sensory function of spinal nerves is to provide feedback to the CNS regarding skin sensation. A defined area of skin supplied by a specific spinal nerve is known as a **dermatome**. Regarding motor function of spinal nerves, a **myotome** is defined as a muscle or group of muscles supplied by a specific spinal nerve. Certain spinal nerves are also responsible for reflexes. See Table 2.3 regarding the specific spinal nerve functions.

TABLE 2.3 • **Spinal nerve root dermatomes, myotomes, reflexes, and functional application**

	Nerve root	Dermatome afferent (sensory)	Myotome efferent (motor)	Reflexes	Functional application
Cervical plexus	C1	Touch: Vertex of skull	Upper neck muscles	None	None
	C2	Touch: Temple, forehead, occiput	Upper neck muscles	None	None
	C3	Touch: Entire neck, posterior cheek, temporal area, under mandible	Trapezius, splenius, capitis	None	Scapula retraction, neck extension Sensation to cheek and side of neck
	C4	Touch: Shoulder area, clavicular area, upper scapular area	Trapezius, levator scapulae	None	Scapula retraction and elevation Sensation to clavicle and upper scapula
Brachial plexus	C5	Touch: Deltoid area, anterior aspect of entire arm to base of thumb	Supraspinatus, infraspinatus, deltoid, biceps brachii	Biceps brachii	Shoulder abduction Sensation to lateral side of arm and elbow
	C6	Touch: Anterior arm, radial side of hand to thumb and index finger	Biceps, supinator, wrist extensors	Biceps brachii, brachioradialis	Elbow flexion, wrist extension Sensation to lateral side of forearm including thumb and index fingers
	C7	Touch: Lateral arm and forearm to index, long, and ring fingers	Triceps brachii, wrist flexors	Triceps brachii	Elbow extension, wrist flexion Sensation to middle of anterior forearm and long finger
	C8	Touch: Medial side of forearm to ring and little fingers	Ulnar deviators, thumb extensors, thumb adductors (rarely triceps)	None	Wrist ulnar deviation, thumb extension Sensation to posterior elbow and medial forearm to little fingers
	T1	Touch: Medial arm and forearm to wrist	Intrinsic muscles of the hand except for opponens pollicis and abductor pollicis brevis	None	Abduction and adduction of fingers Sensation to medial arm and elbow

TABLE 2.3 (continued) • Spinal nerve root dermatomes, myotomes, reflexes, and functional application

	Nerve root	Dermatome afferent (sensory)	Myotome efferent (motor)	Reflexes	Functional application
	T2	Touch: Medial side of upper arm to medial elbow, pectoral and midscapular areas	Intercostal muscles	None	Sensation to medial upper arm, upper chest, and midscapular area
	T3–T12	Touch: T3–T6, upper thorax; T5–T7, coastal margin; T8–T12, abdomen and lumbar region	Intercostal muscles, abdominal muscles	None	Sensation to chest, abdomen, and low back
Lumbosacral plexus	L1	Touch: Lower abdomen, groin, lumbar region from 2nd to 4th vertebrae, upper and outer aspect of buttocks	Quadratus Lumborum	None	Sensation to low back, over trochanter and groin
	L2	Touch: Lower lumbar region, upper buttock, anterior aspect of thigh	Iliopsoas, quadriceps	None	Hip flexion Sensation to back, front of thigh to knee
	L3	Touch: Medial aspect of thigh to knee, anterior aspect of lower 1/3 of the thigh to just below patella	Psoas, quadriceps	Patella or knee extensors	Hip flexion and knee extension Sensation to back, upper buttock, anterior thigh and knee, medial lower leg
	L4	Touch: Medial aspect of lower leg and foot, inner border of foot, great toe	Tibialis anterior, extensor hallucis and digitorum longus, peroneals	Patella or knee extensors	Ankle dorsiflexion, transverse tarsal/subtalar inversion Sensation to medial buttock, lateral thigh, medial leg, dorsum of foot, big toe
	L5	Touch: Lateral border of leg, anterior surface of the lower leg, top of foot to middle three toes	Extensor hallucis and digitorum longus, peroneals, gluteus maximus and medius, and dorsiflexors	None	Great toe extension, transverse talar/subtalar eversion Sensation to upper lateral leg, anterior surface of the lower leg, middle three toes
	S1	Touch: Posterior aspect of the lower 1/4 of the leg, posterior aspect of the foot, including the heel, lateral border of the foot and sole	Gastrocnemius, soleus, gluteus maximus and medius, hamstrings, peroneals	Achilles reflex	Ankle plantarflexion, knee flexion, transverse talar/subtalar eversion Sensation to lateral leg, lateral foot, lateral two toes, plantar aspect of foot
	S2	Touch: Posterior central strip of the leg from below the gluteal fold to 3/4 of the way down the leg	Gastrocnemius, soleus, gluteus maximus, hamstrings	None	Ankle plantarflexion and toe flexion Sensation to posterior thigh and upper posterior leg
	S3	Touch: Groin, medial thigh to knee	Intrinsic foot muscles	None	Sensation to groin and adductor region
	S4	Touch: Perineum, genitals, lower sacrum	Bladder, rectum	None	Urinary and bowel control Sensation to saddle area, genitals, anus

The basic functional units of the nervous system responsible for generating and transmitting impulses are nerve cells known as **neurons**. Neurons consist of a **neuron cell body**, one or more branching projections known as **dendrites**, which transmit impulses to the neuron and cell body, and an **axon**, which is an elongated projection that transmits impulses away from neuron cell bodies. Neurons are classified into three types, according to the direction in which they transmit impulses. **Sensory neurons** transmit impulses to the spinal cord and brain from all parts of the body, whereas **motor neurons** transmit impulses away from the brain and spinal cord to muscle and glandular tissue. **Interneurons** are central or connecting neurons that conduct impulses from sensory neurons to motor neurons.

Proprioception and kinesthesis

The performance of various activities is significantly dependent upon neurological feedback from the body. Very simply, we use the various senses to determine a response to our environment, as when we use sight to know when to lift our hand to catch a fly ball. We are all familiar with the senses of smell, touch, sight, hearing, and taste. We are also aware of other sensations, such as pain, pressure, heat, and cold, but we often take for granted the sensory feedback provided by proprioceptors during neuromuscular activity. Proprioceptors are internal receptors located in the skin, joints, muscles, and tendons that provide us feedback relative to the tension, length, and contraction state of muscle, the position of the body and limbs, and movements of the joints. These proprioceptors in combination with the other sense organs of the body are vital in **kinesthesis**, which is the conscious awareness of the position and movement of the body in space. For example, if standing on one leg with the other knee flexed, you do not have to look at your non–weight-bearing leg to know the approximate number of degrees that you may have it flexed. The proprioceptors in and around the knee provide you information so that you are kinesthetically aware of your knee position. Proprioceptors specific to the muscles are muscle spindles and Golgi tendon organs (GTO), whereas Meissner's corpuscles, Ruffini's corpuscles, Pacinian corpuscles, and Krause's end-bulbs are proprioceptors specific to the joints and skin.

While kinethesis is concerned with the conscious awareness of the body's position, **proprioception** is the subconscious mechanism by which the body is able to regulate posture and movement by responding to stimuli originating in the proprioceptors imbedded in the joints, tendons, muscles, and inner ear. When we unexpectedly step on an unlevel or unstable surface, the muscles in and about our lower extremity may respond very quickly by contracting appropriately to prevent a fall or injury if we have good proprioception. This is a protective response of the body that occurs without us having time to make a conscious decision on how to respond.

Muscle spindles (Fig. 2.7), concentrated primarily in the muscle belly between the fibers, are sensitive to stretch and rate of stretch. Specifically, they insert into the connective tissue within the muscle and run parallel with the muscle fibers. The number of spindles in a particular muscle varies depending upon the level of control needed for the area. Consequently, the concentration of muscle spindles in the hands is much greater than in the thigh.

When rapid stretch occurs, an impulse is sent to the CNS. The CNS then activates the motor neurons of the muscle and causes it to contract. All muscles possess this **myotatic** or **stretch reflex**, but it is more remarkable in the extensor muscles of the extremities. The knee jerk or patella tendon reflex can be used as an example as shown in Fig. 2.8. When the reflex hammer strikes the patella tendon, it causes a quick stretch to the musculotendinous unit of the quadriceps. In response, the quadriceps fires and the knee extends. To an extent, the more sudden the tap of the hammer, the more significant the reflexive contraction. A more practical example is seen in maintaining posture as in the case when a student begins to doze off in class. As the head starts to nod forward, there is a sudden stretch placed on the neck extensors, which activates the muscle spindles and ultimately results in a sudden jerk back to an extended position.

The stretch reflex provided by the muscle spindle may be utilized to facilitate a greater response, as in the case of a quick short squat before attempting a jump. The quick stretch placed upon the muscles in the squat enables the same muscles to generate more force in subsequently jumping off the floor.

The Golgi tendon organs (Fig. 2.7), serially located in the tendon close to the muscle–tendon junction, are continuously sensitive to both muscle tension and active contraction. The GTO is much less sensitive to stretch than muscle spindles are and requires a greater stretch to be activated. Tension in tendons and consequently in the GTO increases as the muscle contracts, which in turn activates the GTO. When the GTO stretch threshold is reached, an impulse is sent to the CNS, which in turn causes the muscle to relax and facilitates activation of the

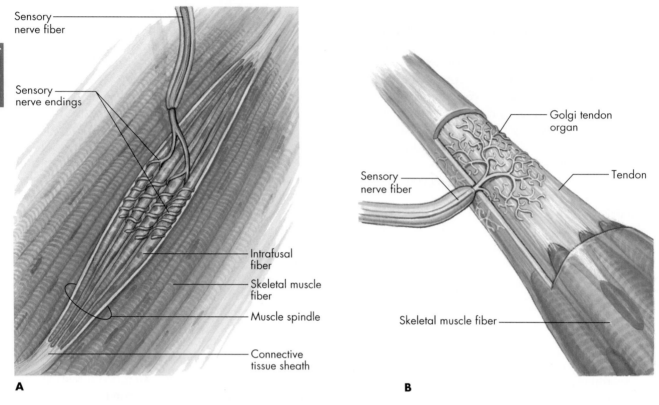

FIG. 2.7 ● Muscles spindles and Golgi tendon organs. **A,** Increased muscle length stimulates muscle spindles, which stimulate muscle contraction; **B,** Golgi tendon organs occupy tendons, where they inhibit muscle contraction.

From Shier D, Butler J, Lewis R: *Hole's human anatomy & physiology,* ed 9, New York, 2002, McGraw-Hill.

antagonists as a protective mechanism. That is, the GTO, through this inverse stretch reflex, protects us from an excessive contraction by causing the muscle it supplies to relax. As an example, when a weight lifter attempts a very heavy resistance in the biceps curl and reaches the point of extreme overload, the GTO is activated, the biceps suddenly relaxes, and the triceps contracts, which is why it appears as if the lifter is throwing the weight down.

Pacinian corpuscles, concentrated around joint capsules, ligaments, and tendon sheaths and beneath the skin, are activated by rapid changes in the joint angle and by pressure changes affecting the capsule. This activation lasts only briefly and is not effective in detecting constant pressure. Pacinian corpuscles are helpful in providing feedback regarding the location of a body part in space following quick movements such as running or jumping.

Ruffini's corpuscles, located in deep layers of the skin and the joint capsule, are activated by strong and sudden joint movements as well as pressure changes. Unlike Pacinian corpuscles, their reaction to pressure changes are slower to

develop, but their activation is continued as long as pressure is maintained. They are essential in detecting even minute joint position changes and providing information as to the exact joint angle.

Meissner's corpuscles and Krause's end-bulbs are located in the skin and in subcutaneous tissues. While they are important in receiving stimuli from touch, they are not so relevant to our discussion of kinesthesis. See Table 2.4 for further comparisons of sensory receptors.

The quality of movement and how we react to position change is significantly dependent upon the proprioceptive feedback from the muscles and joints. Like the other factors involving body movement, proprioception may be enhanced through specific training that utilizes the proprioceptors to a high degree, such as balancing and functional activities. Attempting to maintain your balance on one leg, first with the eyes open on a level surface, may serve as an initial low-level proprioceptive activity, which may eventually progress to a much higher level such as balancing on an unlevel, unstable surface with your eyes closed.

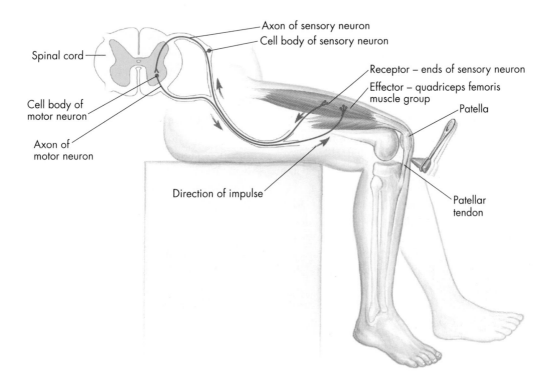

FIG. 2.8 ● Knee jerk or patellar tendon reflex. A sudden tap on the patella tendon causes a quick stretch to the quadriceps, which activates the muscle spindle. The information regarding the stretch is sent via the axon of the sensory neuron to the spinal cord where it synapses with a motor neuron, which, in turn, carries via its axon a motor response for the quadriceps to contract.

From Shier D, Butler J, Lewis R: *Hole's human anatomy & physiology,* ed 9, New York, 2002, McGraw-Hill.

TABLE 2.4 • **Sensory receptors**

Receptors	Sensitivity	Location	Response
Muscle spindles	Subconscious muscle sense, muscle length changes	In skeletal muscles among muscle fibers in parallel with fibers	Initiates rapid contraction of stretched muscle Inhibits development of tension in antagonistic muscles
Golgi tendon organs	Subconscious muscle sense, muscle tension changes	In tendons, near muscle–tendon junction in series with muscle fibers	Inhibits development of tension in stretched muscles Initiates development of tension in antagonistic muscles
Pacinian corpuscles	Rapid changes in joint angles, pressure, vibration	Subcutaneous, submucosa, and subserous tissues around joints and external genitals, mammary glands	Provide feedback regarding location of body part in space following quick movements
Ruffini's corpuscles	Strong, sudden joint movements, touch, pressure	Skin and subcutaneous tissue of fingers, collagenous fibers of the joint capsule	Provide feedback as to exact joint angle
Meissner's corpuscles	Fine touch, vibration	In skin	Provide feedback regarding touch, two-point discrimination
Krause's end-bulbs	Touch, thermal change	Skin, subcutaneous tissue, lip and eyelid mucosa, external genitals	Provide feedback regarding touch

There are numerous additional proprioceptive training activities that are really only limited by the imagination and the level of proprioception.

Neuromuscular concepts

Motor units and the all or none principle

When a particular muscle contracts, the contraction actually occurs at the muscle fiber level within a particular motor unit. A **motor unit** consists of a single motor neuron and all of the muscle fibers it innervates. Motor units function as a single unit. In a typical muscle contraction, the number of motor units responding and consequently the number of muscle fibers contracting within the muscle may vary significantly, from relatively few to virtually all of the muscle fibers, depending on the number of muscle fibers within each activated motor unit and the number of motor units activated. Regardless of the number involved, the individual muscle fibers within a given motor unit will fire and contract either maximally or not at all. This is referred to as the **all or none principle**.

Factors affecting muscle tension development

The difference between a particular muscle contracting to lift a minimal resistance versus the same muscle contracting to lift a maximal resistance is the number of muscle fibers recruited. The number of muscle fibers recruited may be increased by activating those motor units containing a greater number of muscle fibers, by activating more motor units, or by increasing the frequency of motor unit activation. The number of muscle fibers per motor unit varies significantly from less than 10 in muscles requiring a very precise and detailed response such as the muscles of the eye to as many as a few thousand in large muscle groups such as the quadriceps that perform less complex activities.

For the muscle fibers in a particular motor unit to contract, the motor unit must first receive a stimulus via an electrical signal known as an **action potential** from the brain and spinal cord through its axons. If the stimulus is not strong enough to cause an action potential, it is known as a **subthreshold stimulus** and does not result in a contraction. When the stimulus becomes strong enough to produce an action potential in a single motor unit axon, it is known as a **threshold stimulus** and all of the muscle fibers in the motor unit contract. Stimuli that are stronger to the point of producing action potentials in additional motor units are known as **submaximal stimuli**. For action potentials to be produced in all of the motor units of a particular muscle, a **maximal**

stimulus would be required. As the strength of the stimulus increases from threshold up to maximal, more motor units are recruited and the overall force of the muscle contraction increases in a graded fashion. Increasing the stimulus beyond maximal has no effect. The effect of increasing the number of motor units activated is detailed in Fig. 2.9.

Greater contraction forces may also be achieved by increasing the frequency of motor unit activation. To simplify the phases of a single muscle fiber contraction or twitch, a stimulus is provided and followed by a brief **latent period** of a few milliseconds. Then the second phase known as the **contraction phase** begins in which the muscle fiber begins shortening. The contraction phase lasts about 40 milliseconds and is followed by the **relaxation phase**, which lasts approximately 50 milliseconds. This is illustrated in Fig. 2.10. When successive stimuli are provided before the relaxation phase of the first twitch has completed, the subsequent twitches combine with the first to produce a sustained contraction. This **summation** of contractions generates a greater amount of tension than a single contraction would produce individually. As the frequency of stimuli increases, the resultant summation increases accordingly, producing increasingly greater total muscle tension. If the stimuli are provided at a frequency high enough that no relaxation can occur between contractions, then **tetanus** results. Fig. 2.11 illustrates the effect of increasing the rate of stimulation to gain increased muscle tension.

Treppe is another phenomenon of muscle contraction that occurs when multiple maximal stimuli are provided at a low enough frequency to allow complete relaxation between contractions to rested muscle. Slightly greater tension is produced by the second stimulus than with the first. A third stimulus produces even greater tension than the second. This staircase effect as illustrated in Fig. 2.12 occurs only with the first few stimuli and the resultant contractions after the initial ones result in equal tension being produced.

Muscle length–tension relationship

The maximal ability of a muscle to develop tension and exert force varies depending upon the length of the muscle during contraction. Generally, depending upon the particular muscle involved, the greatest amount of tension can be developed when a muscle is stretched between 100% and 130% of its resting length. As a muscle is stretched beyond this point, the amount of force it can exert significantly decreases. Likewise, a proportional decrease in the ability to develop

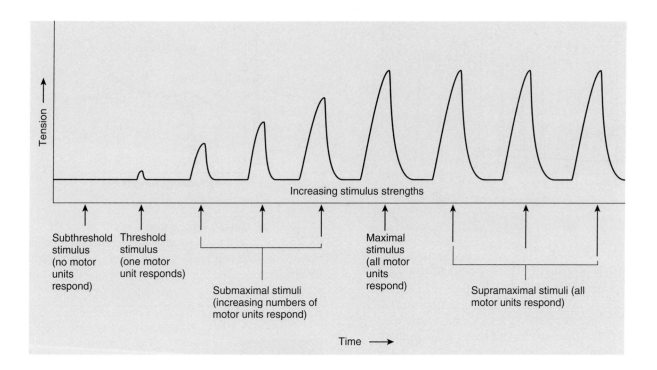

Increasing stimulus strengths

Subthreshold stimulus (no motor units respond)

Threshold stimulus (one motor unit responds)

Submaximal stimuli (increasing numbers of motor units respond)

Maximal stimulus (all motor units respond)

Supramaximal stimuli (all motor units respond)

Time →

FIG. 2.9 ● Achieving threshold stimulus and the effect of recruiting more motor units on increasing tension. If the stimulus does not reach threshold, there is no motor unit response. As the stimulus strength is increased, more motor units are recruited until eventually all motor units are recruited and maximal tension of the muscle is generated. Increasing stimulus strength beyond this has no effect.

From Seeley RR, Stephens TD, Tate P: *Anatomy & physiology,* ed 7, New York, 2006, McGraw-Hill.

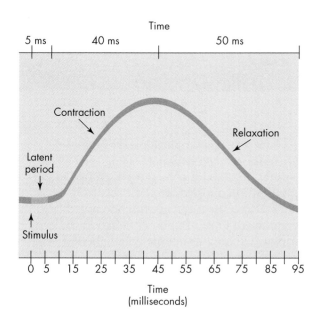

FIG. 2.10 ● A recording of a simple twitch. Note the three time periods (latent period, contraction, and relaxation) following the stimulus.

From Powers SK, Howley ET: *Exercise physiology: theory and application to fitness and performance,* ed 4, New York, 2001, McGraw-Hill.

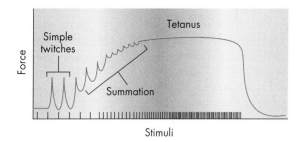

FIG. 2.11 ● Recording showing the change from simple twitches to summation, and finally tetanus. Peaks to the left represent simple twitches, while increasing the frequency of the stimulus results in a summation of the twitches and finally tetanus.

From Powers SK, Howley ET: *Exercise physiology: theory and application to fitness and performance,* ed 4, New York, 2001, McGraw-Hill.

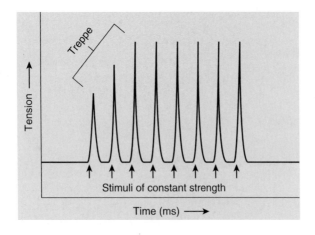

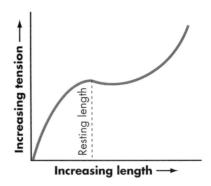

FIG. 2.12 ● Treppe. When a rested muscle is stimulated repeatedly with a maximal stimulus at a frequency that allows complete relaxation between stimuli, the second contraction produces a slightly greater tension than the first, and the third contraction produces greater tension than the second. After a few contractions, the tension produced by all contractions is equal.

From Seeley RR, Stephens TD, Tate P: *Anatomy & physiology,* ed 7, New York, 2006, McGraw-Hill.

tension occurs as a muscle is shortened. When a muscle is shortened to around 50% to 60% of resting length, its ability to develop contractile tension is essentially reduced to zero.

In most sporting activities we generally place the muscles on optimum stretch in the preparatory phase that we intend to contract forcefully in the subsequent movement or action phase of the skill. The various phases of performing a movement skill will be discussed in much greater detail in Chapter 8. This principle may be seen at work when we squat slightly to stretch the calf, hamstrings, and quadriceps before contracting them concentrically to jump. If we do not first lengthen these muscles through squatting slightly, they are unable to generate enough contractile force to jump very high. If we squat fully and lengthen the muscles too much, we lose the ability to generate as much force and as a result we cannot jump as high.

We may take advantage of this principle by effectively reducing the contribution of some muscles in a group by placing them in a shortened state so that we can isolate the work to the muscle(s) remaining in the lengthened state. For example, in hip extension, we may isolate the work of the gluteus maximus as a hip extensor by maximally shortening the hamstrings with flexion of the knee to reduce their ability to act as hip extensors. See Fig. 2.13 and Fig. 2.14.

FIG. 2.13 ● Muscle length–tension relationship. As the muscle's length is shorter to the left, the amount of tension that can be developed is less, and as the length increases to the right, the amount of tension increases.

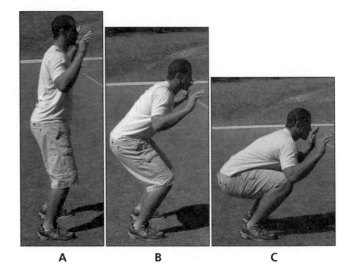

FIG. 2.14 ● Practical application of the muscle length–tension relationship involving the calf, hamstrings, and quadriceps muscle groups in jumping. **A,** The muscles are in a relatively shortened position and consequently not able to generate much tension upon contraction; **B,** The muscles are in a more optimally lengthened position to generate significant tension to jump high; **C,** The muscles are lengthened too much and are not able to generate as much force as in *B*.

Muscle force-velocity relationship

When the muscle is either concentrically or eccentrically contracting, the rate of length change is significantly related to the amount of force potential. When contracting concentrically against a light resistance, the muscle is able to contract at a high velocity. As the resistance increases, the maximal velocity at which the muscle is able to contract decreases. Eventually, as the load increases, the velocity decreases to zero; this results in an isometric contraction.

As the load increases even further beyond that which the muscle can maintain with an isometric contraction, the muscle begins to lengthen, resulting in an eccentric contraction or action. A slight increase in the load will result in relatively low velocity of lengthening. As the load increases even further, the velocity of lengthening will increase as well. Eventually, the load may increase to the point where the muscle can no longer resist. This will result in uncontrollable lengthening, or more likely, dropping of the load.

From this explanation you can see that there is an inverse relationship between concentric velocity and force production. As the force needed to cause movement of an object increases, the velocity of concentric contraction decreases. Furthermore, there is a somewhat proportional relationship between eccentric velocity and force production. As the force needed to control the movement of an object increases, the velocity of eccentric lengthening increases, at least until the point at which control is lost. This is illustrated in Fig. 2.15.

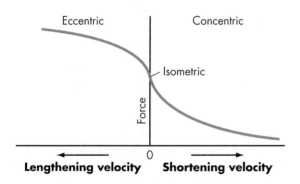

FIG. 2.15 ● Muscle–force velocity relationship. From right to left; when little force is needed to move relatively light loads, the muscles may contract concentrically at a relatively high velocity. As the amount of force needed increases with greater loads, the velocity of the concentric contraction decreases proportionally until eventually reaching zero velocity or an isometric contraction with a very large load. When the muscle can no longer generate the amount of force needed to maintain the load in a static position, the muscle begins eccentrically contracting to control the velocity, and it can do so at a relatively slow velocity. As the amount of force needed increases to control greater loads, the velocity increases proportionally.

Angle of pull

Another factor of considerable importance in using the leverage system is the angle of pull of the muscles on the bone. The angle of pull may be defined as the angle between the line of pull of the muscle and the bone on which it inserts. For the sake of clarity and consistency, we need to specify that the actual angle referred to is the angle toward the joint. With every degree of joint motion, the angle of pull changes. Joint movements and insertion angles involve mostly small angles of pull. The angle of pull decreases as the bone moves away from its anatomical position through the contraction of the local muscle group. This range of movement depends on the type of joint and bony structure.

Most muscles work at a small angle of pull, generally less than 50 degrees. The amount of muscular force needed to cause joint movement is affected by the angle of pull. Three components of muscular force are involved. The **rotary component**, also referred to as the vertical component, is the component of muscular force that acts perpendicular to the long axis of the bone (lever). When the line of muscular force is at 90 degrees to the bone on which it attaches, all of the muscular force is rotary force; therefore, 100% of the force is contributing to the movement. That is, all of the force is being used to rotate the lever about its axis. The closer the angle of pull to 90 degrees, the greater the rotary component. At all other degrees of the angle of pull, one of the other two components of force is operating in addition to the rotary component. The same rotary component is continuing, although with less force, to rotate the lever about its axis. The second force component is the horizontal or **nonrotary component** and is either a **stabilizing component** or a **dislocating component**, depending on whether the angle of pull is less than or greater than 90 degrees. If the angle is less than 90 degrees, the force is a stabilizing force, because its pull directs the bone toward the joint axis. If the angle is greater than 90 degrees, the force is dislocating, because its pull directs the bone away from the joint axis (Fig. 2.16).

In some activities, it is desirable to have a person begin a movement when the angle of pull is at 90 degrees. Many boys and girls are unable to do a chin-up (pull-up) unless they start with the elbow in a position to allow the elbow flexor muscle group to approximate a 90-degree angle with the forearm.

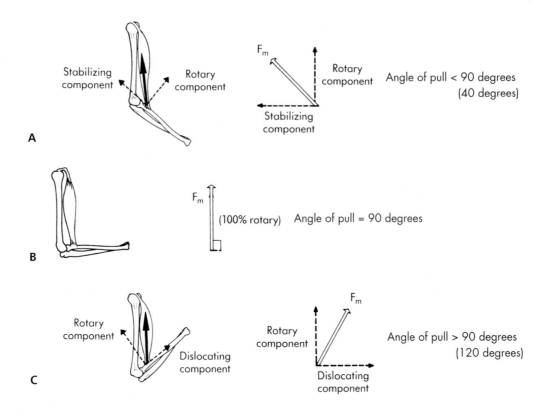

FIG. 2.16 ● **A** to **C,** Components of force due to the angle of pull.

Modified from Hall SJ: *Basic biomechanics*, New York, 2003, McGraw-Hill.

This angle makes the chin-up easier because of the more advantageous angle of pull. The application of this fact can compensate for lack of sufficient strength. In its range of motion, a muscle pulls a lever through a range characteristic of itself, but it is most effective when approaching and going beyond 90 degrees. An increase in strength is the only solution for muscles that operate at disadvantageous angles of pull and require a greater force to operate efficiently.

Uniarticular, biarticular, and multiarticular muscles

Uniarticular muscles are those that cross and act directly only on the joint that they cross. The brachialis of the elbow is an example in that it can only pull the humerus and ulna closer to each other upon concentric contraction. When the humerus is relatively stabilized as in an elbow curl, the brachialis contracts to flex the elbow and pulls the ulna closer to the humerus. However, when the ulna is relatively stabilized as in a pull-up, the brachialis indirectly causes motion at the shoulder even though it does not cross it. In this example the brachialis contracts and pulls the humerus closer to the ulna as an elbow flexor. Correspondingly, the shoulder has to move from flexion into extension for the pull-up to be accomplished.

Biarticular muscles are those that cross and act directly on two different joints. Depending on a variety of factors, a biarticular muscle may contract to cause, control, or prevent motion at either one or both of its joints. Biarticular muscles have two advantages over uniarticular muscles. They can cause, control, and/or prevent motion at more than one joint, and they may be able to maintain a relatively constant length due to "shortening" at one joint and "lengthening" at another joint. The muscle does not actually shorten at one joint and lengthen at the other; instead, the concentric shortening of the muscle to move one joint is offset by motion of the other joint, which moves its attachment of the muscle farther away. This maintenance of a relatively constant length results in the muscle's ability to continue its exertion of force. In the pull-up example, the biceps brachii acts as a flexor at the elbow. In the initial stage of the pull-up, the biceps brachii is in a relatively lengthened state at the elbow due to its extended position and in a relatively shortened state at the shoulder due to its flexed position. To accomplish the pull-up, the biceps brachii contracts concentrically to flex the elbow so it effectively "shortens" at the elbow. Simultaneously, the shoulder is extending during the pull-up, which effectively "lengthens" the biceps brachii at the shoulder.

The biarticular muscles of the hip and knee provide excellent examples of two different patterns of action. An example of a **concurrent** movement pattern occurs when both the knee and the hip extend at the same time. If only the knee were to extend, the rectus femoris would shorten and lose tension as do the other quadriceps muscles, but its relative length and subsequent tension may be maintained due to its relative lengthening at the hip joint during extension. When a ball is kicked, an example of a **countercurrent** movement pattern may be observed. During the forward movement phase of the lower extremity, the rectus femoris is concentrically contracted to both flex the hip and extend the knee. These two movements, when combined, increase the tension or stretch on the hamstring muscles at both the knee and the hip.

Multiarticular muscles act on three or more joints due to the line of pull between their origin and insertion crossing multiple joints. The principles discussed relative to biarticular muscles apply in a similar fashion to multiarticular muscles.

Reciprocal inhibition or innervation

As stated earlier, antagonist muscle groups must relax and lengthen when the agonist muscle group contracts. This effect, reciprocal innervation, occurs through reciprocal inhibition of the antagonists. Activation of the motor units of the agonists causes a reciprocal neural inhibition of the motor units of the antagonists. This reduction in neural activity of the antagonists allows them to subsequently lengthen under less tension. This may be demonstrated by comparing the ease at which one can stretch the hamstrings when simultaneously contracting the quadriceps versus attempting to stretch the hamstrings without the quadriceps contracted.

Active and passive insufficiency

As a muscle shortens, its ability to exert force diminishes, as discussed earlier. When the muscle becomes shortened to the point at which it cannot generate or maintain active tension, **active insufficiency** is reached. If the opposing muscle becomes stretched to the point at which it can no longer lengthen and allow movement, **passive insufficiency** is reached. These principles are most easily observed in either biarticular or multiarticular muscles when the full range of motion is attempted in all of the joints crossed by the muscle.

An example occurs when the rectus femoris contracts concentrically to both flex the hip and extend the knee, as in Fig. 2.17, *A*. It may completely perform either action one at a time but is actively insufficient to obtain full range at both joints simultaneously, as in Fig. 2.17, *B*. Likewise, the hamstrings will not usually stretch enough to allow both maximal hip flexion and maximal knee extension; hence, they are passively insufficient. As a result of these phenomena, it is virtually impossible to actively extend the knee fully when beginning with the hip fully flexed, or vice versa.

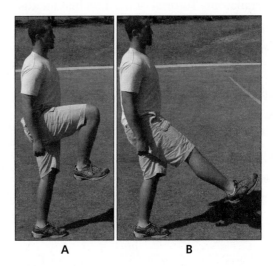

A   **B**

FIG. 2.17 ● Active and passive insufficiency. **A,** The rectus femoris is easily able to actively flex the hip or extend the knee through their respective full ranges of motion individually without fully stretching the hamstrings; **B,** When attempting to both actively flex the hip and extend the knee simultaneously, active insufficiency is reached in the rectus femoris and passive insufficiency is reached in the hamstrings, resulting in the inability to reach full range of motion in both joints.

Web sites

Neurologic Exam: An anatomical approach

http://medlib.med.utah.edu/neurologicexam/home_exam.html

A very thorough site regarding neurological exam including numerous movies with both normal and pathological results.

Cranial Nerves: Review info

www.gwc.maricopa.edu/class/bio201/cn/cranial.htm

A good resource on the cranial nerves.

University of Arkansas for Medical Sciences Nerve tables

http://anatomy.uams.edu/anatomyhtml/nerves.html

Numerous tables of all nerves throughout the body.

A Cyberanatomy Tutorial of the Brachial Plexus and Its Associated Injuries

http://anatome.ncl.ac.uk/tutorials/brachial1/text

A hands-on guide to the brachial plexus.

Dermatomes

www.meddean.luc.edu/lumen/MedEd/GrossAnatomy/learnem/dermat/main_der.htm

An interactive review of the body's dermatomes.

Loyola University Medical Education Network Master Muscle List

www.meddean.luc.edu/lumen/MedEd/GrossAnatomy/dissector/mml

An interactive and graphical review of the muscles indexed alphabetically and by region.

Spinal Cord and Nervous System

www.driesen.com/spine_and_spinal_cord.htm

A review of the spinal cord and nervous system.

Proprioception Exercises Can Improve Balance

http://sportsmedicine.about.com/library/weekly/aa062200.htm

Proprioception.

Meds 1 Neurophysiology 2003

www.med.uwo.ca/physiology/courses/medsweb

A site with Flash animation regarding neurophysiology.

Muscular System

www.bio.psu.edu/faculty/strauss/anatomy/musc/muscular.htm

Cadavaric photos of the cat muscular system.

BBC Science & Nature

http://bbc.co.uk/science/humanbody/body/interactives/3djigsaw_02/index.shtml

An interactive site allowing you to select the innervation for the muscles.

Human Anatomy Online

http://innerbody.com/image/musfov.html

An interactive site with details on the muscular and nervous systems.

Functions of the Muscular System

http://training.seer.cancer.gov/module_anatomy/unit41_muscle_functions.html

Several pages with information on skeletal muscle structure, types, and groups along with unit review and quizzes.

Functions of the Nervous System

http://training.seer.cancer.gov/module_anatomy/unit51_nerve_functions.html

Several pages with information on the nervous system organization, nerve structure, unit review, and quizzes.

Training for Proprioception & Function

www.coachr.org/proprio.htm

Information on improving body awareness and movement efficiency.

Fitter International

www.fitter1.com/article_proprioception.html?mtcPromotion=Articles%3EProprioception

Products and articles on proprioception.

The Physician and Sportsmedicine

www.physsportsmed.com/issues/1997/10oct/laskow.htm

Refining rehabilitation with proprioception training: expediting return to play.

U.S. National Library of Medicine's Visible Human Project AnatQuest Anatomic Images Online

http://anatline.nlm.nih.gov/AnatQuest/AwtCsViewer/aq-cutaway.html

A great resource using a cut-away viewer to see cadaver images throughout the body with the ability to identify the relevant anatomy.

LABORATORY AND REVIEW EXERCISES

1. Observe on a fellow student some of the muscles found in Figs. 2.1 and 2.2.
2. Complete the muscle nomenclature chart by writing in the distinctive characteristics for which each of the muscles is named such as shape, size, number of divisions, fiber direction, location, action. Some muscles have more than one.

Muscle nomenclature chart

Muscle name	Distinctive characteristics for which it is named
Adductor magnus	
Biceps brachii	
Biceps femoris	
Brachialis	
Brachioradialis	
Coracobrachialis	
Deltoid	
Extensor carpi radialis brevis	
Extensor carpi ulnaris	
Extensor digiti minimi	
Extensor digitorum	
Extensor hallucis longus	
Extensor indicis	
Extensor pollicis brevis	
External oblique	
Fibularis brevis	
Flexor carpi radialis	
Flexor digitorum longus	
Flexor digitorum profundus	
Flexor digitorum superficialis	
Flexor pollicus longus	
Gastrocnemius	
Gluteus maximus	

Muscle name	Distinctive characteristics
Gluteus medius	
Iliacus	
Iliocostalis thoracis	
Infraspinatus	
Latissimus dorsi	
Levator scapulae	
Longissimus lumborum	
Obturator externus	
Palmaris longus	
Pectoralis minor	
Peroneus tertius	
Plantaris	
Pronator quadratus	
Pronator teres	
Psoas major	
Rectus abdominis	
Rectus femoris	
Rhomboid	
Semimembranous	
Semitendinosus	
Serratus anterior	
Spinalis cervicis	
Sternocleidomastoid	
Subclavius	
Subscapularis	
Supinator	
Supraspinatus	
Tensor fasciae latae	
Teres major	
Tibialis posterior	
Transversus abdominis	
Trapezius	
Triceps brachii	
Vastus intermedius	
Vastus lateralis	
Vastus medialis	

3. With a partner, choose a diarthrodial joint on the body and carry out each of the following exercises:
   a. Familiarize yourself with all of the joint's various movements and list them.
   b. Determine which muscle or muscle groups are responsible for each of the movements you listed in 3a.
   c. For the muscles or muscle groups you listed for each movement in 3b, determine the type of contraction occurring.
   d. Determine how to change the parameters of gravity and/or resistance so that the opposite muscles contract to control the same movements in 3c. Name the type of contraction occurring.
   e. Determine how to change the parameters of movement, gravity, and/or resistance so that the same muscles listed in 3c contract differently to control the opposite movement.

4. For each of the muscles listed in the muscle shape and fiber arrangement chart, determine first whether it should be classified as parallel or pennate. Complete the chart by writing in flat, fusiform, strap, radiate, or sphincter under those you classify as parallel. Write in unipennate, bipennate, or multipennate for those you classify as pennate.

Muscle shape and fiber arrangement chart

Muscle	Parallel	Pennate
Adductor longus		
Adductor magnus		
Brachioradialis		
Extensor digitorum		
Flexor carpi ulnaris		
Flexor digitorum longus		
Gastrocnemius		
Gluteus maximus		
Iliopsoas		
Infraspinatus		
Latissimus dorsi		
Levator scapulae		
Palmaris longus		
Pronator quadratus		
Pronator teres		
Rhomboid		
Serratus anterior		
Subscapularis		
Triceps brachii		
Vastus intermedius		
Vastus medialis		

5. For each of the following exercises in the muscle contraction typing chart, write the type of contraction (isometric, concentric, or eccentric), if any, in the cell of the muscle group that is contracting. Place a dash in the cell if there is no contraction occurring. Hint: In some instances you may have more than one type of contraction in the same muscle groups throughout varying portions of the exercises. If so, list them in the order of occurrence.

Muscle contraction typing chart

Exercise	Quadriceps	Hamstrings
a. Lay prone on a table with your knee in full extension.		
Maintain your knee in full extension.		
Very slowly flex your knee maximally.		
Maintain your knee in full flexion.		
From the fully flexed position, extend your knee fully as fast as possible but stop immediately before reaching maximal extension.		
From the fully flexed position, very slowly extend your knee fully.		
b. Begin sitting on the edge of the table with your knee in full extension.		
Maintain your knee in full extension.		
Very slowly flex your knee maximally.		
Maintain your knee in full flexion.		
Maintain your knee in approximately 90 degrees of flexion.		
From the fully flexed position, slowly extend your knee fully.		
c. Stand on one leg and move the other knee as directed.		
Maintain your knee in full extension.		
Very slowly flex your knee maximally.		
From the fully flexed position, slowly extend your knee fully.		
From the fully flexed position, extend your knee fully as fast as possible.		

6. Choose a particular sport skill and determine type of muscle contractions occurring in various major muscle groups about the body at different phases of the skill.

7. Utilizing a reflex hammer or the flexed knuckle of your long finger PIP joint, compare the patellar reflex among several subjects.

8. Request a partner to stand with eyes closed while you position his/her arms in an odd position at the shoulders, elbows, and wrists. Ask your partner to describe the exact position of each joint while keeping the eyes closed. Then have your partner begin in the anatomical position, close the eyes, and subsequently reassume the position in which you had previously placed him/her. Explain the neuromechanisms involved in your partner's being able to both sense the joint position you placed him/her in and then reassume the same position.

9. Stand up straight on one leg on a flat surface with the other knee flexed slightly and not in contact with anything. Look straight ahead and attempt to maintain your balance in this position for up to 5 minutes. What do you notice happening about the muscles in your lower leg? Try this again with the knee of the leg you are standing slightly flexed. What differences do you notice? Try it again standing on a piece of thick foam. Try it in the original position with your eyes closed. Elaborate on the differences between the various attempts.

10. Hold a heavy book in your hand with your forearm supinated and elbow flexed approximately 90 degrees while standing. Have a partner suddenly place another heavy book upon the one you are holding. What is the immediate result regarding the angle of flexion in your elbow? Explain why this result occurs.

11. Sit up very straight on a table with the knees flexed 90 degrees and the feet hanging free. Maintain this position while then flexing the right hip and attempting to cross your legs to place the right leg across the left knee. Is this difficult? What tends to happen to the low back and trunk? How can you modify this activity to make it easier?

12. Determine your one-repetition maximum for a biceps curl beginning in full extension and ending in full flexion. Carry out each of the following exercises with adequate periods for recovery in between.

a. Begin with your elbow flexed 45 degrees and have a partner hand you a weight slightly heavier than your one-repetition maximum (approximately 5 pounds). Attempt to lift this through the remaining range of flexion. Can you reach full flexion? Explain your results.

b. Begin with your elbow in 90 degrees of flexion. Have your partner hand you an even slightly heavier weight than in 12a. Attempt to hold the elbow flexed in this position for 10 seconds. Can you do so? Explain.

c. Begin with your elbow in full flexion. Have your partner hand you an even slightly heavier weight than in 12b. Attempt to slowly lower the weight under control till you reach full extension. Can you do this? Explain.

13. With the wrist in neutral, extend the fingers maximally and attempt to maintain the position and then extend the wrist maximally. What happens to the fingers and why?

14. Maximally flex your fingers around a pencil with your wrist in neutral. Maintain the maximal finger flexion while you allow a partner to grasp your forearm with one hand and use his/her other hand to push your wrist into maximal flexion. Can you maintain control of the pencil? Explain.

15. You are walking in a straight line down the street when a stranger bumps into you. You stumble but "catch" your balance. Using the information from this chapter and other resources, explain what just happened.

16. Drinking a glass of water is a normal daily activity in which the mind and body are involved in the controlled task. Explain how the movements happen once your mind indicates that you are thirsty, regarding the nerve roots, muscle contractions, and angle of pull.

References

Bernier MR: Perturbation and agility training in the rehabilitation for soccer athletes, *Athletic Therapy Today* 8(3):20–22, 2003.

Blackburn T, Guskiewicz KM, Petschauer MA, Prentice WE: Balance and joint stability: the relative contributions of proprioception and muscular strength, *Journal of Sport Rehabilitation* 9(4):315–328, 2000.

Carter AM, Kinzey SJ, Chitwood LF, Cole JL: Proprioceptive neuromuscular facilitation decreases muscle activity during the stretch reflex in selected posterior thigh muscles, *Journal of Sport Rehabilitation* 9(4):269–278, 2000.

Dover G, Powers ME: Reliability of joint position sense and force-reproduction measures during internal and external rotation of the shoulder, *Journal of Athletic Training* 38(4):304–310, 2003.

Hall SJ: *Basic biomechanics,* ed 4, New York, 2003, McGraw-Hill.

Hamill J, Knutzen KM: *Biomechanical basis of human movement,* ed 2, Baltimore, 2003, Lippincott Williams & Wilkins.

Hamilton N, Luttgens K: *Kinesiology: scientific basis of human motion,* ed 10, Boston, 2002, McGraw-Hill.

Knight KL, Ingersoll CD, Bartholomew J: Isotonic contractions might be more effective than isokinetic contractions in developing muscle strength, *Journal of Sport Rehabilitation* 10(2):124–131, 2001.

Kreighbaum E, Barthels KM: *Biomechanics: a qualitative approach for studying human movement,* ed 4, Boston, 1996, Allyn & Bacon.

Lindsay DT: *Functional human anatomy,* St. Louis, 1996, Mosby.

Logan GA, McKinney WC: *Anatomic kinesiology,* ed 3, Dubuque, IA, 1982, Brown.

McArdle WD, Katch FI, Katch VI: *Exercise physiology: energy nutrition, and human performance,* ed 5, Baltimore, 2001, Lippincott Williams & Wilkins.

McCrady BJ, Amato HK: Functional strength and proprioception testing of the lower extremity, *Athletic Therapy Today* 9(5):60–61, 2005.

Myers JB, Guskiewicz KM, Schneider, RA, Prentice WE: Proprioception and neuromuscular control of the shoulder after muscle fatigue, *Journal of Athletic Training* 34(4):362–367, 1999.

National Strength and Conditioning Association; Baechle TR, Earle RW: *Essentials of strength training and conditioning,* ed 2, Champaign, IL, 2000, Human Kinetics.

Neumann DA: *Kinesiology of the musculoskeletal system: foundations for physical rehabilitation,* St. Louis, 2002, Mosby.

Norkin CC, Levangie PK: *Joint structure and function—a comprehensive analysis,* Philadelphia, 1983, Davis.

Northrip JW, Logan GA, McKinney WC: *Analysis of sport motion,* ed 3, Dubuque, IA, 1983, Brown.

Olmsted-Kramer LC, Hertel J: Preventing recurrent lateral ankle sprains: an evidence-based approach, *Athletic Therapy Today* 9(2):19–22, 2004.

Powers ME, Buckley BD, Kaminski TW, Hubbard TJ, Ortiz C: Six weeks of strength and proprioception training does not affect muscle fatigue and static balance in functional ankle stability, *Journal of Sport Rehabilitation* 13(3):201–227, 2004.

Powers SK, Howley ET: *Exercise physiology: theory and application of fitness and performance,* ed 5, New York, 2004, McGraw-Hill.

Rasch PJ: *Kinesiology and applied anatomy,* ed 7, 1989, Philadelphia, Lea & Febiger.

Riemann BL, Lephart SM: The sensorimotor system, part I: the physiological basis of functional joint stability, *Journal of Athletic Training* 37(1):71–79, 2002.

Riemann BL, Lephart SM: The sensorimotor system, part II: the role of proprioception in motor control and functional joint stability, *Journal of Athletic Training* 37(1):80–84, 2002.

Riemann BL, Myers JB, Lephart SM: Sensorimotor system measurement techniques, *Journal of Athletic Training* 37(1):85–98, 2002.

Riemann BL, Tray NC, Lephart SM: Unilateral multiaxial coordination training and ankle kinesthesia, muscle strength, and postural control, *Journal of Sport Rehabilitation* 12(1):13–30, 2003.

Ross S, Guskiewicz K, Prentice W, Schneider R, Yu B: Comparison of biomechanical factors between the kicking and stance limbs, *Journal of Sport Rehabilitation* 13(2):135–150, 2004.

Sandrey MA: Using eccentric exercise in the treatment of lower extremity tendinopathies, *Athletic Therapy Today* 9(1):58–59, 2004.

Seeley RR, Stephens TD, Tate P: *Anatomy & physiology,* ed 7, New York, 2006, McGraw-Hill.

Shier D, Butler J, Lewis R: *Hole's essentials of human anatomy and physiology,* ed 9, New York, 2006, McGraw-Hill.

Shier D, Butler J, Lewis R: *Hole's human anatomy and physiology,* ed 9, New York, 2002, McGraw-Hill.

Van De Graaff KM: *Human anatomy,* ed 6, New York, 2002, McGraw-Hill.

Van De Graaf KM, Fox SI, LaFleur KM: *Synopsis of human anatomy & physiology,* Dubuque, IA, 1997, Brown.

Yaggie J, Armstrong WJ: Effects on lower extremity fatigue on indices of balance, *Journal of Sport Rehabilitation* 10(2):124–131, 2004.

Chapter 3

Basic Biomechanical Factors and Concepts

Objectives

- To know and understand how knowledge of levers can help improve physical performance

- To know and understand how the musculoskeletal system functions as a series of simple machines

- To know and understand how knowledge of torque and lever arm lengths can help improve physical performance

- To know and understand how knowledge of Newton's laws of motion can help improve physical performance

- To know and understand how knowledge of balance, equilibrium, and stability can help improve physical performance

- To know and understand how knowledge of force and momentum can help improve physical performance

- To know and understand the basic effects of mechanical loading on the tissues of the body

In Chapter 1 we defined kinesiology, very simply, as the study of muscles as they are involved in the science of movement. From this general definition we can go into greater depth into the science of body movement, which primarily includes anatomy, physiology, and mechanics. To truly understand movement, a vast amount of knowledge is needed in all three areas. The focus of this text is primarily that of structural and functional anatomy, and we have only very minimally touched on some physiology in the first two chapters. A much greater study of physiology as it relates to movement should be addressed in an exercise physiology course for which there are many excellent texts and resources. Likewise, the study of mechanics as it relates to the functional and anatomical analysis of biological systems is known as **biomechanics** and should be addressed to a greater degree in a separate course. Human movement is quite complex. In order to make recommendations for its improvement, we need to study movements from a biomechanical perspective, both qualitatively and quantitatively. This chapter is included to serve as an introduction to some of the basic biomechanical factors and concepts with the understanding that many readers will subsequently study these more in depth in a dedicated course utilizing much more thorough resources.

Many students in kinesiology classes have some knowledge, from a college or high school physics course, of the laws that affect motion. These principles and others will be discussed briefly in this chapter, which should prepare you as you begin to apply them to motion in the human body. The more you can put these principles and concepts into practical applications, the easier it will be to understand them.

Mechanics, the study of physical actions of forces, can be subdivided into **statics** and **dynamics**. Statics involves the study of systems that are in a constant state of motion, whether at rest with no motion or moving at a constant velocity without acceleration. Statics involves all forces acting on the body being in balance, resulting in the body being in equilibrium. Dynamics involves the study of systems in motion with acceleration. A system in acceleration is unbalanced due to unequal forces acting on the body. Additional components of biomechanical study include **kinematics** and **kinetics**. Kinematics is concerned with the description of motion and includes consideration of time, displacement, velocity, acceleration, and space factors of a system's motion. Kinetics is the study of forces associated with the motion of a body.

Types of machines found in the body

As discussed in Chapter 2, we utilized muscles to apply force to the bones upon which they attach to cause, control, or prevent movement in the joints that they cross. As is often the case, we utilize the bones such as those in the hand to either hold, push, or pull on an object while we use a series of bones and joints throughout the body to apply force via the muscles to affect the position of the object. In doing so we are using a series of simple machines to accomplish the tasks. Machines are used to increase or multiply the applied force in performing a task or to provide a **mechanical advantage**. The mechanical advantage provided by machines enables us to apply a relatively small force, or effort, to move a much greater resistance. We can determine mechanical advantage by dividing the load by the effort. The mechanical aspect of each component should be considered in analysis with respect to the components' machinelike function.

Another way of thinking about machines is that they convert smaller amounts of force over a longer distance to large amounts of force exerted over a shorter distance. This may be turned around so that a larger amount of force exerted over a shorter distance converts to a smaller amount of force over a greater distance. Machines function in four ways:

1. To balance multiple forces
2. To enhance force in an attempt to reduce the total force needed to overcome a resistance
3. To enhance range of motion and speed of movement so that resistance may be moved farther or faster than the applied force
4. To alter the resulting direction of the applied force

Simple machines are the lever, wheel and axle, pulley, inclined plane, screw, and the wedge. The arrangement of the musculoskeletal system provides for three types of machines in producing movement: levers, wheel/axles, and pulleys. Each of these involves a balancing of rotational forces about an axis. The lever is the most common form of simple machine found in the human body.

Levers

It may be difficult for a person to visualize his or her body as a system of levers, but this is actually the case. Human movement occurs through the organized use of a system of levers. The anatomical levers of the body cannot be changed, but, when the system is properly understood, it can be used more efficiently to maximize the muscular efforts of the body.

A lever is defined as a rigid bar that turns about an **axis** of rotation, or fulcrum. The axis is the point of rotation about which the lever moves.

The lever rotates about the axis as a result of **force** (sometimes referred to as effort, E) being applied to it to cause its movement against a **resistance** or weight. In the body, the bones represent the bars, the joints are the axes, and the muscles contract to apply the force. The amount of resistance can vary from maximal to minimal. In fact, the bones themselves or weight of the body segment may be the only resistance applied. All lever systems have each of these three components in one of three possible arrangements.

The arrangement or location of these three points in relation to one another determines the type of lever and the application for which they are best suited. These points are the axis, the point of force application (usually the muscle insertion), and the point of resistance application (sometimes the center of gravity of the lever and sometimes the location of an external resistance). When the axis (A) is placed between the force (F) and the resistance (R), a first-class lever is produced (Figs. 3.1 and 3.2). In second-class levers, the resistance is between the axis and the force (Figs. 3.1 and 3.3). If the force is placed between the axis and the resistance, a third-class lever is created (Figs. 3.1 and 3.4). Table 3.1 provides a summary of the three classes of levers and the characteristics of each.

The mechanical advantage of levers may be determined using the following equations:

$$\text{Mechanical advantage} = \frac{\text{resistance}}{\text{force}}$$

or

$$\text{Mechanical advantage} = \frac{\text{length of force arm}}{\text{length of resistance arm}}$$

First-class levers

Typical examples of a first-class lever are the crowbar, the seesaw, pliers, oars, and the triceps in overhead elbow extension. An example of this type of lever in the body is seen with the triceps applying the force to the olecranon (F) in extending the nonsupported forearm (R) at the elbow

(A). Other examples of this type of lever may be seen in the body when the agonist and antagonist muscle groups on either side of a joint axis are contracting simultaneously, with the agonist producing force and the antagonist supplying the resistance. A first-class lever (Fig. 3.2) is designed basically to produce balanced movements when the axis is midway between the force and the resistance (e.g., a seesaw). When the axis is close to the force, the lever produces speed and range of motion (e.g., the triceps in elbow extension). When the axis is close to the resistance, the lever produces force motion (e.g., a crowbar).

In applying the principle of levers to the body, it is important to remember that the force

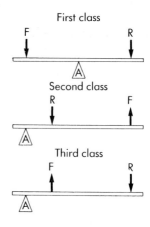

FIG. 3.1 ● Classification of levers.

Modified from Hall SJ: *Basic biomechanics,* ed 4, 2003, McGraw-Hill.

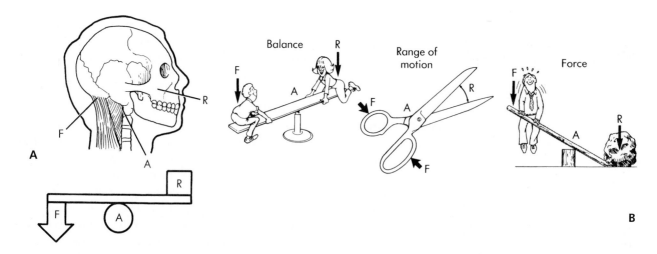

FIG. 3.2 ● **A** and **B,** First-class levers.

A modified from Booher JM, Thibodeau GA: *Athletic injury assessment,* ed 4, 2000, McGraw-Hill;
B modified from Hall SJ: *Basic biomechanics,* ed 4, 2003, McGraw-Hill.

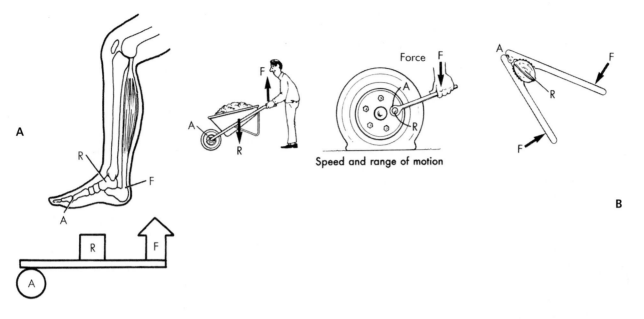

FIG. 3.3 ● **A** and **B,** Second-class levers.

A modified from Booher JM, Thibodeau GA: *Athletic injury assessment,* ed 4, 2000, McGraw-Hill;
B modified from Hall SJ: *Basic biomechanics,* ed 4, 2003, McGraw-Hill.

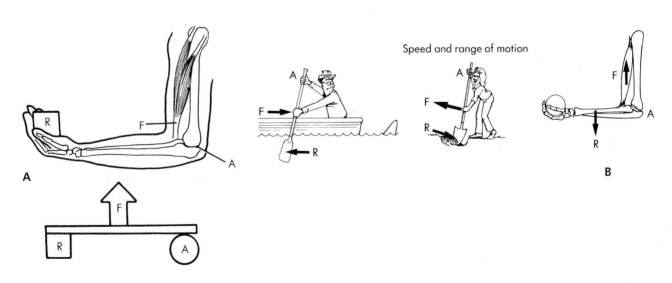

FIG. 3.4 ● **A** and **B,** Third-class levers.

A modified from Booher JM, Thibodeau GA: *Athletic injury assessment,* ed 4, 2000, McGraw-Hill;
B modified from Hall SJ: *Basic biomechanics,* ed 4, 2003, McGraw-Hill.

TABLE 3.1 • Classification of levers and characteristics of each

Class	Arrangement	Arm movement	Functional design	Relationship to axis	Practical example	Human example
1st	**F–A–R** Axis between force and resistance	Resistance arm and force arm move in opposite directions	Balanced movements	Axis near middle	Seesaw	Erector spinae extending the head on cervical spine
			Speed and range of motion	Axis near force	Scissors	Triceps brachii in extending the elbow
			Force motion	Axis near resistance	Crowbar	
2nd	**A–R–F** Resistance between axis and force	Resistance arm and force arm move in the same direction	Force motion (large resistance can be moved with relatively small force)	Axis near resistance	Wheelbarrow, nutcracker	Gastrocnemius and soleus in plantar flexing the foot to raise the body on the toes
3rd	**A–F–R** Force between axis and resistance	Resistance arm and force arm move in the same direction	Speed and range of motion (requires large force to move a relatively small resistance)	Axis near force	Shoveling dirt, catapult	Biceps brachii and brachialis in flexing the elbow

is applied where the muscle inserts in the bone, not in the belly of the muscle. For example, in elbow extension with the shoulder fully flexed and the arm beside the ear, the triceps applies the force to the olecranon of the ulna behind the axis of the elbow joint. As the applied force exceeds the amount of forearm resistance, the elbow extends.

The type of lever may be changed for a given joint and muscle, depending on whether the body segment is in contact with a surface such as a floor or wall. For example, we have demonstrated that the triceps in elbow extension is a first-class lever with the hand free in space where the arm is pushed away from the body. By placing the hand in contact with the floor, as in performing a push-up to push the body away from the floor, the same muscle action at this joint now changes the lever to second class, because the axis is at the hand and the resistance is the body weight at the elbow joint.

Second-class levers

A second-class lever (Fig. 3.3) is designed to produce force movements, since a large resistance can be moved by a relatively small force. Examples of second-class levers include a bottle opener, wheelbarrow, and a nutcracker. Besides the example given previously of the triceps ex-

tending the elbow in a push-up, a similar example of a second-class lever in the body is plantar flexion of the ankle to raise the body up on the toes. The ball (A) of the foot serves as the axis of rotation as the ankle plantar flexors apply force to the calcaneus (F) to lift the resistance of the body at the tibiofibular articulation (R) with the talus. Opening the mouth against resistance provides another example of a second-class lever. There are relatively few other occurrences of second-class levers in the body.

Third-class levers

Third-class levers (Fig. 3.4), with the force being applied between the axis and the resistance, are designed to produce speed and range-of-motion movements. Most of the levers in the human body are of this type, which require a great deal of force to move even a small resistance. Examples include a catapult, a screen door operated by a short spring, and application of lifting force to a shovel handle with the lower hand while the upper hand on the shovel handle serves as the axis of rotation. The biceps brachii is a typical example in the body. Using the elbow joint (A) as the axis, the biceps brachii applies force at its insertion on the radial tuberosity (F) to rotate the forearm up, with its center of gravity (R) serving as the point of resistance application.

The brachialis is an example of true third-class leverage. It pulls on the ulna just below the elbow, and, since the ulna cannot rotate, the pull is direct and true. The biceps brachii, on the other hand, supinates the forearm as it flexes, so that the third-class leverage applies to flexion only.

Other examples include the hamstrings contracting to flex the leg at the knee while in a standing position and the iliopsoas when it is used to flex the thigh at the hip.

Factors in use of anatomical levers

Our anatomical leverage system can be used to gain a mechanical advantage that will improve simple or complex physical movements. Some individuals unconsciously develop habits of using human levers properly, but frequently this is not the case.

Torque and length of lever arms

To understand the leverage system, the concept of torque must be understood. **Torque**, or moment of force, is the turning effect of an eccentric force. **Eccentric force** is a force that is applied in a direction not in line with the center of rotation of an object with a fixed axis. In objects without a fixed axis, it is an applied force that is not in line with the object's center of gravity. For rotation to occur, an eccentric force must be applied. In the human body, the contracting muscle applies an eccentric force (not to be confused with eccentric contraction) to the bone on which it attaches and causes the bone to rotate about an axis at the joint. The amount of torque can be determined by multiplying the amount of force (**force magnitude**) by the **force arm**. The perpendicular distance between the location of force application and the axis is known as the force arm, moment arm, or torque arm. The force arm may be best understood as the shortest distance from the axis of rotation to the line of action of the force. The greater the distance of the force arm, the more torque produced by the force. A frequent practical application of torque and levers occurs when we purposely increase the force arm length in order to increase the torque so that we can more easily move a relatively large resistance. This is commonly referred to as increasing our leverage.

It is also important to note the **resistance arm**, which may be defined as the distance between the axis and the point of resistance application. In discussing the application of levers, it is necessary to understand the length relationship between the two lever arms. There is an inverse relationship between force and the force arm, just as there is between resistance and the resistance arm. The longer the force arm, the less force required to move the lever if the resistance and resistance arm remain constant as shown graphically in Fig. 3.5. In addition, if the force and force arm remain constant, a greater resistance may be moved by shortening the resistance arm.

Also, there is a proportional relationship between the force components and the resistance components. That is, for movement to occur when either of the resistance components increases, there must be an increase in one or both of the force components. See Figs. 3.6, 3.7, and 3.8 as they relate to first-, second-, and third-class levers, respectively. Even slight variations in the location of the force and the resistance are important in determining the mechanical advantage

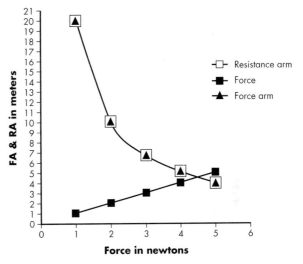

Relationship between Force, Force Arm, & Resistance Arm with Constant Resistance of 20kg

FIG. 3.5 ● Relationship between forces, force arms, and resistance arms. (The graph assumes a constant resistance of 20 Newtons and as a result the graphical representation of the resistance arm and force arm lie directly over one another.) **A,** With the resistance being held constant at 20 Newtons and a resistance arm of 1 meter, the product of the (force) × (force arm) must equal 20 Newton-meters. As a result there is an inverse relationship between the force and the force arm. As the force increases in Newtons, the force arm length decreases in meters and vice versa. **B,** If the force increases and the force arm is held constant, then the resistance arm decreases for a constant resistance of 20 Newtons.

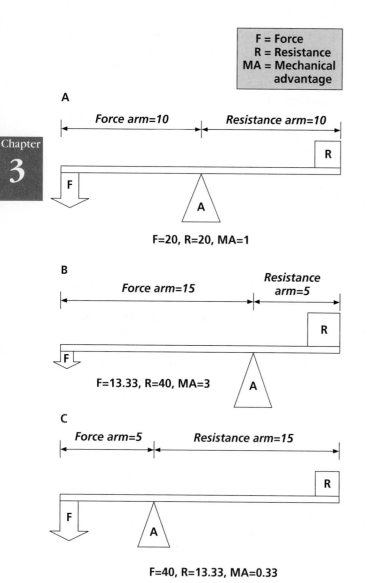

FIG. 3.6 ● First-class levers. **A,** If the force arm and resistance arm are equal in length, a force equal to the resistance is required to balance it. **B,** As the force arm becomes longer, a decreasing amount of force is required to move a relatively larger resistance. **C,** As the force arm becomes shorter, an increasing amount of force is required to move a relatively smaller resistance, but the speed and range of motion that the resistance can be moved are increased.

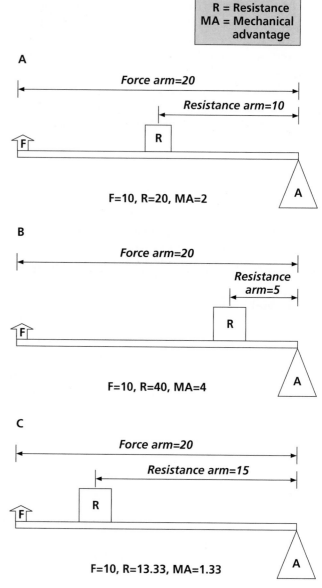

FIG. 3.7 ● Second-class levers have a positive mechanical advantage due to the force arm always being longer than the resistance arm and are well suited for moving larger resistances with smaller forces. **A,** Placing the resistance halfway between the axis and the point of force application provides a mechanical advantage of 2. **B,** Moving the resistance closer to the axis increases the mechanical advantage but decreases the distance that the resistance is moved. **C,** The closer the resistance is positioned to the point of force application, the less of a mechanical advantage, but the greater the distance it is moved.

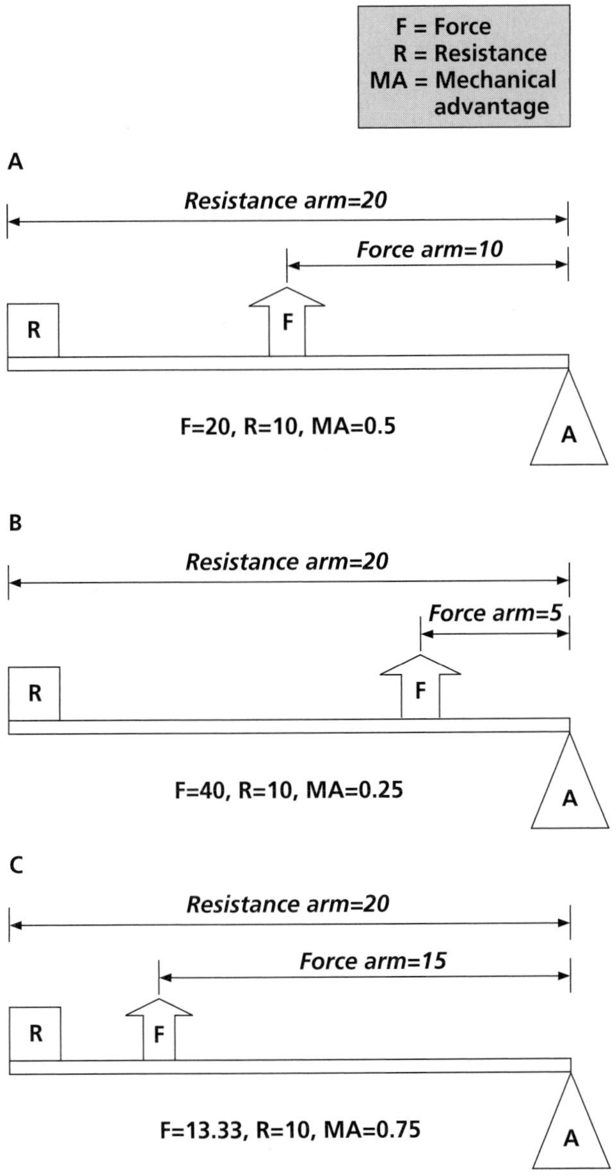

| F = Force |
| R = Resistance |
| MA = Mechanical advantage |

A

Resistance arm=20

Force arm=10

R | F

F=20, R=10, MA=0.5

A

B

Resistance arm=20

Force arm=5

R | F

F=40, R=10, MA=0.25

A

C

Resistance arm=20

Force arm=15

R | F

F=13.33, R=10, MA=0.75

A

FIG. 3.8 ● Third-class levers. **A,** A force greater than the resistance, regardless of the point of force application, is required due to the resistance arm always being longer. **B,** Moving the point of force application closer to the axis increases the range of motion and speed, but requires more force. **C,** Moving the point of force application closer to the resistance decreases the force needed, but also decreases the speed and range of motion.

and the effective force of the muscle. This point can be illustrated in the following simple formula, using the biceps brachii muscle in each example, as shown in Fig. 3.9:

Lever equation

| F | x | FA | = | R | x | RA |
| (Force) | x | (Force arm) | = | (Resistance) | x | (Resistance) |

Initial Example

$$F \times 0.1 = 45 \text{ Newtons} \times 0.25 \text{ meters}$$
$$F \times 0.1 = 11.25 \text{ Newton-meters}$$
$$F = 112.5 \text{ Newton-meters}$$

Example A – Lengthening the force arm

Increase the (FA) by moving the insertion distally 0.05 meters:

$$F \times 0.15 = 45 \text{ Newtons} \times 0.25 \text{ meters}$$
$$F \times 0.15 = 11.25 \text{ Newton-meters}$$
$$F = 75 \text{ Newton-meters}$$

An increase in the insertion from the axis by 0.05 meters results in a substantial reduction in the force necessary to move the resistance.

Example B – Shortening the resistance arm

Reduce the (RA) by moving the point of resistance application proximally by 0.05 meters:

$$F \times 0.1 = 45 \text{ Newtons} \times 0.2 \text{ meters}$$
$$F \times 0.1 = 9 \text{ Newton-meters}$$
$$F = 90 \text{ Newton-meters}$$

A decrease in the resistance application from the axis by 0.05 meters results in a considerable reduction in the force necessary to move the resistance.

Example C – Reducing the resistance

Reduce the (R) amount by reducing the resistance 1 Newton:

$$F \times 0.1 = 44 \text{ Newtons} \times 0.25 \text{ meters}$$
$$F \times 0.1 = 11 \text{ Newton-meters}$$
$$F = 110 \text{ Newton-meters}$$

Decreasing the amount of resistance can decrease the amount of force needed to move the lever.

FIG. 3.9 ● Torque calculations with examples of modifications in force arms, resistance arms, and resistance.

In Example A, we can only move the insertion of the biceps brachii with surgery, so this is not practical. In some orthopaedic conditions, the attachments of tendons are surgically relocated in an attempt to change the dynamic forces of the muscles on the joints. In Example B, we can and often do shorten the resistance arm to enhance our ability to move an object. When attempting a maximal weight in a biceps curl exercise, we may flex our wrist to move the weight just a little closer, which shortens the resistance arm. Example C is straightforward in that we can obviously reduce the force needed by reducing the resistance.

The system of leverage in the human body is built for speed and range of movement at the expense of force. Short force arms and long resistance arms require great muscular strength to produce movement. In the forearm, the attachments of the biceps and triceps muscles clearly illustrate this point, since the force arm of the biceps is 1 to 2 inches and that of the triceps less than 1 inch. Many other similar examples are found all over the body. From a practical point of view, this means that the muscular system should be strong to supply the necessary force for body movements, especially in strenuous sports activities.

When we speak of human leverage in relation to sport skills, we are generally referring to several levers. For example, in throwing a ball, there are levers at the shoulder, elbow, and wrist joints as well as from the ground up through the lower extremities and the trunk.

In fact, it can be said that there is one long lever from the feet to the hand.

The longer the lever, the more effective it is in imparting velocity. A tennis player can hit a tennis ball harder (deliver more force to it) with a straight-arm drive than with a bent elbow because the lever (including the racket) is longer and moves at a faster speed.

Fig. 3.10 indicates that a longer lever (Z') travels faster than a shorter lever (S') in traveling the same number of degrees. In sports activities in which it is possible to increase the length of a lever with a racket or bat, the same principle applies.

In baseball, hockey, golf, field hockey, and other sports, long levers similarly produce more linear force and thus better performance. However, to be able to fully execute the movement in as short a time as possible, it is sometimes desirable to have a short lever arm. For example, a baseball catcher attempting to throw a runner out

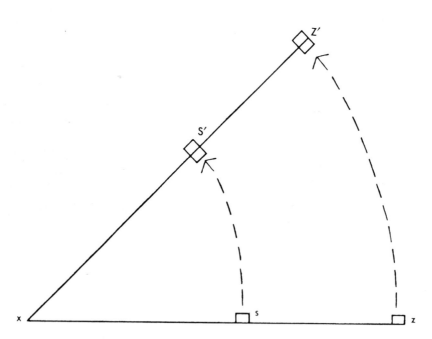

FIG. 3.10 ● Length of levers. The end of a longer lever (Z') travels faster than a shorter lever (S') when moved over the same degrees in the same amount of time. The farther the resistance is from the axis, the farther it is moved and the more force delivered to it, resulting in the resistance moving at a greater velocity. In sports activities in which it is possible to increase the length of a lever with a racket or bat, the same principle applies.

at second base does not have to throw the ball so that it travels as fast as it may when the pitcher is attempting to throw a strike. In the catcher's case, it is more important to initiate and complete the throw as soon as possible rather than deliver as much velocity to the ball as possible. The pitcher, when attempting to throw a ball at 90-plus miles per hour, will utilize his body as a much longer lever system throughout a greater range of motion to impart velocity to the ball.

Wheels and axles

Wheels and axles are used primarily to enhance range of motion and speed of movement in the musculoskeletal system. A wheel and an axle essentially function as a form of a lever. When either the wheel or the axle turns, the other must turn as well. Both complete one turn at the same time. The center of the wheel and the axle both correspond to the fulcrum. Both the radius of the wheel and the radius of the axle correspond to the force arms. If the radius of the wheel is greater than the radius of the axle, then, due to the longer force arm, the wheel has a mechanical advantage over the axle. That is, a relatively smaller force may be applied to the wheel to move a relatively greater resistance applied to the axle. Very simply, if the radius of the wheel is three times the radius of the axle, then the wheel has a 3 to 1 mechanical advantage over the axle, as in Fig. 3.11. The mechanical advantage of a wheel and axle for this scenario may be calculated by considering the radius of the wheel over that of the axle.

$$\text{Mechanical advantage} = \frac{\text{radius of the wheel}}{\text{radius of the axle}}$$

In this case the mechanical advantage is always more than 1. An application of this example is using the outer portion of an automobile steering wheel to turn the steering mechanism. Previously, before the development of power steering, steering wheels had a much larger diameter than today in order to enable the driver to have more of a mechanical advantage. An example of applying force to the wheel in the body is when we attempt to manually force a person's shoulder into internal rotation while they hold it in external rotation isometrically. The humerus acts as the axle and the person's hand and wrist are located near the outside of the wheel when the elbow is flexed approximately 90 degrees. If we unsuccessfully attempt to break the force of the contraction of their external rotators by pushing internally at their mid-forearm, we can increase our leverage or mechanical advantage and our likelihood of success by applying our force nearer to their hand and wrist.

If the application of force is reversed so that it is applied to the axle, then the mechanical advantage results from the wheel's turning a greater distance and speed. Using the same example, if the wheel radius is three times greater than the radius of the axle, the outside of the wheel will turn at a speed three times that of the axle. Additionally, the distance that the outside of the wheel turns will be three times that of the outside of the axle. The mechanical advantage of a wheel and axle for this scenario may be calculated by considering the radius of the axle over the radius of the wheel.

$$\text{Mechanical advantage} = \frac{\text{radius of the axle}}{\text{radius of the wheel}}$$

In this case the mechanical advantage is always less than 1. This is the principle utilized in the drive train of an automobile to turn the axle, which subsequently turns the tire around one revolution for every turn of the axle. We utilize the powerful engine of the automobile to supply the force to increase the speed of the tire and subsequently carry us great distances. An example of the muscles applying force to the axle to result in greater range of motion and speed may again be seen in the upper extremity, in the case of the internal rotators attaching to the humerus. With the humerus acting as the axle and the hand and wrist located at the outside of the wheel (when the elbow is flexed approximately 90 degrees), the internal rotators apply force to the humerus. With the internal rotators concentrically internally rotating the humerus a relatively small amount, the hand and wrist travel a great distance. Using the wheel and axle in this manner allows us to significantly increase the speed at which we can throw objects.

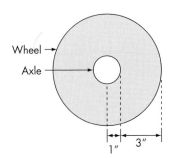

FIG. 3.11 ● Wheel and axle.

Pulleys

Single pulleys have a fixed axle and function to change the effective direction of force application. Single pulleys have a mechanical advantage of 1 as shown in Fig. 3.12, *A*. Numerous weight machines utilize pulleys to alter the direction of the resistive force. Pulleys may be movable and combined to form compound pulleys to further increase the mechanical advantage. Every additional rope connecting to movable pulleys increases the mechanical advantage by 1 as shown in Fig. 3.12, *B*.

In the human body, an excellent example is provided by the lateral malleolus, acting as a pulley around which the tendon of the peroneus longus runs. As this muscle contracts, it pulls toward its belly, which is toward the knee. Due to its use of the lateral malleolus as a pulley (Fig. 3.13), the force is transmitted to the plantar aspect of the foot; this results in downward and outward movement of the foot. Other examples in the human body include pulleys on the volar aspect of the phalanges to redirect the force of the flexor tendons.

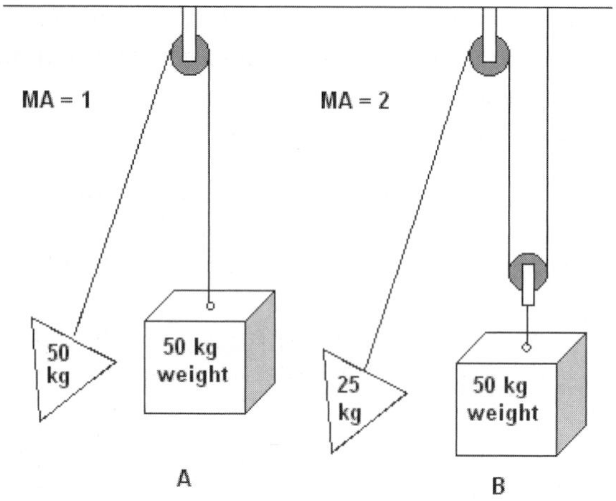

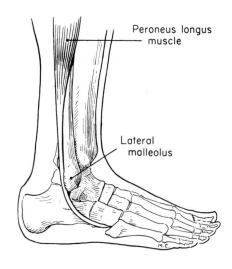

FIG. 3.12 ● **A,** Single pulley; **B,** Compound movable pulley.

FIG. 3.13 ● Pulley. The lateral malleolus serving as a pulley for the peroneus longus tendon.

From Hamilton N, Luttgens K: *Kinesiology: scientific basis of human motion,* ed 10, 2002, McGraw-Hill.

Laws of motion and physical activities

Motion is fundamental in physical education and sports activity. Body motion is generally produced, or at least started, by some action of the muscular system. Motion cannot occur without a force, and the muscular system is the source of force in the human body. Thus, development of the muscular system is indispensable to movement.

Basically, there are two types of motion: **linear motion** and **angular motion**. Linear motion, also referred to as translatory motion, is motion along a line. If the motion is along a straight line, it is **rectilinear** motion, whereas motion along a curved line is known as **curvilinear** motion. Angular motion, also known as rotary motion, involves rotation around an axis. In the human body, the axis of rotation is provided by the various joints. In a sense, these two types of motion are related, since angular motion of the joints can produce the linear motion of walking.

For example, in many sports activities, the cumulative angular motion of the joints of the body imparts linear motion to a thrown object (ball, shot) or to an object struck with an instrument (bat, racket).

Displacement refers to a change in position or location of an object from its original point of reference, whereas **distance**, or the path of movement, refers to the actual sum length of measurement traveled. An object may have traveled a distance of 10 meters along a linear path in two or more directions but be displaced from its original reference point by only 6 meters. Fig. 3.14 provides an example. **Angular displacement** refers to the change in location of a rotating body. **Linear displacement** is the distance that a system moves in a straight line.

We are sometimes concerned about the time it takes for displacement to occur. **Speed** is how fast an object is moving, or the distance an object travels in a specific amount of time. **Velocity** includes the direction and describes the rate of displacement.

A brief review of Newton's laws of motion will indicate the many applications of these laws to physical education activities and sports. Newton's laws explain all the characteristics of motion, and they are fundamental to understanding human movement.

Law of inertia

A body in motion tends to remain in motion at the same speed in a straight line unless acted on by a force; a body at rest tends to remain at rest unless acted on by a force.

Inertia may be described as the resistance to action or change. In terms of human movement, *inertia* refers to resistance to acceleration or deceleration. Inertia is the tendency for the current state of motion to be maintained, regardless of whether the body segment is moving at a particular velocity or is motionless.

Muscles produce the force necessary to start motion, stop motion, accelerate motion, decelerate motion, or change the direction of motion. Put another way, inertia is the reluctance to change status; only force can do it. The greater the mass of an object, the greater its inertia. Therefore, the greater the mass, the more force needed to significantly change an object's inertia. Numerous examples of this law are found in physical education activities. A sprinter in the starting blocks must apply considerable force to overcome resting inertia. A runner on an indoor track must apply considerable force to overcome moving inertia and stop before hitting the wall. Fig. 3.15 provides an example of how a skier in motion remains in motion, even though airborne after skiing off a hill. Balls and other objects that are thrown or struck require force to stop them. Starting, stopping, and changing direction—a part of many physical activities—provide many other examples of the law of inertia applied to body motion.

Since force is required to change inertia, it is obvious that any activity that is carried out at a steady pace in a consistent direction will conserve

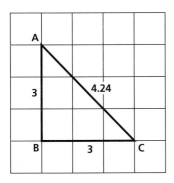

FIG. 3.14 ● Displacement. If the path of movement is from A to B, then from B to C, the distance covered is AB + BC, but the displacement would be AC. If each cell is 1 square meter, then AB is 3 meters and BC is 3 meters, so the distance covered would be 6 meters, but the displacement, AC, would be 4.24 meters.

FIG. 3.15 ● An example of Newton's first law of motion.

From Hamilton N, Luttgens K: *Kinesiology: scientific basis of human motion,* ed 10, 2002, McGraw-Hill.

energy, and that any irregularly paced or directed activity will be very costly to energy reserves. This explains in part why activities such as handball and basketball are so much more fatiguing than jogging or dancing.

Law of acceleration

A change in the acceleration of a body occurs in the same direction as the force that caused it. The change in acceleration is directly proportional to the force causing it and inversely proportional to the mass of the body.

Acceleration may be defined as the rate of change in velocity. To attain speed in moving the body, a strong muscular force is generally necessary. **Mass**, the amount of matter in a body, affects the speed and acceleration in physical movements. A much greater force is required from the muscles to accelerate an 80-kilogram man than to accelerate a 58-kilogram man to the same running speed. Also, it is possible to accelerate a baseball faster than a shot because of the difference in weight. The force required to run at half speed is less than the force required to run at top speed. To impart speed to a ball or an object, it is necessary to rapidly accelerate the part of the body holding the object. Football, basketball, track, and field hockey are a few sports that demand speed and acceleration.

Law of reaction

For every action there is an opposite and equal reaction.

As we place force on a supporting surface by walking over it, the surface provides an equal resistance back in the opposite direction to the soles of our feet. Our feet push down and back, while the surface pushes up and forward. The force of the surface reacting to the force we place on it is referred to as **ground reaction force**. We provide the action force, while the surface provides the reaction force. It is easier to run on a hard track than on a sandy beach because of the difference in the ground reaction forces of the two surfaces. The track resists the runner's propulsion force, and the reaction drives the runner ahead. The sand dissipates the runner's force, and the reaction force is correspondingly reduced, with the apparent loss in forward force and speed (Fig. 3.16). A sprinter applies a force in excess of 1335 Newtons on the starting blocks, which resist with an equal force. When a body is in flight, as it is in jumping, movement of one part of the body produces a reaction in another part. This occurs because there is no resistive surface to supply a reaction force.

FIG. 3.16 ● An example of Newton's third law of motion. To accelerate forward, a walker must push backward. Notice how the front part of the footprint in the sand is more deeply depressed than the rear.

From Hamilton N, Luttgens K: *Kinesiology: scientific basis of human motion,* ed 10, 2002, McGraw-Hill.

Friction

Friction is the force that results from the resistance between the surfaces of two objects moving upon one another. Depending upon the activity involved, we may desire increased or decreased friction. In running, we depend upon friction forces between our feet and the ground so that we may exert force against the ground and propel ourselves forward. When friction is reduced due to a slick ground or shoe surface, we are more likely to slip. In skating, we desire decreased friction so that we may slide across the ice with less resistance. Friction may be further characterized as either static friction or kinetic friction. **Static friction** refers to the amount of friction between two objects that have not yet begun to move, whereas **kinetic friction** is the friction occurring between two objects that are sliding upon one another. Static friction is always greater than kinetic friction. As a result, it is always more difficult to initiate dragging an object across a surface than it is to continue. Static friction may be increased by increasing the normal or perpendicular forces pressing the two objects together, as in adding more weight to one object sitting on the other object. To determine the amount of friction forces, we must consider both the forces pressing the two objects together and the **coefficient of friction**, which depends upon the hardness and roughness of the surface textures. The coefficient of friction is the ratio between the force needed to overcome the friction over the force holding the surfaces together. **Rolling friction** is the resistance to an object rolling across a surface, such as a ball rolling across a court or a tire rolling across the ground. Rolling friction is always much less than static or kinetic friction.

Balance, equilibrium, and stability

Balance is the ability to control equilibrium, either static or dynamic. In relation to human movement, **equilibrium** refers to a state of zero acceleration where there is no change in the speed or direction of the body. Equilibrium may be either static or dynamic. If the body is at rest or completely motionless, it is in **static equilibrium**. **Dynamic equilibrium** occurs when all of the applied and inertial forces acting on the moving body are in balance, resulting in movement with unchanging speed or direction. For us to control equilibrium and, hence, achieve balance, we need to maximize **stability**. Stability is the resistance to a change in the body's acceleration or, more appropriately, the resistance to a disturbance of the body's equilibrium. Stability may be enhanced by determining the body's **center of gravity** and changing it appropriately. The center of gravity is the point at which all of the body's mass and weight is equally balanced or equally distributed in all directions.

Balance is important for the resting body, as well as for the moving body. Generally, balance is to be desired, but there are circumstances in which movement is improved when the body tends to be unbalanced. Following are certain general factors that apply toward enhancing equilibrium, maximizing stability, and ultimately achieving balance:

1. A person has balance when the center of gravity falls within the base of support.
2. A person has balance in direct proportion to the size of the base. The larger the base of support, the more balance.
3. A person has balance depending on the weight (mass). The greater the weight, the more balance.
4. A person has balance depending on the height of the center of gravity. The lower the center of gravity, the more balance.
5. A person has balance depending on where the center of gravity is in relation to the base of support. The balance is less if the center of gravity is near the edge of the base. However, when anticipating an oncoming force, stability may be improved by placing the center of gravity nearer the side of the base of support expected to receive the force.
6. In anticipation of an oncoming force, stability may be increased by enlarging the size of the base of support in the direction of the anticipated force.
7. Equilibrium may be enhanced by increasing the friction between the body and the surfaces it contacts.
8. Rotation about an axis aids balance. A moving bike is easier to balance than a stationary bike.
9. Kinesthetic physiological functions contribute to balance. The semicircular canals of the inner ear, vision, touch (pressure), and kinesthetic sense all provide balance information to the performer. Balance and its components of equilibrium and stability are essential in all movements. They are all affected by the constant force of gravity, as well as by inertia. Walking has been described as an activity in which a person throws the body in and out of balance with each step. In rapid running movements in which moving inertia is

high, the individual has to lower the center of gravity to maintain balance when stopping or changing direction. Conversely, in jumping activities, the individual attempts to raise the center of gravity as high as possible.

Force

Muscles are the main source of force that produces or changes movement of a body segment, the entire body, or an object thrown, struck, or stopped. Strong muscles are able to produce more force than weak muscles. This refers to both maximum and sustained exertion over a period of time.

Forces either push or pull on an object in an attempt to affect motion or shape. Without forces acting on an object, there is no motion. Force is the product of mass times acceleration. The mass of a body segment or the entire body times the speed of acceleration determines the force. Obviously, in football this is very important, yet it is just as important in other activities that use only a part of the human body. When one throws a ball, the force applied to the ball is equal to the mass of the arm times the arm's speed of acceleration. Also, as previously discussed, leverage is important.

$$\text{Force} = \text{mass} \times \text{acceleration}$$
$$F = m \times a$$

The quantity of motion or, more scientifically stated, the **momentum**, which is equal to mass times velocity, is important in skill activities. The greater the momentum, the greater the resistance to change in the inertia or state of motion.

It is not necessary to apply maximum force and thus increase the momentum of a ball or an object being struck in all situations. In skillful performance, regulation of the amount of force is necessary. Judgment as to the amount of force required to throw a softball a given distance, hit a golf ball 200 yards, or hit a tennis ball across the net and into the court is important.

In activities involving movement of various joints, as in throwing a ball or putting a shot, there should be a summation of forces from the beginning of movement in the lower segment of the body to the twisting of the trunk and movement at the shoulder, elbow, and wrist joints. The speed at which a golf club strikes the ball is the result of a summation of forces of the lower extremeties, trunk, shoulders, arms, and wrists. Shot-putting and discus and javelin throwing are other good examples that show that summation of forces is essential.

Mechanical loading basics

As we utilize the musculoskeletal system to exert force on the body to move and to interact with the ground and other objects or people, significant mechanical loads are generated and absorbed by the tissues of the body. The forces causing these loads may be internal or external. Only muscles can actively generate internal force, but tension in tendons, connective tissues, ligaments, and joints capsules may generate passive internal forces. External forces are produced from outside the body and originate from gravity, inertia, or direct contact. All tissues, in varying degrees, resist changes in their shape. Obviously, tissue deformation may result from external forces, but it must be realized that we also have the ability to generate internal forces large enough to fracture bones, dislocate joints, and disrupt muscles and connective tissues. To prevent injury or damage from tissue deformation, we must use the body to absorb energy from both internal and external forces. Along this line, it is to our advantage to absorb this force over larger aspects of our body rather than smaller ones, and to spread the absorption rate over a greater period of time. Additionally, the stronger and healthier we are, the more likely we are able to withstand excessive mechanical loading and the resultant excessive tissue deformation. Tension (stretching or strain), compression, shear, bending, and torsion (twisting) are all forces that act individually or in combination to provide mechanical loading that may result in excessive tissue deformation. Fig. 3.17 illustrates the mechanical forces that act on tissues of the body.

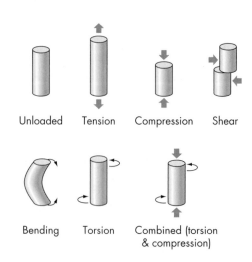

Unloaded   Tension   Compression   Shear

Bending   Torsion   Combined (torsion & compression)

FIG. 3.17 ● Mechanical loading forces.

Throwing

In the performance of various sport skills, many applications of the laws of leverage, motion, and balance may be found. A skill common to many activities is throwing. The object thrown may be some type of ball, but it is frequently an object of another size or shape, such as a rock, beanbag, Frisbee, discus, or javelin. A brief analysis of some of the basic mechanical principles involved in the skill of throwing will help indicate the importance of understanding the applications of these principles. Many activities involve these and often other mechanical principles. Motion is basic to throwing when the angular motion (Fig. 2.16) of the levers (bones) of the body (trunk, shoulder, elbow, and wrist) is used to give linear motion to the ball when it is released.

Newton's laws of motion apply in throwing because the individual's inertia and the ball's inertia (p. 77) must be overcome by the application of force. The muscles of the body provide the force to move the body parts and the ball held in the hand. The **law of acceleration** (Newton's second law) comes into operation with the muscular force necessary to accelerate the arm, wrist, and hand. The greater the force (mass times acceleration) that a person can produce, the faster the arm will move and, thus, the greater the speed that will be imparted to the ball. The reaction of the feet against the surface on which the person stands indicates the application of the **law of reaction**.

The leverage factor is very important in throwing a ball or an object. The longer the lever, the greater the speed that can be imparted to the ball. For all practical purposes, the body from the feet to the fingers can be considered as one long lever. The longer the lever, either from natural body length or from the movements of the body to the extended backward position (as in throwing a softball, with extension of the shoulder and the elbow joints), the greater will be the arc through which it accelerates, and thus the greater the speed imparted to the thrown object.

In certain circumstances, when the ball is to be thrown only a short distance, as in baseball when it is thrown by the catcher to the bases, the short lever is advantageous because it takes less total time to release the ball.

Balance, or equilibrium, is a factor in throwing when the body is rotated to the rear in the beginning of the throw. This motion moves the body nearly out of balance to the rear, and then balance changes again in the body with the forward movement. Balance is again established with the follow-through, when the feet are spread and the knees and trunk are flexed to lower the center of gravity.

Summary

The preceding discussion has been a brief overview of some of the factors affecting motion. Analysis of human motion in light of the laws of physics poses a problem: How comprehensive is the analysis to be? It can become very complex, particularly when body motion is combined with the manipulation of an object in the hand involved in throwing, kicking, striking, or catching.

These factors become involved when an analysis is attempted of the activities common to our physical education programs—football, baseball, basketball, track and field, field hockey, and swimming, to mention a few. However, to have a complete view of which factors control human movement one must have a working knowledge of both the physiological and the biomechanical principles of kinesiology.

It is beyond the scope of this book to make a detailed analysis of other activities. Some sources that consider these problems in detail are listed in the references and Web sites.

Web sites

Biomechanics World Wide

www.per.ualberta.ca/biomechanics

This site enables the reader to search the biomechanics journals for recent information regarding mechanism of injury.

Biomechanics: The Magazine of Body Movement and Medicine

www.biomech.com

COSI Hands-On Science Centers

www.cosi.org/onlineExhibits/simpMach/sm1.swf

A Flash site demonstrating simple machine explanations.

Edquest

www.edquest.ca/pdf/sia84notes.pdf

Text, pictures, and illustrations on simple and complex machines.

GRD Training Corporation

www.physchem.co.za/Motion/Motion%20Index.htm

Explanations of physics principles in motion with quizzes.

Integrated Publishing

http://www.tpub.com/content/engine/14037/index.htm

Engine mechanics.

International Society of Biomechanics

www.isbweb.org

Software, data, information, resources, yellow pages, conferences.

James Madison Memorial High School

www.madison.k12.wi.us/jmm/isp/U7PDF08.pdf

A pdf file explaining the six types of simple machines.

Kinesiology Biomechanics Classes

www.uoregon.edu/~karduna/biomechanics/kinesiology.htm

A listing of numerous biomechanics and kinesiology class sites on the Web with many downloadable presentations and notes.

Marsden State High School

www.marsdenshs.qld.edu.au/subjects/science/juniorscience/quizzes/physics-simple.html

Physics quiz—simple machines.

Optusnet.com

www.members.optusnet.com.au/ncrick/converters/moment.html

Conversion formulas for physics variables.

The Physics Classroom

http://www.glenbrook.k12.il.us/qbssci/phys/Class/BBoard.html

Numerous topics including the laws of motion and other physics principles.

Physics concepts—Simple Machines

http://www.ceeo.tufts.edu/robolabatceeo/K12/classroom/lever.asp

An overview of physics concepts involved in the study of biomechanics.

Simple Simon

http://geo.hdsb.ca/grassroots2003/grassroots/MichalopoulosPd1/Lever/web%20page.htm

Basic physics of machines.

Sports Coach—Levers

www.brianmac.demon.co.uk/levers.htm

A basic review of levers with excellent links to the study of muscle training and function.

Worsley School

www.wcsscience.com/simple/machines.html

Detailed explanations and illustrations on simple machines.

LABORATORY AND REVIEW

1. In the lever component identification chart below, list two levers for each class other than those illustrated in this chapter. For each identify the force, axis, and resistance. Also explain the advantage of using each lever as to whether it is to achieve balance, force motion, speed, or range of motion.

Lever Component Identification Chart

Lever class	Example	Force	Axis	Resistance	Advantage provided
1st					
1st					
2nd					
2nd					
3rd					
3rd					

2. Anatomical levers can improve physical performance. Explain how this occurs using the information you have learned in relation to throwing.

3. What class of lever is an automobile steering wheel?

4. If your biceps muscle inserts to your forearm 2 inches distal to your elbow axis, the distance from the elbow axis to the palm of your hand is 18 inches, and you lift a 20-pound weight, how much pull must your muscle exert to achieve elbow flexion?

5. If the weight of an object is 50 kg and your mechanical advantage is 4, how much force would you need to exert to lift the object with a lever system?

6. Utilizing Fig. 3.18, assume the components are arranged as listed in the lever variable calculation chart and determine the value for each of the variables. In Fig. 3.18, each vertical line on the lever bar represents the point at which the component is to be arranged with the left end point representing 0 and the right end point representing 20.

7. List two different wheel and axles in which the force is applied to the wheel. From your observation, estimate which of the two has the greater mechanical advantage.

8. List two different wheel and axles in which the force is applied to the axle. From your observation, estimate which of the two has the greater mechanical advantage.

9. When attempting to remove a screw, is it easier when using a screw driver with a larger grip on the handle? Why?

10. If a pulley set-up has four supporting ropes, what would be the mechanical advantage of the set-up?

11. What would be the amount of force needed to lift an object in a pulley system if the weight of an object being lifted is 200 kg and the number of supporting ropes is four?

12. Identify a practical example of Newton's law of inertia. Explain how the example illustrates the law.

13. Identify a practical example of Newton's law of acceleration. Explain how the example illustrates the law.

14. Identify a practical example of Newton's law of reaction. Explain how the example illustrates the law.

FIG. 3.18 ● Components of a lever system.

Lever Variable Calculation Chart

	Lever components			Variables				
	Force applied at	Axis placed at	25 Newtons of resistance placed at	Lever class	FA	RA	Force	MA
a.	0	2	20					
b.	0	9	15					
c.	3	17	13					
d.	8	4	19					
e.	12	0	18					
f.	19	9	3					
g.	16	2	7					
h.	13	20	4					
i.	8	17	1					
j.	20	4	11					

Laws of Motion Task Comparison Chart

	Column A	Column B
a.	Throw a baseball 60 mph.	Throw a shot put 60 mph.
Ex.		
b.	Kick a bowling ball 40 yards.	Kick a soccer ball 40 yards.
Ex.		
c.	Bat a whiffle ball over a 320-yard fence.	Bat a baseball over a 320-yard fence.
Ex.		
d.	Catch a shot put that was thrown 60 mph.	Catch a softball that was thrown 60 mph.
Ex.		
e.	Tackle a 240-lb running back sprinting toward you at full speed.	Tackle a 200-lb running back sprinting toward you at full speed.
Ex.		
f.	Run a 40-yard dash in 4.5 seconds on a wet field.	Run a 40-yard dash in 4.5 seconds on a dry field.
Ex.		

15. For the laws of motion task comparison chart, assume that you possess the skill, strength, etc. to be able to perform each of the paired tasks. Circle the task in either column A or column B that would be the easiest to accomplish based upon Newton's laws of motion and explain why.

16. If a baseball player hit a triple and ran around the bases to third base, what would be his displacement? Hint: There are 90 feet between each base.

17. Select a sporting activity and explain how the presence of too much friction becomes a problem in the activity.

18. Select a sporting activity and explain how the presence of too little friction becomes a problem in the activity.

19. Using the mechanical loading basic forces of compression, torsion, and shear, describe each force by using examples from soccer or volleyball.

20. Develop special projects and class reports by individuals or small groups of students on the mechanical analysis of all the skills involved in the following:
    a. Basketball g. Golf
    b. Baseball   h. Gymnastics
    c. Dancing   i. Soccer
    d. Diving   j. Swimming
    e. Football   k. Tennis
    f. Field hockey l. Wrestling

21. Develop term projects and special class reports by individuals or small groups of students about the following factors in motion:
    a. Balance   m. Spin
    b. Force   n. Rebound angle
    c. Gravity   o. Momentum
    d. Motion   p. Center of gravity
    e. Torque   q. Equilibrium
    f. Leverage   r. Stability
    g. Projectiles   s. Base of support
    h. Friction   t. Inertia
    i. Buoyancy   u. Linear displacement
    j. Aerodynamics   v. Angular displacement
    k. Hydrodynamics   w. Speed
    l. Restitution   x. Velocity

22. Develop demonstrations, term projects, or special reports by individuals or small groups of students on the following activities:
    a. Lifting
    b. Throwing
    c. Standing
    d. Walking
    e. Running
    f. Jumping
    g. Falling
    h. Sitting
    i. Pushing and pulling
    j. Striking

References

Adrian MJ, Cooper JM: *The biomechanics of human movement,* Indianapolis, IN, 1989, Benchmark.

American Academy of Orthopaedic Surgeons; Schenck RC, ed.: *Athletic training and sports medicine,* ed 3, Rosemont, IL, 1999, American Academy of Orthopaedic Surgeons.

Barham JN: *Mechanical kinesiology,* St. Louis, 1978, Mosby.

Broer MR: *An introduction to kinesiology,* Englewood Cliffs, NJ, 1968, Prentice-Hall.

Broer MR, Zernicke RF: *Efficiency of human movement,* ed 3, Philadelphia, 1979, Saunders.

Bunn JW: *Scientific principles of coaching,* ed 2, Englewood Cliffs, NJ, 1972, Prentice-Hall.

Cooper JM, Adrian M, Glassow RB: *Kinesiology,* ed 5, St. Louis, 1982, Mosby.

Donatelli R, Wolf SL: *The biomechanics of the foot and ankle,* Philadelphia, 1990, Davis.

Hall SJ: *Basic biomechanics,* ed 4, New York, 2000, McGraw-Hill.

Hamill J, Knutzen KM: *Biomechanical basis of human movement,* ed 2, Baltimore, 2003, Lippincott Williams & Wilkins.

Hinson M: *Kinesiology,* ed 4, New York, 1981, McGraw-Hill.

Kegerreis S, Jenkins WL, Malone TR: Throwing injuries, *Sports Injury Management* 2:4, 1989.

Kelley DL: *Kinesiology: fundamentals of motion description,* Englewood Cliffs, NJ, 1971, Prentice-Hall.

Kreighbaum E, Barthels KM: *Biomechanics: a qualitative approach for studying human movement,* ed 3, New York, 1990, Macmillan.

Logan GA, McKinney WC: *Anatomic kinesiology,* ed 3, New York, 1982, McGraw-Hill.

Luttgens K, Hamilton N: *Kinesiology: scientific basis of human motion,* ed 10, New York, 2002, McGraw-Hill.

McCreary EK, Kendall FP, Rodgers MM, Provance PG, Romani WA: *Muscles: testing and function with posture and pain,* ed 5, Philadelphia, 2005, Lippincott Williams & Wilkins.

McGinnis PM: *Biomechanics of sport and exercise,* ed 2, Champaign, IL, 2005, Human Kinetics.

Neumann, DA: *Kinesiology of the musculoskeletal system: foundations for physical rehabilitation,* St. Louis, 2002, Mosby.

Nordin M, Frankel VH: *Basic biomechanics of the musculoskeletal system,* ed 2, Philadelphia, 1989, Lea & Febiger.

Norkin CC, Levangie PK: *Joint structure and function: a comprehensive analysis,* Philadelphia, 1983, Davis.

Northrip JW, Logan GA, McKinney WC: *Analysis of sport motion: anatomic and biomechanic perspectives,* ed 3, New York, 1983, McGraw-Hill.

Piscopo J, Baley J: *Kinesiology: the science of movement,* New York, 1981, Wiley.

Prentice WE: *Arheim's principles of athletic training,* ed 11, New York, 2003, McGraw-Hill.

Rasch PJ: *Kinesiology and applied anatomy,* ed 7, Philadelphia, 1989, Lea & Febiger.

Scott MG: *Analysis of human motion,* ed 2, New York, 1963, Appleton-Century-Crofts.

Segedy A: Braces' joint effects spur research surge, *Biomechanics* February, 2005.

Weineck J: *Functional anatomy in sports,* ed 2, St. Louis, 1990, Mosby.

Wirhed R: *Athletic ability and the anatomy of motion,* London, 1984, Wolfe.

Whiting WC: *Biomechanics of musculoskeletal injury,* Champaign, IL, 1998, Human Kinetics.

Chapter 4

The Shoulder Girdle

Objectives

- To identify on the skeleton important bony features of the shoulder girdle

- To label on a skeletal chart the important bony features of the shoulder girdle

- To draw on a skeletal chart the muscles of the shoulder girdle and indicate shoulder girdle movements using arrows

- To demonstrate, using a human subject, all of the movements of the shoulder girdle and list their respective planes of movement and axes of rotation

- To palpate the muscles of the shoulder girdle on a human subject and list their antagonists

- To palpate the joints of the shoulder girdle on a human subject during each movement through the full range of motion

---

Online Learning Center Resources

Visit *Manual of Structural Kinesiology's* Online Learning Center at **www.mhhe.com/floyd16e** for additional information and study material for this chapter, including:

- *Self-grading quizzes*
- *Anatomy flashcards*
- *Animations*

Brief descriptions of the most important bones in the shoulder region will help you understand the skeletal structure and its relationship to the muscular system.

Bones

Two bones are primarily involved in movements of the shoulder girdle (Figs. 4.1 and 4.2). They are the scapula and clavicle, which generally move as a unit. Their only bony link to the axial skeleton is provided by the clavicle's articulation with the sternum. Key bony landmarks for studying the shoulder girdle are the manubrium, clavicle, coracoid process, acromion process, glenoid fossa, lateral border, inferior angle, medial border, superior angle, and spine of the scapula (Figs. 4.1, 4.2, 4.3).

Joints

When analyzing shoulder girdle (scapulothoracic) movements, it is important to realize that the scapula moves on the rib cage due to the joint motion actually occurring at the sternoclavicular joint and to a lesser extent at the acromioclavicular joint (see Figs. 4.1 and 4.2).

Sternoclavicular (SC)

This is classified as a (multiaxial) arthrodial joint. It moves anteriorly 15 degrees with protraction and posteriorly 15 degrees with retraction. It moves superiorly 45 degrees with elevation and inferiorly 5 degrees with depression. Some rotation of the clavicle along its axis during various movements of the shoulder girdle result in slight rotary gliding movement at the sternoclavicular joint. It

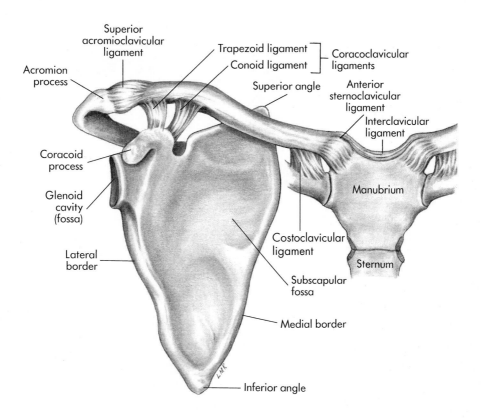

FIG. 4.1 ● Right shoulder girdle, anterior view.

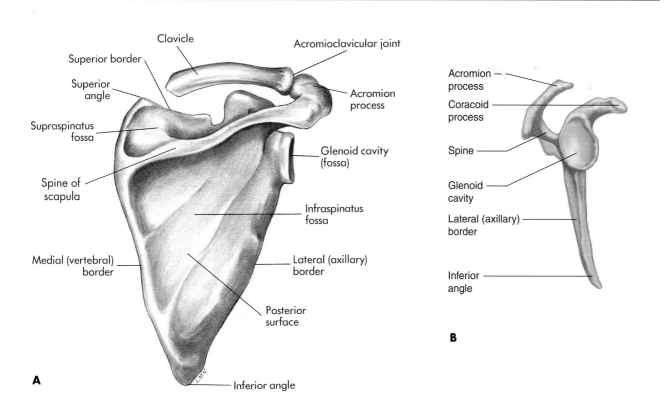

FIG. 4.2 ● Right scapula. **A,** Posterior view; **B,** Lateral view.

A from Seeley RR, Stephens TD, Tate P; *Anatomy and physiology,* ed 7, New York, 2006, McGraw-Hill; **B** from Shier D, Butler J, Lewis R: *Hole's human anatomy and physiology,* ed 9, New York, 2002, McGraw-Hill.

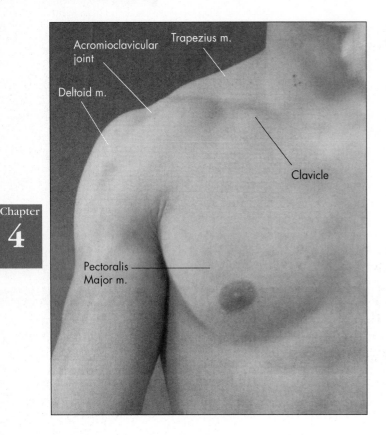

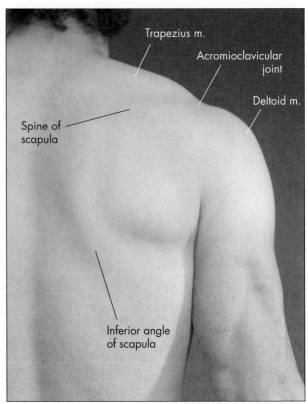

FIG. 4.3 ● Right shoulder girdle surface anatomy, anterior and posterior views.

is supported anteriorly by the anterior sternoclavicular ligament and posteriorly by the posterior ligament. Additionally, the costoclavicular and interclavicular ligaments provide stability against superior displacement.

Acromioclavicular (AC)

This is classified as an arthrodial joint. It has a 20- to 30-degree total gliding and rotational motion accompanying other shoulder girdle and shoulder joint motions. In addition to the strong support provided by the coracoclavicular ligaments (trapezoid and conoid), the superior and inferior acromioclavicular ligaments provide stability to this often injured joint. The coracoclavicular joint, classified as a syndesmotic-type joint, functions through its ligaments to greatly increase the stability of the acromioclavicular joint.

Scapulothoracic

This is not a true synovial joint, due to its not having regular synovial features and due to the fact that its movement is totally dependent on the sternoclavicular and acromioclavicular joints. Even though scapula movement occurs as a result of motion at the SC and AC joints, the scapula can be described as having a total range of 25-degree abduction-adduction movement, 60-degree upward-downward rotation, and 55-degree elevation-depression. The scapulothoracic joint is supported dynamically by its muscles and lacks ligamentous support, since it has no synovial features.

There is not a typical articulation between the anterior scapula and the posterior rib cage. Between these two osseous structures are the serratus anterior muscle originating off the upper nine ribs laterally and running just behind the rib cage posteriorly to insert on the medial border of the scapula. Immediately posterior to the serratus anterior is the subscapularis muscle (see Chapter 5) on the anterior scapula.

Movements   FIGS. 4.4, 4.5

In analyzing shoulder girdle movements, it is often helpful to focus on a specific scapular bony landmark, such as the inferior angle (posteriorly), the glenoid fossa (laterally), and the acromion process (anteriorly). All of these movements have their pivotal point where the clavicle joins the sternum at the sternoclavicular joint.

Movements of the shoulder girdle can be described as movements of the scapula. See Figs. 4.4 and 4.5 for movements of the shoulder girdle.

Abduction (protraction)

Movement of the scapula laterally away from the spinal column, as in reaching for an object in front of the body.

Adduction (retraction)

Movement of the scapula medially toward the spinal column, as in pinching the shoulder blades together.

Upward rotation

Turning the glenoid fossa upward and moving the inferior angle superiorly and laterally away from the spinal column.

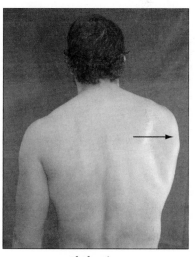

Abduction
(protraction)
A

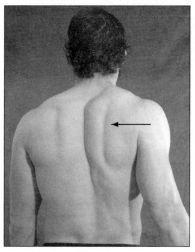

Adduction
(retraction)
B

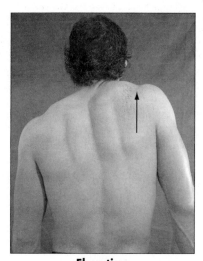

Elevation
C

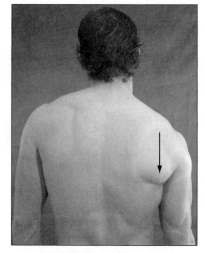

Depression
D

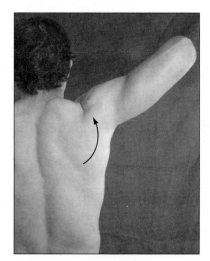

Upward rotation
E

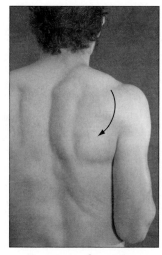

Downward rotation
F

FIG. 4.4 ● Movements of the shoulder girdle.

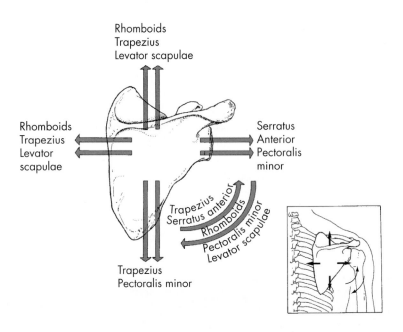

Rhomboids
Trapezius
Levator scapulae

Rhomboids
Trapezius
Levator
scapulae

Serratus
Anterior
Pectoralis
minor

Trapezius
Serratus anterior
Rhomboids
Pectoralis minor
Levator scapulae

Trapezius
Pectoralis minor

FIG. 4.5 ● Actions of the scapular muscles.

Downward rotation

Returning the inferior angle medially and inferiorly toward the spinal column and the glenoid fossa to its normal position. (Once the scapula has returned to its anatomical position, further downward rotation actually results in the superior angle moving slightly superomedial.)

Elevation

Upward or superior movement, as in shrugging the shoulders.

Depression

Downward or inferior movement, as in returning to a normal position from a shoulder shrug.

To accomplish some of the previously listed shoulder girdle movements, the scapula must rotate or tilt on its axis. While these are not primary movements of the shoulder girdle, they are necessary for the scapula to move normally throughout its range of motion during shoulder girdle movements.

Lateral tilt (outward tilt)

Consequential movement during abduction in which the scapula rotates about its vertical axis, resulting in posterior movement of the medial border and anterior movement of the lateral border.

Medial tilt (inward tilt)

Return from lateral tilt; consequential movement during extreme adduction in which the scapula rotates about its vertical axis, resulting in anterior movement of the medial border and posterior movement of the lateral border.

Anterior tilt (upward tilt)

Consequential rotational movement of the scapula about the frontal axis occurring during hyperextension of the glenohumeral joint, resulting in the superior border moving anteroinferiorly and the inferior angle moving posterosuperiorly.

Posterior tilt (downward tilt)

Consequential rotational movement of the scapula about the frontal axis occurring during hyperflexion of the glenohumeral joint, resulting in the superior border moving posteroinferiorly and the inferior angle moving anterosuperiorly.

Synergy with the muscles of the glenohumeral joint

The shoulder joint and shoulder girdle work together in carrying out upper extremity activities. It is critical to understand that movement of the shoulder girdle is not dependent on the shoulder joint and its muscles. However, the muscles of the shoulder girdle are essential in providing a scapula-stabilizing effect, so that the muscles of the shoulder joint will have a stable base from which to exert force for powerful movement involving the humerus. Consequently, the shoulder girdle muscles contract to maintain the scapula in a relatively static position during many shoulder joint actions.

As the shoulder joint goes through more extreme ranges of motion, the scapular muscles contract to move the shoulder girdle so that its glenoid fossa will be in a more appropriate position from which the humerus can move. Without the accompanying movement of the scapula, we can only raise our humerus into approximately 90 degrees of total shoulder abduction and flexion. This works through the appropriate muscles of both joints working in synergy to accomplish the desired action of the entire upper extremity. For example, if we desire to raise our hand out to the side laterally as high as possible, the serratus anterior and trapezius (middle and lower fibers) muscles upwardly rotate the scapula as the supraspinatus and deltoid initiate glenohumeral abduction. This synergy between the scapula and shoulder joint muscles enhances movement of the entire upper extremity. Further discussion of the interaction and teamwork between these joints is provided at the beginning of Chapter 5 with Table 5.1 listing the shoulder girdle movements that usually accompany shoulder joint movements.

Muscles

There are five muscles primarily involved in shoulder girdle movements. To avoid confusion, it is helpful to group muscles of the shoulder girdle separately from those of the shoulder joint. The subclavius muscle is also included in this group, but it is not regarded as a primary mover in any actions of the shoulder girdle. All five shoulder girdle muscles have their origin on the axial skeleton, with their insertion located on the scapula and/or the clavicle. Shoulder girdle muscles do not attach to the humerus, nor do they cause actions of the shoulder joint. The pecto-

ralis minor and subclavius are located anteriorly in relation to the trunk. The serratus anterior is located anteriorly to the scapula, but posteriorly and laterally to the trunk. Located posteriorly to the trunk are the trapezius, rhomboid, and levator scapula.

The shoulder girdle muscles are essential in providing dynamic stability of the scapula, so that it can serve as a relative base of support for shoulder joint activities such as throwing, batting, and blocking.

Shoulder girdle muscles—location and action
Anterior
    Pectoralis minor—abduction, downward rotation, and depression
    Subclavius—depression
Posterior and laterally
    Serratus anterior—abduction and upward rotation
Posterior
    Trapezius
        Upper fibers—elevation and extension of the head
        Middle fibers—elevation, adduction, and upward rotation
        Lower fibers—adduction, depression, and upward rotation
    Rhomboid—adduction, downward rotation, and elevation
    Levator scapula—elevation

It is important to understand that muscles may not necessarily be active throughout the absolute full range of motion for which they are noted as being agonists.

Table 4.1 provides a detailed breakdown of the muscles responsible for primary movements of the shoulder girdle.

TABLE 4.1 • Agonist muscles of the shoulder girdle

	Muscle	Origin	Insertion	Action	Plane of motion	Palpation	Innervation
Anterior muscle	Pectoralis minor	Anterior surfaces of the 3rd to 5th ribs	Coracoid process of the scapula	Abduction	Transverse	Difficult, but under the pectoralis major muscle and just inferior to the coracoid process during resisted depression; enhanced by placing the subject's hand behind the back and having him/her actively lift the hand away	Medial pectoral nerve (C8 and T1)
				Downward rotation	Frontal		
				Depression	Frontal		
Posterior and lateral muscle	Serratus anterior	Surface of the upper 9 ribs at the side of the chest	Anterior aspect of the whole length of the medial border of the scapula	Abduction	Transverse	Frontal and lateral side of the chest below the 5th and 6th ribs just proximal to the origin during abduction; best accomplished with the glenohumeral joint flexed 90 degrees; in the same position palpate the upper fibers between the lateral borders of the pectoralis major and latissimus dorsi in the axilla	Long thoracic nerve (C5–C7)
				Upward rotation	Frontal		
Posterior muscles	Trapezius upper fibers	Base of skull, occipital protuberance, and posterior ligaments of neck	Posterior aspect of the lateral 3rd of the clavicle	Elevation	Frontal	Halfway between occipital protuberance to C6 and laterally to acromion, particularly during elevation and extension of the head at the neck	Spinal accessory nerve and branches of C3 and C4
				Extension of head at neck	Sagittal		
				Rotation of head at neck	Transverse		
	Trapezius middle fibers	Spinous process of 7th cervical and upper 3 thoracic vertebrae	Medial border of the acromion process and upper border of the scapular spine	Elevation	Frontal	From C7 to T3 and laterally to acromion process and scapula spine, particularly during adduction	
				Adduction	Transverse		
				Upward rotation	Frontal		
	Trapezius lower fibers	Spinous process of 4th–12th thoracic vertebrae	Triangular space at the base of the scapular spine	Adduction	Transverse	From T4 to T12 and medial aspect of scapula spine, particularly during depression and adduction	
				Depression	Frontal		
				Upward rotation	Frontal		
	Rhomboids	Spinous processes of the last cervical and the first 5 thoracic vertebrae	Medial border of the scapula, below the scapula spine	Adduction	Transverse	Difficult due to being deep to trapezius, but may be palpated through it during adduction; best accomplished with subject's ipsilateral hand behind the back to relax the trapezius and bring the rhomboid into action when the subject lifts the hand away from the back	Dorsal scapula nerve (C5)
				Downward rotation	Frontal		
				Elevation	Frontal		
	Levator scapulae	Transverse processes of the upper 4 cervical vertebrae	Medial border of the scapula	Elevation	Frontal	Difficult to palpate due to being deep to trapezius; best palpated at insertion just medial to the superior angle of scapula, particularly during slight elevation	Dorsal scapula nerve C5 and branches of C3 and C4

Note: The subclavius is not listed since it is not a prime mover in shoulder girdle movements.

Nerves

The muscles of the shoulder girdle are innervated primarily from the nerves of the cervical plexus and brachial plexus as illustrated in Figs. 4.6 and 4.7. The trapezius is innervated by the spinal accessory nerve and from branches of C3 and C4. In addition to supplying the trapezius, C3 and C4 also innervate the levator scapula. The levator scapula receives further innervation from the dorsal scapula nerve originating from C5. The dorsal scapula nerve also innervates the rhomboid. The long thoracic nerve originates from C5, C6, and C7 and innervates the serratus anterior. The medial pectoral nerve arises from C8 and T1 to innervate the pectoralis minor.

Roots: C5, C6, C7, C8, T1
Trunks: upper, middle, lower
Anterior divisions
Posterior divisions
Cords: posterior, lateral, medial
Branches: Axillary nerve
Radial nerve
Musculocutaneous nerve
Median nerve
Ulnar nerve

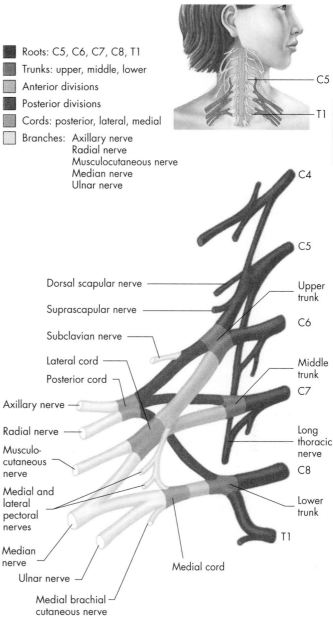

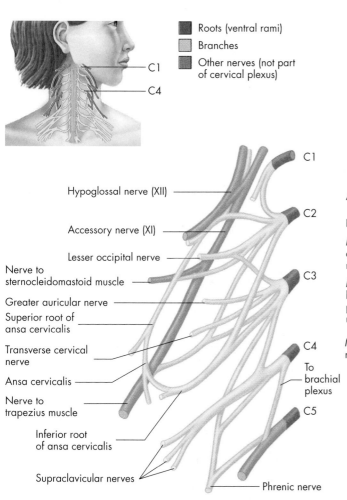

Roots (ventral rami)
Branches
Other nerves (not part of cervical plexus)

FIG. 4.6 ● Cervical plexus, anterior view. The roots of the plexus are formed by the ventral rami of the spinal nerves C1–C4.

FIG. 4.7 ● Brachial plexus, anterior view. The roots of the plexus are formed by the ventral rami of the spinal nerves C5–T1 and join to form an upper, middle, and lower trunk. Each trunk divides into anterior and posterior divisions. The divisions join together to form the posterior, lateral, and medial cords from which the major brachial plexus nerves arise.

From Seeley RR, Stephens TD, Tate P: *Anatomy & physiology*, ed 6, Dubuque, IA, 2003, McGraw-Hill.

From Seeley RR, Stephens TD, Tate P: *Anatomy & physiology*, ed 6, Dubuque, IA, 2003, McGraw-Hill.

Trapezius muscle   FIG. 4.8

(tra-pe´zi-us)

Origin

Upper fibers: base of skull, occipital protuberance, and posterior ligaments of neck

Middle fibers: spinous processes of seventh cervical and upper three thoracic vertebrae

Lower fibers: spinous processes of fourth through twelfth thoracic vertebrae

Insertion

Upper fibers: posterior aspect of the lateral third of the clavicle

Middle fibers: medial border of the acromion process and upper border of the scapular spine

Lower fibers: triangular space at the base of the scapular spine

Action

Upper fibers: elevation of the scapula; extension and rotation of the head at the neck

Middle fibers: elevation, upward rotation, and adduction (retraction) of the scapula

Lower fibers: depression, adduction (retraction), and upward rotation of the scapula

Palpation

Upper fibers: halfway between occipital protuberance to C6 and laterally to acromion, particularly during elevation and extension of the head at the neck

Middle fibers: fro C7 to T3 and laterally to acromion process and scapula spine, particularly during adduction

Lower fibers: from T4 to T12 and medial aspect of scapula spine, particularly during depression and adduction

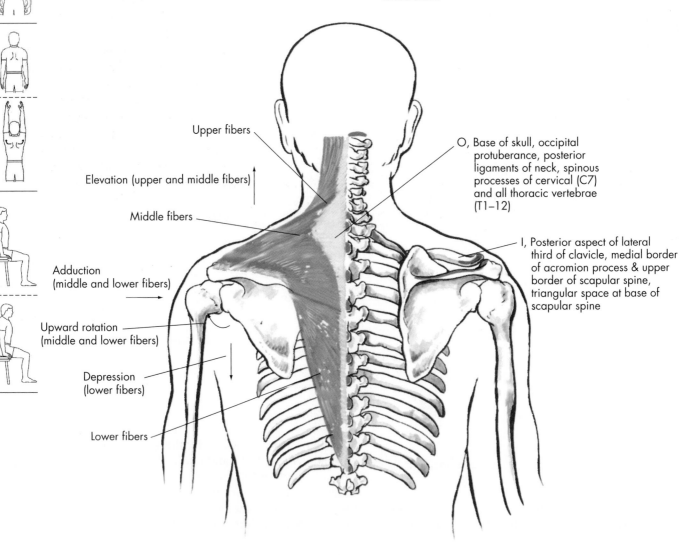

FIG. 4.8 ● Trapezius muscle. *O,* Origin; *I,* Insertion; posterior view.

Innervation

Spinal accessory nerve (cranial nerve XI) and branches of C3, C4

Application, strengthening, and flexibility

The upper fibers are a thin and relatively weak part of the muscle. They provide some elevation of the clavicle. Due to their origin on the base of the skull, they assist in extension of the head.

The middle fibers are stronger and thicker and provide strong elevation, upward rotation, and adduction (retraction) of the scapula. Rarely is this portion of the muscle weak due to it being so active in positioning the shoulder for function and posture. As a result it is often a source of tenderness and discomfort due to chronic tension.

The lower fibers assist in adduction (retraction) and rotate the scapula upward. It is typically weak, particularly in individuals whose activities demand a significant amount of scapula abduction.

When all the parts of the trapezius are working together, they tend to pull upward and adduct at the same time. Typical action of the trapezius muscle is fixation of the scapula for deltoid action. Continuous action in upward rotation of the scapula permits the arms to be raised over the head. The muscle is always used in preventing the glenoid fossae from being pulled down during the lifting of objects with the arms. It is also typically seen in action during the holding of an object overhead. Holding the arm at the side horizontally shows typ-

ical fixation of the scapula by the trapezius muscle, while the deltoid muscle holds the arm in that position. The muscle is used strenuously when lifting with the hands, as in picking up a heavy wheelbarrow. The trapezius must prevent the scapula from being pulled downward. Carrying objects on the tip of the shoulder also calls this muscle into play. Strengthening of the upper and middle fibers can be accomplished through shoulder-shrugging exercises. The middle and lower fibers can be strengthened through bent-over rowing and shoulder joint horizontal abduction exercises from a prone position. The lower fibers can be emphasized with a chest-proud shoulder retraction exercise attempting to place the elbows in the back pants pockets with depression.

To stretch the trapezius, each portion needs to be specifically addressed. The upper fibers may be stretched by using one hand to pull the head and neck forward into flexion or slight lateral flexion to the opposite side while the ipsilateral hand is hooked under a table edge to maintain the scapula in depression. The middle fibers are stretched to some extent with the procedure used for the upper fibers, but they may be stretched further by using a partner to passively pull the scapula into full protraction. The lower fibers are perhaps best stretched with the subject in a side-lying position while a partner grasps the lateral border and inferior angle of the scapula and moves it passively into maximal elevation and protraction.

Levator scapulae muscle   FIG. 4.9

(le-va´tor scap´u-lae)

Origin

Transverse processes of the upper four cervical vertebrae

Insertion

Medial border of the scapula above the base of the scapular spine

Action

Elevates the medial margin of the scapula

Palpation

Difficult to palpate due to being deep to trapezius; best palpated at insertion just medial to the superior angle of scapula, particularly during slight elevation

Innervation

Dorsal scapula nerve C5 and branches of C3 and C4

Application, strengthening, and flexibility

Shrugging the shoulders calls the levator scapulae muscle into play, along with the upper trapezius muscle. Fixation of the scapula by the pectoralis minor muscle allows the levator scapulae muscles on both sides to extend the neck or to flex laterally if used on one side only.

The levator scapula is perhaps best stretched by rotating the head approximately 45 degrees to the opposite side and flexing the cervical spine actively while maintaining the scapula in a relaxed, depressed position.

In addition to the trapezius, the levator scapulae is a very common site for tightness, tenderness, and discomfort secondary to chronic tension and from carrying items with straps over the shoulder.

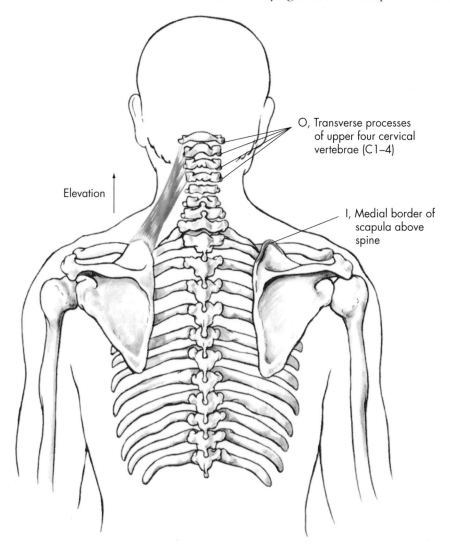

FIG. 4.9 ● Levator scapulae muscle. *O*, Origin; *I*, Insertion; posterior view.

Rhomboid muscles—major and minor FIG. 4.10

(rom´boyd)

Origin

Spinous processes of the last cervical and the first five thoracic vertebrae

Insertion

Medial border of the scapula, below the spine of the scapula

Action

The rhomboid major and minor muscles work together
Adduction (retraction): draw the scapula toward the spinal column
Downward rotation: from the upward rotated position; they draw the scapula into downward rotation
Elevation: slight upward movement accompanying adduction

Palpation

Difficult to palpate due to being deep to trapezius, but may be palpated through the relaxed trapezius during adduction. This is best accomplished by placing the subject's ipsilateral hand behind the back (glenohumeral internal rotation and scapula downward rotation), which relaxes the trapezius and brings the rhomboid into action when the subject lifts the hand away from the back.

Innervation

Dorsal scapula nerve (C5)

Application, strengthening, and flexibility

The rhomboid muscles fix the scapula in adduction (retraction) when the muscles of the shoulder joint adduct or extend the arm. These muscles are used powerfully in chinning. As one hangs from the horizontal bar, suspended by the hands, the scapula tends to be pulled away from the top of the chest. When the chinning movement begins, it is the rhomboid muscles that rotate the medial border of the scapula down and back toward the spinal column. Note their favorable position to do this. Related to this, the rhomboid works in a similar manner to prevent scapula winging.

The trapezius and rhomboid muscles working together produce adduction with slight elevation of the scapula. To prevent this elevation, the latissimus dorsi muscle is called into play.

Chin-ups and dips are excellent exercises for developing strength in this muscle. The rhomboids may be stretched by passively moving the scapula into full protraction while maintaining depression. Upward rotation may assist in this stretch as well.

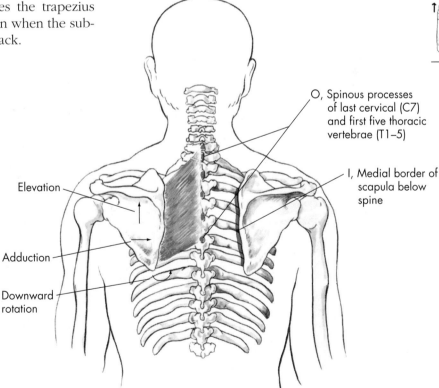

O, Spinous processes of last cervical (C7) and first five thoracic vertebrae (T1–5)

I, Medial border of scapula below spine

Elevation

Adduction

Downward rotation

FIG. 4.10 ● Rhomboid muscles (major and minor). *O*, Origin; *I*, Insertion; posterior view.

Serratus anterior muscle   FIG. 4.11

(ser-a´tus an-tir´e-or)

Origin

Surface of the upper nine ribs at the side of the chest

Insertion

Anterior aspect of the whole length of the medial border of the scapula

Action

Abduction (protraction): draws the medial border of the scapula away from the vertebrae

Upward rotation: longer, lower fibers tend to draw the inferior angle of the scapula farther away from the vertebrae, thus rotating the scapula upward slightly

Palpation

Front and lateral side of the chest below the fifth and sixth ribs just proximal to their origin during abduction, which is best accomplished from a supine position with the glenohumeral joint in 90 degrees of flexion. The upper fibers may be palpated in the same position between the lateral borders of the pectoralis major and latissimus dorsi in the axilla.

Innervation

Long thoracic nerve (C5–C7)

Application, strengthening, and flexibility

The serratus anterior muscle is used commonly in movements drawing the scapula forward with slight upward rotation, such as throwing a baseball, punching in boxing, shooting and guarding in basketball, and tackling in football. It works along with the pectoralis major muscle in typical action, such as throwing a baseball.

The serratus anterior muscle is used strongly in doing push-ups, especially in the last 5 to 10 degrees of motion. The bench press and overhead press are good exercises for this muscle. A winged scapula condition usually results from weakness of the rhomboid and/or the serratus anterior. Serratus anterior weakness may result from an injury to the long thoracic nerve.

The serratus anterior can be stretched by standing, facing a corner and placing each hand at shoulder level on the two walls. As you lean in and attempt to place your nose in the corner, both scapula are pushed into an adducted position, which stretches the serratus anterior.

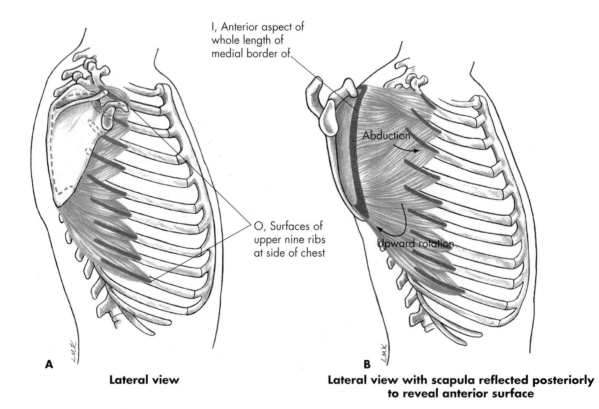

A
Lateral view

B
Lateral view with scapula reflected posteriorly to reveal anterior surface

FIG. 4.11 ● Serratus anterior muscle. *O,* Origin; *I,* Insertion; lateral view.

Pectoralis minor muscle FIG. 4.12

(pek-to-ra´lis mi´nor)

Origin
Anterior surfaces of the third to fifth ribs

Insertion
Coracoid process of the scapula

Action
Abduction (protraction): draws the scapula forward and
tends to tilt the lower border away from the ribs

Downward rotation: as it abducts, it draws the scapula downward

Depression: when the scapula is rotated upward, it
assists in depression

Palpation
Difficult to palpate, but can be palpated under the
pectoralis major muscle and just inferior to the coracoid process during resisted depression. This may
be enhanced by placing the subject's hand behind
the back and having him/her actively lift the hand
away, which causes downward rotation.

Innervation
Medial pectoral nerve (C8–T1)

Application, strengthening, and flexibility

The pectoralis minor muscle is used, along with
the serratus anterior muscle, in true abduction
(protraction) without rotation. This is seen particularly in movements such as push-ups. True abduction of the scapula is necessary. Therefore, the
serratus anterior draws the scapula forward with
a tendency toward upward rotation, the pectoralis minor pulls forward with a tendency toward
downward rotation, and the two pulling together
give true abduction, which is necessary in push-ups. These muscles will be seen working together
in most movements of pushing with the hands.

The pectoralis minor is most used in depressing and rotating the scapula downward from an
upwardly rotated position, as in pushing the body
upward on dip bars.

The pectoralis minor is often tight due to being
overused in activities involving abduction, which
may lead to forward and rounded shoulders. As a
result stretching may be indicated, which can be accomplished with a wall push-up in the corner as
used for stretching the serratus anterior. Additionally, lying supine with a rolled towel directly under
the thoracic spine while a partner pushes each scapula into retraction places this muscle on stretch.

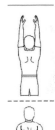

Chapter 4

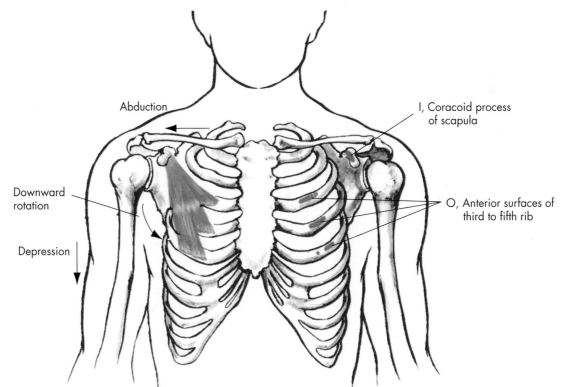

FIG. 4.12 ● Pectoralis minor muscle. *O,* Origin; *I,* Insertion; anterior view.

Subclavius muscle   FIG. 4.13

(sub-klá-ve-us)

Origin

Superior aspect of first rib at its junction with its costal cartilage

Insertion

Inferior groove in the midportion of the clavicle

Action

Stabilization and protection of the sternoclavicular joint
Depression
Abduction (protraction)

Palpation

Difficult to distinguish from the pectoralis major, but may be palpated just inferior to the middle third of the clavicle with the subject side-lying and in a somewhat upwardly rotated position and the humerus supported in a partially passively flexed position. Slight active depression and abduction of the scapula may enhance palpation.

Innervation

Nerve fibers from C5 and C6

Application, strengthening, and flexibility

The subclavius pulls the clavicle anteriorly and inferiorly toward the sternum. In addition to assisting in abducting and depressing the clavicle and the shoulder girdle, it has a significant role in protecting and stabilizing the sternoclavicular joint during upper extremity movements. It may be strengthened during activities in which there is active depression, such as dips, or active abduction, such as push-ups. Extreme elevation and retraction of the shoulder girdle provides a stretch to the subclavius.

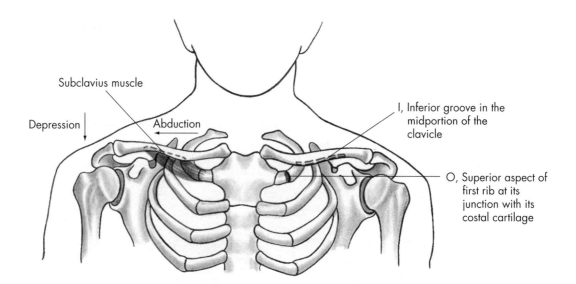

FIG. 4.13 ● Subclavius muscle. *O,* Origin; *I,* Insertion; anterior view.

Web sites

Radiologic Anatomy Browser

http://radlinux1.usuf1.usuhs.mil/rad/iong/ index.html

This site has numerous radiological views of the musculoskeletal system.

University of Arkansas Medical School Gross Anatomy for Medical Students

http://anatomy.uams.edu/anatomyhtml/ grossresources.html

Dissections, anatomy tables, atlas images, links, etc.

Loyola University Medical Center: Structure of the Human Body

www.meddean.luc.edu/lumen/MedEd/ GrossAnatomy/GA.html

An excellent site with many slides, dissections, tutorials, etc., for the study of human anatomy.

Wheeless' Textbook of Orthopaedics

www.wheelessonline.com

This site has an extensive index of links to the fractures, joints, muscles, nerves, trauma, medications, medical topics, lab tests, and links to orthopaedic journals and other orthopaedic and medical news.

Scapulothoracic Positions and Motion

www.orthop.washington.edu/uw/scapulothoracic/ tabID_3376/ItemID_212/Articles/Default.aspx

Scapula movements and movies.

The Physician and Sportsmedicine

www.physsportsmed.com/issues/2003/0703/ depalma.htm

Detecting and treating shoulder impingement syndrome: the role of scapulothoracic dyskinesis.

The Physician and Sportsmedicine

www.physsportsmed.com/issues/2001/11_01/ johnson.htm

Acromioclavicular joint injuries: identifying and treating "separated shoulder" and other conditions.

The Physician and Sportsmedicine

www.physsportsmed.com/issues/1999/02_99/ williams.htm

Posterior sternoclavicular joint dislocation.

Baseball Almanac

www.baseball-almanac.com/chapters/cap-ch8.shtml

Coaching adult pitchers.

Lecture Topics in Kinesiology

http://moon.ouhsc.edu/dthompso/namics/ shoulder.htm

Shoulder articulations, movements, and muscles that are within the shoulder girdle.

Radiologic Anatomy Browser

http://radlinux1.usuf1.usuhs.mil/rad/long/ index.html

This site has numerous radiological views of the musculoskeletal system.

Premiere Medical Search Engine

www.medsite.com/Default.asp?bhcp=1

This site allows the reader to enter any medical condition, and it will search the net to find relevant articles.

Virtual Hospital

www.vh.org

Numerous slides, patient information, etc.

Worksheet exercises

As an aid to learning, for in-class or out-of-class assignments, or for testing, tear-out worksheets are found at the end of the text (pp. 374 and 375).

Skeletal worksheet (no. 1)

Draw and label on the worksheet the following muscles:

a. Trapezius
b. Rhomboid major and minor
c. Serratus anterior
d. Levator scapulae
e. Pectoralis minor
f. Subclavius

Human figure worksheet (no. 2)

Label the circles next to the arrows with the appropriate letter that corresponds to the movements of the shoulder girdle indicated by the arrow:

a. Adduction (retraction)
b. Abduction (protraction)
c. Rotation upward
d. Rotation downward
e. Elevation
f. Depression

LABORATORY AND REVIEW EXERCISES

1. Locate the following prominent skeletal features on a human skeleton and on a subject:
   a. Scapula
      1. Medial border
      2. Inferior angle
      3. Superior angle
      4. Coracoid process
      5. Spine of scapula
      6. Glenoid cavity
      7. Acromion process
      8. Supraspinatus fossa
      9. Infraspinatus fossa
   b. Clavicle
      1. Sternal end
      2. Acromial end
   c. Joints
      1. Sternoclavicular joint
      2. Acromioclavicular joint

2. How and where do you palpate the following muscles on a human subject?
   a. Serratus anterior
   b. Trapezius
   c. Rhomboid major and minor
   d. Levator scapulae
   e. Pectoralis minor
      NOTE: "How" means palpating during active contraction and possibly resisting a primary movement of the muscle. Some muscles have several primary movements, such as the trapezius with rotation upward and adduction. "Where" refers to the location on the body where the muscle can be felt.

3. Palpate the sternoclavicular and acromioclavicular joint movements and the muscles primarily involved while demonstrating the following shoulder girdle movements.
   a. Adduction
   b. Abduction
   c. Rotation upward
   d. Rotation downward
   e. Elevation
   f. Depression

4. List the planes in which each of the following shoulder girdle movements occur. List the respective axis of rotation for each movement in each plane.
   a. Adduction
   b. Abduction
   c. Rotation upward
   d. Rotation downward
   e. Elevation
   f. Depression

5. Fill in the muscle analysis chart below by listing the muscles primarily involved in each movement.

6. Fill in the antagonistic muscle action chart below by listing the muscle(s) or parts of muscles that are antagonist in their actions to the muscles in the left column.

Muscle analysis chart • Shoulder girdle

Shoulder girdle	
Abduction	Adduction
Elevation	Depression
Upward rotation	Downward rotation

Antagonistic Muscle Action Chart • Shoulder girdle

Agonist	Antagonist
Serratus anterior	
Trapezius (upper fibers)	
Trapezius (middle fibers)	
Trapezius (lower fibers)	
Rhomboid	
Levator scapulae	
Pectoralis minor	

7. After analyzing each of the exercises in the shoulder girdle movement analysis chart, break each into two primary movement phases such as a lifting phase and lowering phase. For each phase, determine the shoulder girdle movement occurring, and then list the shoulder girdle muscles primarily responsible for causing/controlling those movements. Beside each muscle in each movement, indicate the type of contraction as follows: I-isometric; C-concentric; E-eccentric.

8. Analyze each skill in the shoulder girdle sport skill analysis chart and list the movements of the right and left shoulder girdle in each phase of the skill. You may prefer to list the initial position the shoulder girdle is in for the stance

Shoulder Girdle Movement Analysis Chart

Exercise	Initial movement phase		Secondary movement phase	
	Movement(s)	Agonist(s)–(contraction type)	Movement(s)	Agonist(s)–(contraction type)
Push-up				
Chin-up				
Bench press				
Dip				
Lat pull				
Overhead press				
Prone row				
Barbell shrugs				

Shoulder Girdle Sport Skill Analysis Chart

Exercise		Stance phase	Preparatory phase	Movement phase	Follow-through phase
Baseball pitch	(R)				
	(L)				
Volleyball serve	(R)				
	(L)				
Tennis serve	(R)				
	(L)				
Softball pitch	(R)				
	(L)				
Tennis backhand	(R)				
	(L)				
Batting	(R)				
	(L)				
Bowling	(R)				
	(L)				
Basketball free throw	(R)				
	(L)				

phase. After each movement, list the shoulder girdle muscle(s) primarily responsible for causing/controlling those movements. Beside each muscle in each movement, indicate the type of contraction as follows: I-isometric; C-concentric; E-eccentric. It may be desirable to review the concepts for analysis in Chapter 8 for the various phases.

References

Andrews JR, Wilk KE: *The athlete's shoulder,* New York, 1994, Churchill Livingstone.

Andrews JR, Zarins B, Wilk KE: *Injuries in baseball,* Philadelphia, 1998, Lippincott-Raven.

DePalma MJ, Johnson EW: Detecting and treating shoulder impingement syndrome: the role of scapulothoracic dyskinesis, *The Physician and Sportsmedicine* 31(7), 2003.

Field D: *Anatomy: palpation and surface markings,* ed 3, Oxford, 2001, Butterworth-Heinemann.

Hislop HJ, Montgomery J: *Daniels and Worthingham's muscle testing: techniques of manual examination,* ed 7, Philadelphia, 2002, Saunders.

Johnson RJ: Acromioclavicular joint injuries: identifying and treating "separated shoulder" and other conditions practice essentials,

Harmon K, Rubin A, eds: *The Physician and Sportsmedicine* 29(11), 2001.

McMurtrie H, Rikel JK: *The coloring review guide to human anatomy,* New York, 1991, McGraw-Hill.

Muscolino JE: *The muscular system manual: the skeletal muscles of the human body,* ed 2, St. Louis, 2003, Elsevier Mosby.

Neumann, DA: *Kinesiology of the musculoskeletal system: foundations for physical rehabilitation,* St. Louis, 2002, Mosby.

Norkin CC, Levangie PK: *Joint structure and functional comprehensive analysis,* Philadelphia, 1983, Davis.

Rasch PJ: *Kinesiology and applied anatomy,* ed 7, Philadelphia, 1989, Lea & Febiger.

Seeley RR, Stephens TD, Tate P: *Anatomy & physiology,* ed 7, Dubuque, IA, 2006, McGraw-Hill.

Smith LK, Weiss EL, Lehmkuhl LD: *Brunnstrom's clinical kinesiology,* ed 5, Philadelphia, 1996, Davis.

Sobush DC, et al: The Lennie test for measuring scapula position in healthy young adult females: a reliability and validity study, *Journal of Orthopedic and Sports Physical Therapy* 23:39, January 1996.

Soderburg GL: *Kinesiology—application to pathological motion,* Baltimore, 1986, Williams & Wilkins.

Van De Graaff KM: *Human anatomy,* ed 6, Dubuque, IA. 2002, McGraw-Hill.

Williams CC: Posterior sternoclavicular joint dislocation emergencies series, Howe WB, ed.: *The Physician and Sportsmedicine* 27(2), 1999.

Chapter

4

The Shoulder Joint

Objectives

- To identify on a human skeleton or human subject selected bony structures of the shoulder joint

- To label on a skeletal chart selected bony structures of the shoulder joint

- To draw on a skeletal chart the muscles of the shoulder joint and indicate, using arrows, shoulder joint movements

- To demonstrate with a fellow student all of the movements of the shoulder joints and list their respective planes and axes of rotation

- To learn and understand how movements of the scapula accompany movements of the humerus in achieving movement of the entire shoulder complex

- To determine and list the muscles of the shoulder joint and their antagonists

- To organize and list the muscles that produce the movements of the shoulder girdle and the shoulder joint

---

Online Learning Center Resources

Visit *Manual of Structural Kinesiology's* Online Learning Center at **www.mhhe.com/floyd16e** for additional information and study material for this chapter, including:

- *Self-grading quizzes*
- *Anatomy flashcards*
- *Animations*

The only attachment of the shoulder joint to the axial skeleton is via the scapula and its attachment through the clavicle at the sternoclavicular joint. Movements of the shoulder joint are many and varied. It is unusual to have movement of the humerus without scapula movement. When the humerus is flexed above shoulder level, the scapula is elevated, rotated upward, and abducted. With glenohumeral abduction above shoulder level, the scapula is rotated upward and elevated. Adduction of the humerus results in rotation downward and depression, whereas extension of the humerus results in depression, rotation downward, and adduction of the scapula. The scapula abducts with humeral internal rotation and horizontal adduction. Scapula adduction accompanies external rotation and horizontal abduction of the humerus. For a summary of these movements and the muscles primarily responsible for them, refer to Table 5.1.

Because the shoulder joint has such a wide range of motion in so many different planes, it also has a significant amount of laxity, which often results in instability problems such as rotator cuff impingement, subluxations, and dislocations. The price of mobility is reduced stability. The concept that the more mobile a joint is, the less stable it is and that the more stable it is, the less mobile it is applies generally throughout the body, but particularly in the shoulder joint.

Bones

The scapula, clavicle, and humerus serve as attachments for most of the muscles of the shoulder joint. Learning the specific location and importance

TABLE 5.1 • Pairing of shoulder girdle and shoulder joint movements. When the muscles of the shoulder joint (second column) perform the actions in the first column through any substantial range of motion, the muscles of the shoulder girdle (fourth column) work in concert by performing the actions in the third column.

Shoulder joint actions	Shoulder joint agonists	Shoulder girdle actions	Shoulder girdle agonists
Abduction	Supraspinatus, deltoid, upper pectoralis major	Upward rotation	Serratus anterior, middle and lower trapezius
Adduction	Latissimus dorsi, teres major, lower pectoralis major	Downward rotation	Pectoralis minor, rhomboids
Flexion	Anterior deltoid, upper pectoralis major	Elevation/upward rotation	Levator scapulae, serratus anterior, upper and middle trapezius, rhomboids
Extension	Latissimus dorsi, teres major, lower pectoralis major, posterior deltoid	Depression/downward rotation	Pectoralis minor, lower trapezius
Internal rotation	Latissimus dorsi, teres major, pectoralis major, subscapularis	Abduction (protraction)	Serratus anterior, pectoralis minor
External rotation	Infraspinatus, teres minor	Adduction (retraction)	Middle and lower trapezius, rhomboids
Horizontal abduction	Middle and posterior deltoid, infraspinatus, teres minor	Adduction (retraction)	Middle and lower trapezius, rhomboids
Horizontal adduction	Pectoralis major, anterior deltoid, coracobrachialis	Abduction (protraction)	Serratus anterior, pectoralis minor
Diagonal abduction (overhand activities)	Posterior deltoid, infraspinatus, teres minor	Adduction (retraction)/upward rotation/elevation	Trapezius, rhomboids, serratus anterior, levator scapulae
Diagonal adduction (overhand activities)	Pectoralis major, anterior deltoid, coracobrachialis	Abduction (protraction)/depression/downward rotation	Serratus anterior, pectoralis minor

of certain bony landmarks is critical to understanding the functions of the shoulder complex. Some of these scapular landmarks are the supraspinatus fossa, infraspinatus fossa, subscapular fossa, spine of the scapula, glenoid cavity, coracoid process, acromion process, and inferior angle. Humeral landmarks are the head, greater tubercle, lesser tubercle, intertubercular groove, and deltoid tuberosity (review Figs. 4.1, 4.2 and see Fig. 5.1).

Joint

The shoulder joint, specifically known as the glenohumeral joint, is a multiaxial ball-and-socket joint classified as enarthrodial (Fig. 5.1). As such, it moves in all planes and is the most movable joint in the body. Its stability is enhanced slightly by the glenoid labrum, a cartilaginous ring that surrounds the glenoid fossa just inside its periphery. It is further stabilized by the glenohumeral ligaments, especially anteriorly and inferiorly. The anterior glenohumeral ligaments become taut as external rotation, extension, abduction, and horizontal abduction occur, whereas the very thin posterior capsular ligaments become taut in internal rotation, flexion, and horizontal adduction. In recent years the importance of the inferior glenohumeral ligament in providing both anterior and posterior stability has come to light (see Figs. 5.2 and 5.3). It is, however, important to note that, due to the wide range of motion involved in the glenohumeral joint, the ligaments are quite lax until the extreme ranges of motion are reached. Stability is sacrificed to gain mobility.

Movement of the humerus from the side position is common in throwing, tackling, and striking

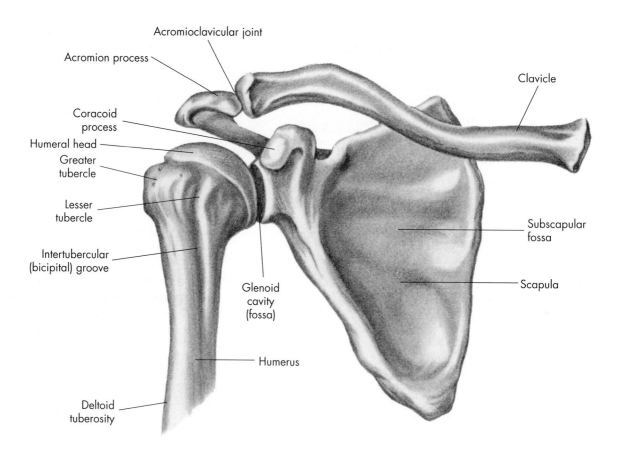

FIG. 5.1 ● Right glenohumeral joint, anterior view.

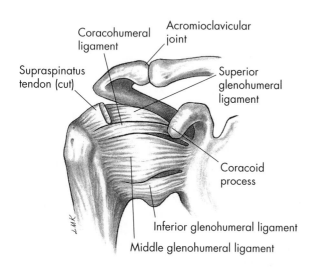

FIG. 5.2 ● Glenohumeral ligaments, anterior view.

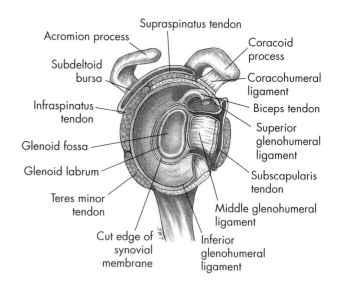

FIG. 5.3 ● Glenohumeral joint, lateral view with humerus removed.

activities. Flexion and extension of the shoulder joint are performed frequently when supporting body weight in a hanging position or in a movement from a prone position on the ground.

Determining the exact range of each movement for the glenohumeral joint is difficult because of the accompanying shoulder girdle movement. However, if the shoulder girdle is prevented from moving, then the glenohumeral joint movements are generally thought to be in the following ranges: 90 to 95 degrees abduction, 0 degrees adduction (prevented by the trunk) or 75 degrees anterior to the trunk, 40 to 60 degrees of extension, 90 to 100 degrees of flexion, 70 to 90 degrees of internal and external rotation, 45 degrees of horizontal abduction, and 135 degrees of horizontal adduction. If the shoulder girdle is free to move, then the total range of the combined joints is 170 to 180 degrees of abduction and 170 to 180 degrees of flexion.

The shoulder joint is frequently injured because of its anatomical design. A number of factors contribute to its injury rate, including the shallowness of the glenoid fossa, the laxity of the ligamentous structures necessary to accommodate its wide range of motion, and the lack of strength and endurance in the muscles, which are essential in providing dynamic stability to the joint. As a result, anterior or anteroinferior glenohumeral subluxations and dislocations are quite common with physical activity. Although posterior dislocations are fairly rare in occurrence, problems about the shoulder due to posterior instability are somewhat commonplace.

Another frequent injury is to the rotator cuff. The subscapularis, supraspinatus, infraspinatus, and teres minor muscles make up the rotator cuff. They are small muscles whose tendons cross the front, top, and rear of the head of the humerus to attach on the lesser and greater tuberosities, respectively. Their point of insertion enables them to rotate the humerus, an essential movement in this freely movable joint. Most important, however, is the vital role that the rotator cuff muscles play in maintaining the humeral head in correct approximation within the glenoid fossa while the more powerful muscles of the joint move the humerus through its wide range of motion.

Movements   FIG. 5.4, & FIG. 5.5

Flexion

Movement of the humerus straight anteriorly from any point in the sagittal plane.

Extension

Movement of the humerus straight posteriorly from any point in the sagittal plane, sometimes referred to as hyperextension.

Abduction

Upward lateral movement of the humerus in the frontal plane out to the side, away from the body.

Adduction

Downward movement of the humerus in the frontal plane medially toward the body from abduction.

External rotation

Movement of the humerus laterally in the transverse plane around its long axis away from the midline.

Internal rotation

Movement of the humerus in the transverse plane medially around its long axis toward the midline.

Horizontal adduction (flexion)

Movement of the humerus in a horizontal or transverse plane toward and across the chest.

Horizontal abduction (extension)

Movement of the humerus in a horizontal or transverse plane away from the chest.

Diagonal abduction

Movement of the humerus in a diagonal plane away from the midline of the body.

Diagonal adduction

Movement of the humerus in a diagonal plane toward the midline of the body.

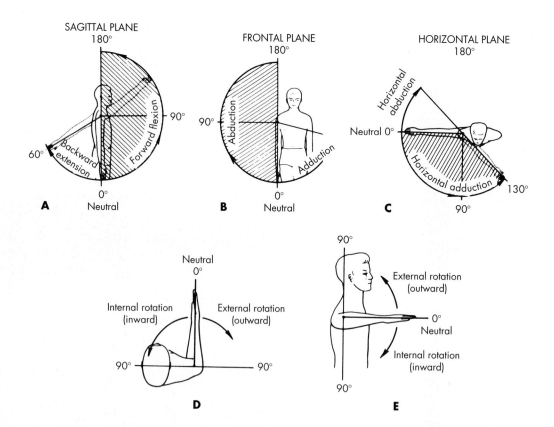

FIG. 5.4 ● Range of motion of the shoulder. **A,** Flexion and extension; **B,** Abduction and adduction; **C,** Horizontal abduction and adduction; **D,** Internal and external rotation with the arm at the side of the body; and **E,** Internal and external rotation with the arm abducted to 90 degrees.

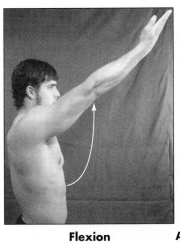

Flexion   A

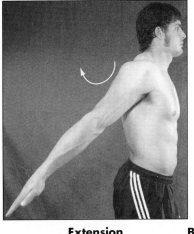

Extension   B

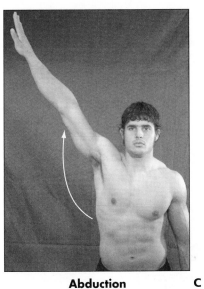

Abduction   C

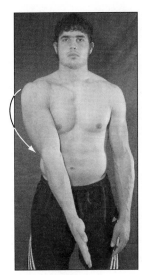

Adduction   D

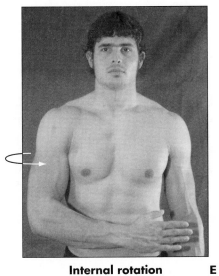

Internal rotation   E

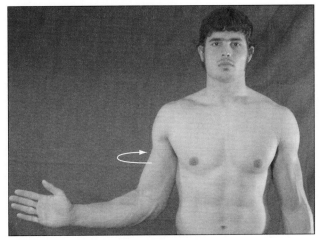

External rotation   F

FIG. 5.5 ● Movements of the shoulder joint.

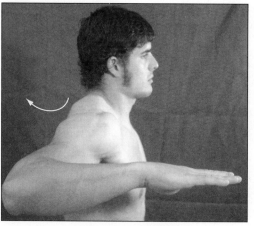

Horizontal abduction   **G**

Horizontal adduction   **H**

Diagonal abduction   **I**

Diagonal adduction   **J**

FIG. 5.5 (continued) ● Movements of the shoulder joint.

Muscles

When attempting to learn and understand the muscles of the glenohumeral joint, it may be helpful to group them according to their location and function. Muscles that originate on the scapula and clavicle and insert on the humerus may be thought of as muscles intrinsic to the glenohumeral joint, whereas muscles originating on the trunk and inserting on the humerus are considered extrinsic to the joint. The intrinsic muscles include the deltoid, the coracobrachialis, the teres major, and the rotator cuff group, which is composed of the subscapularis, the supraspinatus, the infraspinatus, and the teres minor. Extrinsic glenohumeral muscles are the latissimus dorsi and pectoralis major. It may also be

helpful to organize the muscles according to their general location. The pectoralis major, coracobrachialis, and subscapularis are anterior muscles. Located superiorly are the deltoid and supraspinatus. The latissimus dorsi, teres major, infraspinatus, and teres minor are located posteriorly. Table 5.1 lists the glenohumeral joint movements and the muscles primarily responsible for them, and Table 5.2 provides the action of each muscle.

The biceps brachii and triceps brachii (long head) are also involved in glenohumeral movements. Primarily, the biceps brachii assists in flexing and horizontally adducting the shoulder, whereas the long head of the triceps brachii assists in extension and horizontal abduction. Further discussion of these muscles appears in Chapter 6.

Shoulder joint muscles—location

Anterior
> Pectoralis major
> Coracobrachialis
> Subscapularis

Superior
> Deltoid
> Supraspinatus

Posterior
> Latissimus dorsi
> Teres major
> Infraspinatus
> Teres minor

Muscle identification

In Figs. 5.6 and 5.7, identify the anterior and posterior muscles of the shoulder joint and shoulder girdle. Compare Fig. 5.6 with Fig. 5.8 and Fig. 5.7 with Fig. 5.9, and refer to Table 5.2 for a detailed breakdown of the agonist muscles for the gleno-humeral joint.

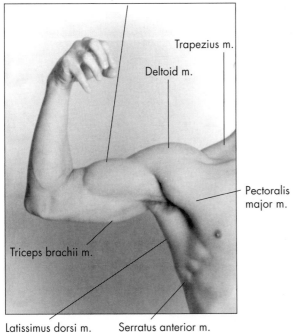

FIG. 5.6 ● Anterior shoulder joint and shoulder girdle muscles.

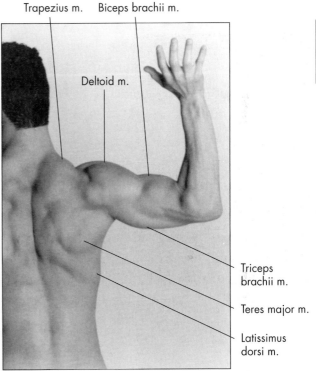

FIG. 5.7 ● Posterior shoulder joint and shoulder girdle muscles.

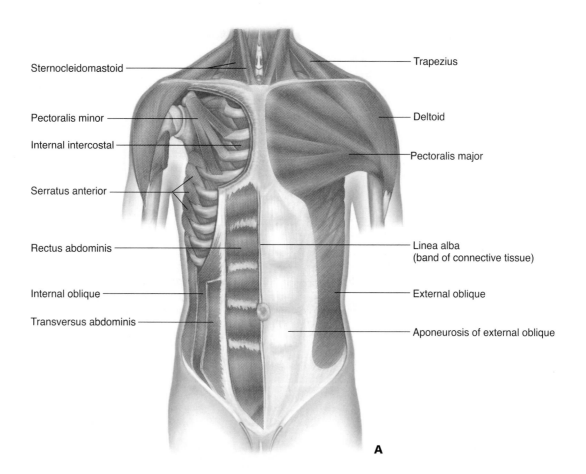

Sternocleidomastoid

Pectoralis minor

Internal intercostal

Serratus anterior

Rectus abdominis

Internal oblique

Transversus abdominis

Trapezius

Deltoid

Pectoralis major

Linea alba
(band of connective tissue)

External oblique

Aponeurosis of external oblique

A

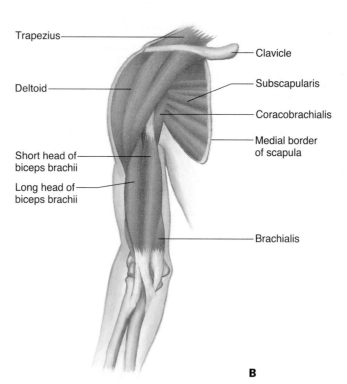

Trapezius

Deltoid

Short head of
biceps brachii

Long head of
biceps brachii

Clavicle

Subscapularis

Coracobrachialis

Medial border
of scapula

Brachialis

B

FIG. 5.8 ● Anterior muscles of the shoulder. **A,** The right pectoralis major is removed to show the pectoralis minor and serratus anterior; **B,** Muscles of the anterior right shoulder and arm, with the rib cage removed.

From Shier D, Butler J, Lewis R: *Hole's essentials of human anatomy and physiology,* ed 9, New York, 2006, McGraw-Hill.

Chapter

5

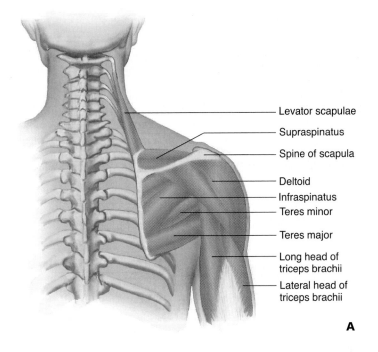

Levator scapulae
Supraspinatus
Spine of scapula
Deltoid
Infraspinatus
Teres minor
Teres major
Long head of triceps brachii
Lateral head of triceps brachii

A

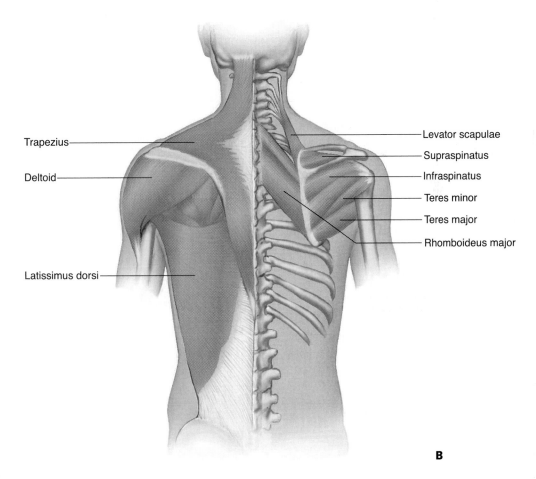

Trapezius
Deltoid
Latissimus dorsi

Levator scapulae
Supraspinatus
Infraspinatus
Teres minor
Teres major
Rhomboideus major

B

FIG. 5.9 ● Posterior muscles of the shoulder. **A,** The right trapezius and deltoid are removed to show the underlying muscles; **B,** Muscles of the posterior surface of the scapula and arm.

From Shier D, Butler J, Lewis R: *Hole's essentials of human anatomy and physiology,* ed 9, New York, 2006, McGraw-Hill.

TABLE 5.2 • Agonist muscles of the glenohumeral joint

	Muscle	Origin	Insertion	Action	Plane of motion	Palpation	Innervation
Anterior muscles	Pectoralis major upper fibers	Medial half of anterior surface of clavicle	Flat tendon 2 or 3 inches wide to lateral lip of intertubercular groove of humerus	Internal rotation	Transverse	From medial end of clavicle to the intertubercular groove of the humerus, during flexion and adduction from the anatomical position	Lateral pectoral nerve (C5, C6, C7)
				Horizontal adduction			
				Diagonal adduction	Diagonal		
				Flexion	Sagittal		
				Abduction	Frontal		
	Pectoralis major lower fibers	Anterior surface of costal cartilages of first six ribs, and adjoining portion of sternum	Flat tendon 2 or 3 inches wide to lateral lip of intertubercular groove of humerus	Internal rotation	Transverse	From the lower ribs and sternum to the intertubercular groove of the humerus, during resisted extension from a flexed position	Medial pectoral nerve (C8, T1)
				Horizontal adduction			
				Diagonal adduction	Diagonal		
				Extension	Sagittal		
				Adduction	Frontal		
	Sub-scapularis	Entire anterior surface of subscapular fossa	Lesser tubercle of humerus	Internal rotation	Transverse	Mostly innaccessible, lateral portion may be palpated on supine subject (arm in slight flexion and adduction with elbow lying across abdomen); pull medial border laterally with one hand while palpating between the scapula and rib cage with other hand (subject actively internally rotates)	Upper and lower subscapular nerve (C5, C6)
	Coraco-brachialis	Coracoid process of scapula	Middle of medial border of humeral shaft	Horizontal adduction	Transverse	The belly may be palpated high up on the medial arm just posterior to the short head of the biceps brachii and toward the coracoid process, particularly with resisted adduction	Musculo-taneous nerve (C5, C6, C7)
				Diagonal adduction	Diagonal		
Superior muscles	Deltoid anterior fibers	Anterior lateral third of clavicle	Deltoid tuberosity on the lateral humerus	Abduction	Frontal	From the clavicle toward the anterior humerus during resisted flexion or horizontal adduction	Axillary nerve (C5, C6)
				Flexion	Sagittal		
				Horizontal adduction	Transverse		
				Diagonal adduction	Diagonal		
	Deltoid middle fibers	Lateral aspect of acromion	Deltoid tuberosity on lateral humerus	Abduction	Frontal	From the lateral border of the acromion down toward the deltoid tuberosity during resisted abduction	Axillary nerve (C5, C6)
				Horizontal abduction	Transverse		
	Deltoid posterior fibers	Inferior edge of spine of scapula	Deltoid tuberosity on lateral humerus	Abduction	Frontal	From the lower lip of the spine of the scapula toward the posterior humerus during resisted extension or horizontal abduction	Axillary nerve (C5, C6)
				Horizontal abduction	Transverse		
				Diagonal abduction	Diagonal		
	Supra-spinatus	Medial 2/3 of supraspinatus fossa	Superiorly on greater tubercle of humerus	Abduction	Frontal	Above the spine of the scapula in supraspinatus fossa during initial abduction in the scapula plane, tendon may be palpated just off acromion on greater tubercle	Suprascapula nerve (C5)

TABLE 5.2 (continued) • Agonist muscles of the glenohumeral joint

	Muscle	Origin	Insertion	Action	Plane of motion	Palpation	Innervation
Posterior muscles	Latissimus dorsi	Posterior crest of ilium, back of sacrum and spinous processes of lumbar and lower six thoracic vertebrae, slips from lower three ribs	Medial side of intertubercular groove of humerus, just anterior to the insertion of the teres major	Extension	Sagittal	The tendon may be palpated as it passes under the teres major at the posterior axillary wall, particularly during resisted extension and internal rotation. The muscle can be palpated in the upper lumbar/lower thoracic area during extension from a flexed position and throughout most of its length during resisted adduction from a slightly abducted position	Thoracodorsal (C6, C7, C8)
				Adduction	Frontal		
				Internal rotation	Transverse		
	Teres major	Posteriorly on inferior third of lateral border of scapula and just superior to inferior angle	Medial lip of intertubercular groove of humerus, just posterior to the insertion of the latissimus dorsi	Extension	Sagittal	Just above the latissimus dorsi and below the teres minor on the posterior scapula surface, moving diagonally upward and laterally from the inferior angle of the scapula during resisted internal rotation	Lower subscapular nerve (C5, C6)
				Adduction	Frontal		
				Internal rotation	Transverse		
	Infraspinatus	Medial aspect of infraspinatus fossa just below spine of scapula	Posteriorly on greater tubercle of humerus	External rotation	Transverse	Just below the spine of the scapula passing upward and laterally to the humerus during resisted external rotation	Suprascapula nerve (C5, C6)
				Horizontal abduction			
				Diagonal abduction	Diagonal		
	Teres minor	Posteriorly on upper and middle aspect of lateral border of scapula	Posteriorly on greater tubercle of humerus	External rotation	Transverse	Just above the teres major on the posterior scapula surface, moving diagonally upward and laterally from the inferior angle of the scapula during resisted external rotation	Axillary nerve (C5, C6)
				Horizontal abduction			
				Diagonal adduction	Diagonal		

Note: The biceps brachii assists in flexion, horizontal adduction, and diagonal adduction, while the long head of the triceps brachii assists in extension, adduction, horizontal abduction, and diagonal abduction. Due to their coverage in Chapter 6, neither is listed above.

Chapter

5

Nerves   FIG. 5.10 & FIG.5.11

The muscles of the shoulder joint are all inner-vated from the nerves of the brachial plexus. The pectoralis major is innervated by the pectoral nerves. Specifically, the lateral pectoral nerve aris-ing from C5, C6, and C7 innervates the clavicu-lar head, while the medial pectoral nerve arising from C8 and T1 innervates the sternal head. The thoracodorsal nerve, arising from C6, C7, and C8, supplies the latissimus dorsi. The axillary nerve (Fig. 5.10), branching from C5 and C6, innervates the deltoid and teres minor. A lateral patch of skin over the deltoid region of the arm is provided sen-sation by the axillary nerve. Both the upper and lower subscapular nerves arising from C5 and C6 innervate the subscapularis, while only the lower subscapular nerve supplies the teres major. The supraspinatus and infraspinatus are innervated by the suprascapula nerve, which originates from C5 and C6. The musculocutaneous nerve, as seen in Fig. 5.11, branches from C5, C6, and C7 and in-nervates the coracobrachialis. It supplies sensa-tion to the radial aspect of the forearm.

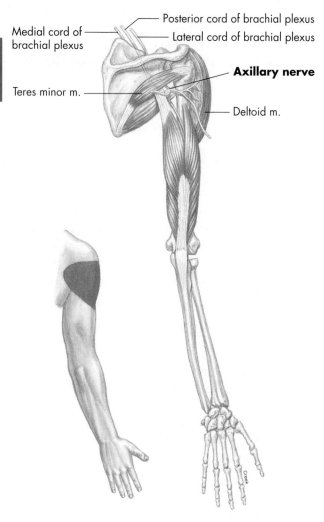

FIG. 5.10 ● Muscular and cutaneous distribution of the axillary nerve.

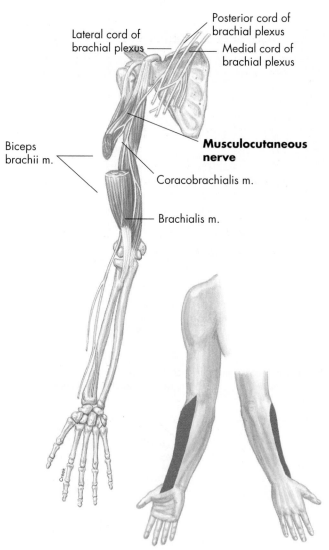

FIG. 5.11 ● Muscular and cutaneous distribution of the musculocutaneous nerve.

Deltoid muscle   FIG. 5.12

(del-toyd´)

Origin

Anterior fibers: anterior lateral third of the clavicle
Middle fibers: lateral aspect of the acromion
Posterior fibers: inferior edge of the spine of the scapula

Insertion

Deltoid tuberosity on the lateral humerus

Action

Anterior fibers: abduction, flexion, horizontal adduction, and internal rotation of the glenohumeral joint
Middle fibers: abduction of the glenohumeral joint
Posterior fibers: abduction, extension, horizontal abduction, and external rotation of the glenohumeral joint

Palpation

Anterior fibers: from the clavicle toward the anterior humerus during resisted flexion or horizontal adduction
Middle fibers: from the lateral border of the acromion down toward the deltoid tuberosity during resisted abduction
Posterior fibers: from the lower lip of the spine of the scapula toward the posterior humerus during resisted extension or horizontal abduction

Innervation

Axillary nerve (C5, C6).

Application, strengthening, and flexibility

The deltoid muscle is used commonly in any lifting movement. The trapezius muscle stabilizes the scapula as the deltoid pulls on the humerus. The anterior fibers of the deltoid muscle flex and internally rotate the humerus. The posterior fibers extend and externally rotate the humerus. The anterior fibers also horizontally adduct the humerus while the posterior fibers horizontally abduct it. This muscle is used in all lifting movements. Any movement of the humerus on the scapula will involve part or all of the deltoid muscle.

Lifting the humerus from the side to the position of abduction is a typical action of the deltoid. Side-arm dumbbell raises are excellent for strengthening the deltoid, especially the middle fibers. By abducting the arm in a slightly horizontally adducted (30 degrees) position, the anterior deltoid fibers can be emphasized. The posterior fibers can be strengthened better by abducting the arm in a slightly horizontally abducted (30 degrees) position.

Stretching the deltoid requires varying positions, depending on the fibers to be stretched. The anterior deltoid is stretched by taking the humerus into extreme horizontal abduction or by extreme extension and adduction. The middle deltoid is stretched by taking the humerus into extreme adduction behind the back. Extreme horizontal adduction stretches the posterior deltoid.

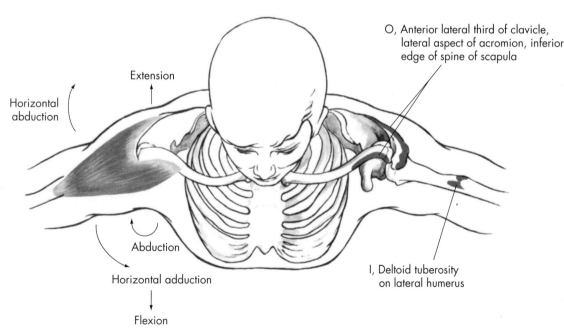

O, Anterior lateral third of clavicle, lateral aspect of acromion, inferior edge of spine of scapula

Extension

Horizontal abduction

Abduction

Horizontal adduction

Flexion

I, Deltoid tuberosity on lateral humerus

FIG. 5.12 ● Deltoid muscle, superior view. *O,* Origin; *I,* insertion.

Pectoralis major muscle FIG. 5.13

(pek-to-ra´lis ma´jor)

Origin

Upper fibers (clavicular head): medial half of the anterior surface of the clavicle

Lower fibers (sternal head): anterior surfaces of the costal cartilage of the first six ribs, and adjacent portion of the sternum

Insertion

Flat tendon 2 or 3 inches wide to the lateral lip of the intertubercular groove of the humerus

Action

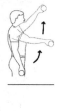

Upper fibers (clavicular head): internal rotation, horizontal adduction, flexion up to about 60 degrees, abduction (once the arm is abducted 90 degrees, the upper fibers assist in further abduction), and adduction (with the arm below 90 degrees of abduction) of the glenohumeral joint

Lower fibers (sternal head): internal rotation, horizontal adduction, adduction and extension of the glenohumeral joint from a flexed position to the anatomical position

Palpation

Upper fibers: from the medial end of the clavicle to the intertubercular groove of the humerus, during flexion and adduction from the anatomical position

Lower fibers: from the ribs and sternum to the intertubercular groove of the humerus, during resisted extension from a flexed position and resisted adduction from the anatomical position

Innervation

Upper fibers: lateral pectoral nerve (C5–C7)

Lower fibers: medial pectoral nerve (C8, T1)

Application, strengthening, and flexibility

The anterior axillary fold is formed primarily by the pectoralis major. It aids the serratus anterior muscle in drawing the scapula forward as it moves the humerus in flexion and internal rotation. Even though the pectoralis major is not attached to the scapula, it is effective in this scapula protraction because of its anterior pull on the humerus, which joins

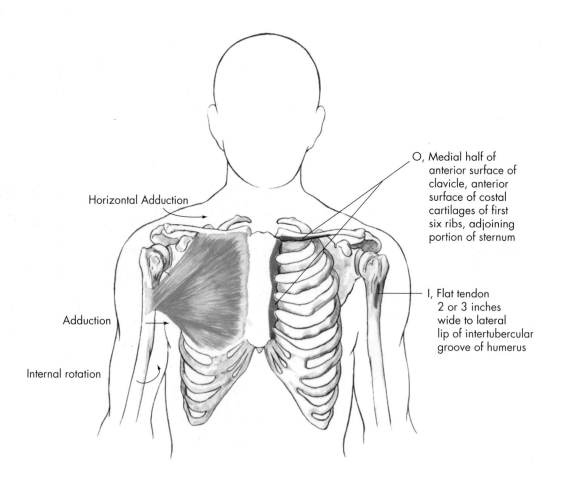

Horizontal Adduction

Adduction

Internal rotation

O, Medial half of anterior surface of clavicle, anterior surface of costal cartilages of first six ribs, adjoining portion of sternum

I, Flat tendon 2 or 3 inches wide to lateral lip of intertubercular groove of humerus

FIG. 5.13 ● Pectoralis major muscle. *O*, Origin; *I*, Insertion.

to the scapula at the glenohumeral joint. Typical action is shown in throwing a baseball. As the glenohumeral joint is flexed, the humerus is internally rotated and the scapula is drawn forward with upward rotation. It also works as a helper of the latissimus dorsi muscle when extending and adducting the humerus from a raised position.

The pectoralis major and the anterior deltoid work closely together. The pectoralis major is used powerfully in push-ups, pull-ups, throwing, and tennis serves. With a barbell, the subject takes a supine position on a bench with the arms at the side and moves the arms to a horizontally adducted position. This exercise, known as bench pressing, is widely used for pectoralis major development.

Due to the popularity of bench pressing and other weight-lifting exercises that emphasize the pectoralis major and its use in most sporting activities, it is often overdeveloped in comparison to its antagonists. As a result, stretching is often needed and can be done by passive external rotation. It is also stretched when the shoulder is horizontally abducted. Extending the shoulder fully provides stretching to the upper pectoralis major, while full abduction stretches the lower pectoralis major.

Latissimus dorsi muscle   FIG. 5.14

(lat-is´i-mus dor´si)

Origin

Posterior crest of the ilium, back of the sacrum and spinous processes of the lumbar and lower six thoracic vertebrae (T6–T12); slips from the lower three ribs

Insertion

Medial lip of the intertubercular groove of the humerus, just anterior to the insertion of the teres major

Action

Adduction of the glenohumeral joint
Extension of the glenohumeral joint
Internal rotation of the glenohumeral joint
Horizontal abduction of the glenohumeral joint

Palpation

The tendon may be palpated as it passes under the teres major at the posterior wall of the axilla, particularly during resisted extension and internal rotation. The muscle can be palpated in the upper lumbar/lower thoracic area during extension from a flexed position. The muscle may be palpated throughout most of its length during resisted adduction from a slightly abducted position.

Innervation

Thoracodorsal nerve (C6–C8)

Application, strengthening, and flexibility

The latissimus dorsi, along with the teres major, forms the posterior axillary fold. It has a strong action in adduction, extension, and internal rotation of the humerus. Due to the upward rotation of the scapula that accompanies glenohumeral abduction, the latissimus effectively downwardly rotates the scapula by way of its action in pulling the entire shoulder girdle downward in active glenohumeral adduction. It is one of the most important extensor muscles of the humerus and contracts powerfully in chinning. The latissimus dorsi is assisted in all of its actions by the teres major and is sometimes referred to as the swimmer's muscle because of its function in pulling the body forward in the water during internal rotation, adduction, and extension.

Exercises in which the arms are pulled down bring the latissimus dorsi muscle into powerful contraction. Chinning, rope climbing, dips on parallel bars, and other uprise movements on the horizontal bar are good examples. In barbell exercises, the basic rowing and pullover exercises are good for developing the "lats." Pulling the bar of an overhead pulley system down toward the shoulders, known as "lat pulls," is a common exercise for this muscle.

The latissimus dorsi is stretched with the teres major when the shoulder is externally rotated while in a 90-degree abducted position. This stretch may be accentuated further by abducting the shoulder fully while maintaining external rotation and then laterally flexing and rotating the trunk to the opposite side.

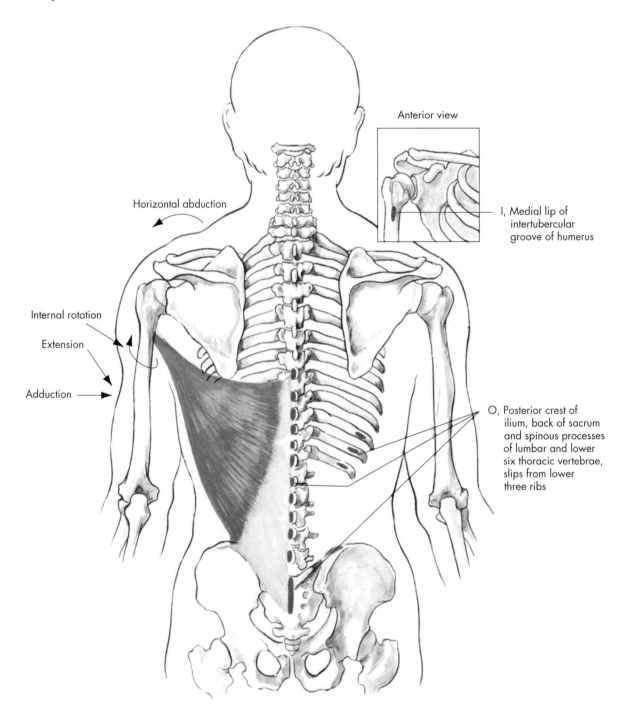

Anterior view

Horizontal abduction

Internal rotation

Extension

Adduction

I, Medial lip of intertubercular groove of humerus

O, Posterior crest of ilium, back of sacrum and spinous processes of lumbar and lower six thoracic vertebrae, slips from lower three ribs

FIG. 5.14 ● Latissimus dorsi muscle. *O*, Origin; *I*, Insertion.

Coracobrachialis muscle FIG. 5.15

(kor-a-ko-bra´ki-a´lis)

Origin
Coracoid process of the scapula

Insertion
Middle of the medial border of the humeral shaft

Action
Flexion of the glenohumeral joint
Adduction of the glenohumeral joint
Horizontal adduction of the glenohumeral joint

Palpation
The belly may be palpated high up on the medial
arm just posterior to the short head of the biceps
brachii and toward the coracoid process, particularly with resisted adduction.

Innervation
Musculocutaneous nerve (C5–C7)

Application, strengthening, and flexibility
The coracobrachialis is not a powerful muscle,
but it does assist in flexion and adduction and is
most functional in moving the arm horizontally
toward and across the chest. It is best strengthened by horizontally adducting the arm against
resistance, as in bench pressing. It may also be
strengthened by performing lat pulls, which are
defined on page 121.

The coracobrachialis is best stretched in extreme horizontal abduction, although extreme extension also stretches this muscle.

Chapter
5

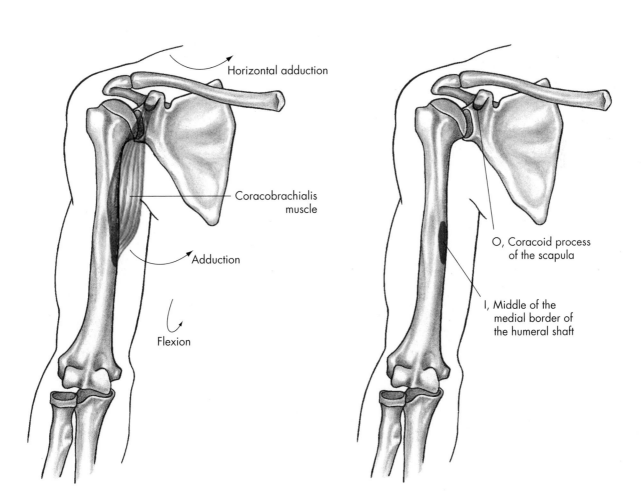

FIG. 5.15 ● Coracobrachialis muscle, anterior view. *O,* Origin; *I,* Insertion.

Rotator cuff muscles

Figs. 5.16 and 5.17 illustrate the rotator cuff muscle group, which, as previously mentioned, is most important in maintaining the humeral head in its proper location within the glenoid cavity. The acronym **SITS** may be used in learning the names of the supraspinatus, infraspinatus, teres minor, and subscapularis. These muscles, which are not very large in comparison with the deltoid and pectoralis major, must possess not only adequate strength but also a significant amount of muscular endurance to ensure their proper functioning, particularly in repetitious overhead activities such as throwing, swimming, and pitching. Quite often when these types of activities are conducted with poor technique, muscle fatigue, or inadequate warm-up and conditioning, the rotator cuff muscle group, particularly the supraspinatus, fails to dynamically stabilize the humeral head in the glenoid cavity, leading to further rotator cuff problems such as tendinitis and rotator cuff impingement within the subacromial space.

FIG. 5.16 ● Rotator cuff muscles, superior view, right shoulder.

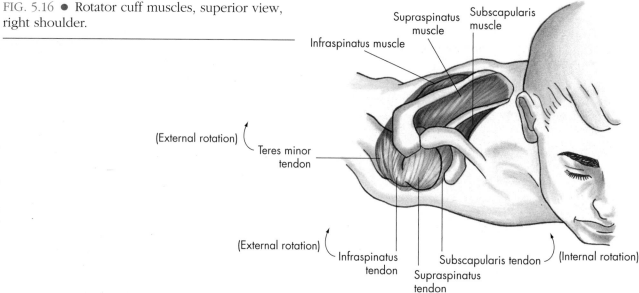

(External rotation)

Teres minor tendon

(External rotation)

Infraspinatus tendon

Supraspinatus tendon

(Abduction)

Subscapularis tendon

(Internal rotation)

Infraspinatus muscle

Supraspinatus muscle

Subscapularis muscle

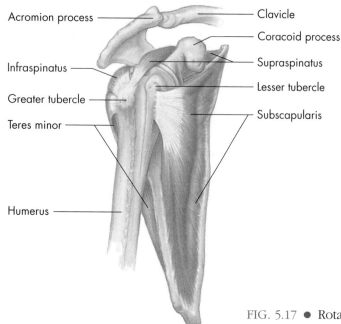

Acromion process

Infraspinatus

Greater tubercle

Teres minor

Humerus

Clavicle

Coracoid process

Supraspinatus

Lesser tubercle

Subscapularis

FIG. 5.17 ● Rotator cuff muscles, anterolateral view, right shoulder.

Subscapularis muscle FIG. 5.18

(sub-skap-u-la´ris)

Origin

Entire anterior surface of the subscapular fossa

Insertion

Lesser tubercle of the humerus

Action

Internal rotation of the glenohumeral joint
Adduction of the glenohumeral joint
Extension of the glenohumeral joint
Stabilization of the humeral head in the glenoid fossa

Palpation

The subscapularis, in conjunction with the latissimus dorsi and teres major, form the posterior axillary fold. Most of the subscapularis is inaccessible on the anterior scapula behind the rib cage. The lateral portion may be palpated with the subject supine and arm in slight flexion and adduction so that the elbow is lying across the abdomen. Use one hand posteriorly to grasp the medial border and pull it laterally while palpating between the scapula and rib cage with the other hand with the

subject actively internally rotating by pressing the forearm against the chest.

Innervation

Upper and lower subscapular nerve (C5, C6)

Application, strengthening, and flexibility

The subscapularis muscle, another rotator cuff muscle, holds the head of the humerus in the glenoid fossa from in front and below. It acts with the latissimus dorsi and teres major muscles in its typical movement but is less powerful in its action because of its proximity to the joint. The muscle also requires the help of the rhomboid in stabilizing the scapula to make it effective in the movements described. The subscapularis is relatively hidden behind the rib cage in its location on the anterior aspect of the scapula in the subscapular fossa. It may be strengthened with exercises similar to those used for the latissimus dorsi and teres major, such as rope climbing and lat pulls. A specific exercise for its development is done by internally rotating the arm against resistance in the beside-the-body position at 0 degrees of glenohumeral abduction.

External rotation with the arm adducted by the side stretches the subscapularis.

FIG. 5.18 ● Subscapularis muscle, anterior view. *O,* Origin; *I,* Insertion.

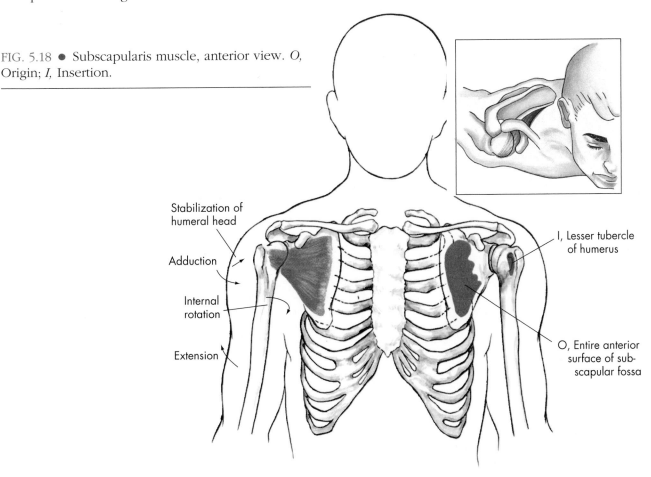

Stabilization of humeral head

Adduction

Internal rotation

Extension

I, Lesser tubercle of humerus

O, Entire anterior surface of subscapular fossa

Supraspinatus muscle   FIG. 5.19

(su´pra-spi-na´tus)

Origin

Medial two-thirds of the supraspinatus fossa

Insertion

Superiorly on the greater tubercle of the humerus

Action

Abduction

Stabilization of the humeral head in the glenoid fossa

Palpation

Anterior and superior to the spine of the scapula in the supraspinatus fossa during initial abduction in the scapula plane. Also, the tendon may be palpated in a seated position just off the acromion on the greater tubercle.

Innervation

Suprascapular nerve (C5)

Application, strengthening, and flexibility

The supraspinatus muscle holds the head of the humerus in the glenoid fossa. In throwing movements, it provides important dynamic stability by maintaining the proper relationship between the humeral head and the glenoid fossa. In the cocking phase of throwing, there is a tendency for the humeral head to subluxate anteriorly. In the

follow-through phase, the humeral head tends to move posteriorly.

The supraspinatus, along with the other rotator cuff muscles, must have excellent strength and endurance to prevent abnormal and excessive movement of the humeral head in the fossa.

The supraspinatus is the most often injured rotator cuff muscle. Acute severe injuries may occur with trauma to the shoulder. However, mild to moderate strains or tears often occur with athletic activity, particularly if the activity involves repetitious overhead movements, such as throwing or swimming.

Injury or weakness in the supraspinatus may be detected when the athlete attempts to substitute the scapula elevators and upward rotators to obtain humeral abduction. An inability to smoothly abduct the arm against resistance is indicative of possible rotator cuff injury.

The supraspinatus muscle may be called into play whenever the middle fibers of the deltoid muscle are used. An "empty-can exercise" may be used to emphasize supraspinatus action. This is performed by internally rotating the humerus, followed by abducting the arm to 90 degrees in a 30- to 45-degree horizontally adducted position, as if one were emptying a can.

Adducting the arm behind the back with the shoulder internally rotated and extended stretches the supraspinatus.

FIG. 5.19 ● Supraspinatus muscle, posterior view. *O*, Origin; *I*, Insertion.

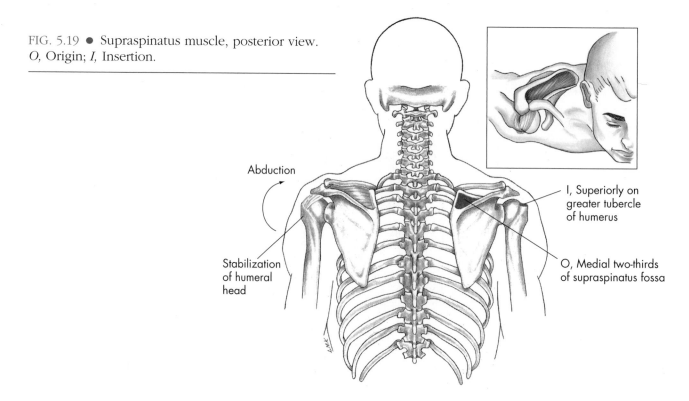

Abduction

Stabilization of humeral head

I, Superiorly on greater tubercle of humerus

O, Medial two-thirds of supraspinatus fossa

Infraspinatus muscle   FIG. 5.20

(in´fra-spi-na´tus)

Origin

Medial aspect of the infraspinatus fossa just below the spine of the scapula

Insertion

Posteriorly on the greater tubercle of the humerus

Action

External rotation of the glenohumeral joint
Horizontal abduction of the glenohumeral joint
Extension of the glenohumeral joint
Stabilization of the humeral head in the glenoid fossa

Palpation

Just below the spine of the scapula passing upward and laterally to the humerus during resisted external rotation

Innervation

Suprascapular nerve (C5, C6)

Application, strengthening, and flexibility

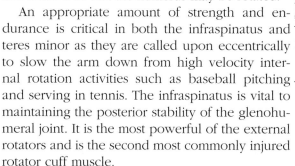

The infraspinatus and teres minor muscles are effective when the rhomboid muscles stabilize the scapula. When the humerus is rotated outward, the rhomboid muscles flatten the scapula to the back and fixate it so that the humerus may be rotated.

An appropriate amount of strength and endurance is critical in both the infraspinatus and teres minor as they are called upon eccentrically to slow the arm down from high velocity internal rotation activities such as baseball pitching and serving in tennis. The infraspinatus is vital to maintaining the posterior stability of the glenohumeral joint. It is the most powerful of the external rotators and is the second most commonly injured rotator cuff muscle.

Both the infraspinatus and the teres minor can best be strengthened by externally rotating the arm against resistance in the 15- to 20-degree abducted position and the 90-degree abducted position.

Stretching of the infraspinatus is accomplished with internal rotation and extreme horizontal adduction.

Chapter 5

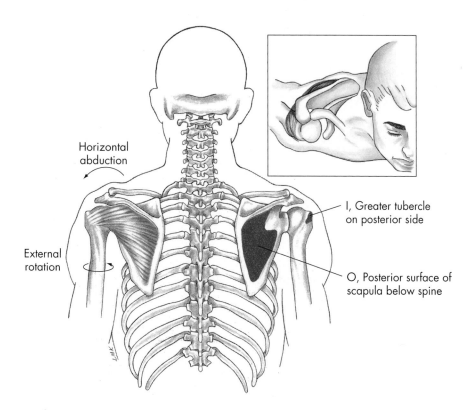

FIG. 5.20 ● Infraspinatus muscle, posterior view. *O,* Origin; *I,* Insertion.

Teres minor muscle   FIG. 5.21

(teˊrez miˊnor)

Origin

Posteriorly on the upper and middle aspect of the lateral border of the scapula

Insertion

Posteriorly on the greater tubercle of the humerus

Action

External rotation of the glenohumeral joint
Horizontal abduction of the glenohumeral joint
Extension of the glenohumeral joint
Stabilization of the humeral head in the glenoid fossa

Palpation

Just above the teres major on the posterior scapula surface, moving diagonally upward and laterally from the inferior angle of the scapula during resisted external rotation

Innervatio

Axillary nerve (C5, C6)

Application, strengthening, and flexibility

The teres minor functions very similarly to the infraspinatus in providing dynamic posterior stability to the glenohumeral joint. Both of these muscles perform the same actions together. The teres minor is strengthened with the same exercises that are used in strengthening the infraspinatus.

The teres minor is stretched similarly to the infraspinatus by internally rotating the shoulder while moving into extreme horizontal adduction.

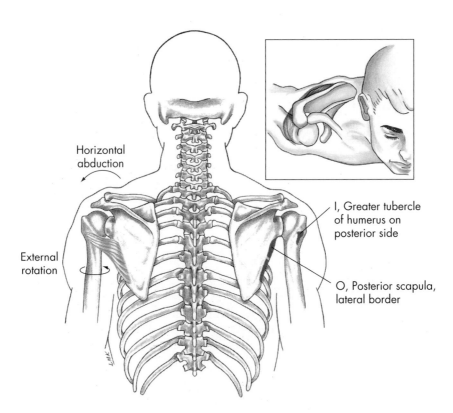

FIG. 5.21 ● Teres minor muscle, posterior view. *O*, Origin; *I*, Insertion.

Teres major muscle   FIG. 5.22

(te´rez ma´jor)

Origin

Posteriorly on the inferior third of the lateral border of the scapula and just superior to the inferior angle

Insertion

Medial lip of the intertubercular groove of the humerus just posterior to the insertion of the latissimus dorsi

Action

Extension of the glenohumeral joint, particularly from the flexed position to the posteriorly extended position

Internal rotation of the glenohumeral joint

Adduction of the glenohumeral joint, particularly from the abducted position down to the side and toward the midline of the body

Palpation

Just above the latissimus dorsi and below the teres minor on the posterior scapula surface, moving diagonally upward and laterally from the inferior angle of the scapula during resisted internal rotation

Innervation

Lower subscapular nerve (C5, C6)

Application, strengthening, and flexibility

The teres major muscle is effective only when the rhomboid muscles stabilize the scapula or move the scapula in downward rotation. Otherwise, the scapula would move forward to meet the arm.

This muscle works effectively with the latissimus dorsi. It assists the latissimus dorsi, pectoralis major, and subscapularis in adducting, internally rotating, and extending the humerus. It is said to be the latissimus dorsi's "little helper." It may be strengthened by lat pulls, rope climbing, and internal rotation exercises against resistance.

Externally rotating the shoulder in a 90-degree abducted position stretches the teres major.

Chapter
5

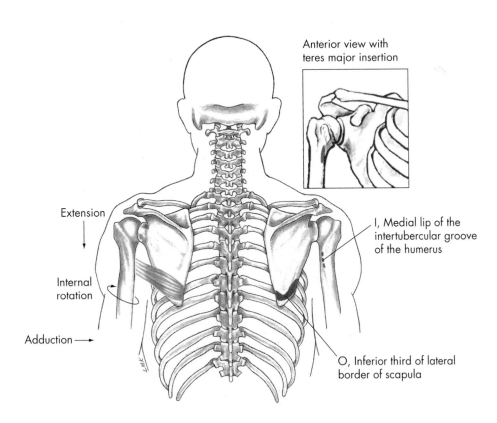

FIG. 5.22 ● Teres major muscle, posterior view. *O*, Origin; *I*, Insertion.

Web sites

Radiologic Anatomy Browser

http://radlinux1.usuf1.usuhs.mil/rad/iong/index.html

This site has numerous radiological views of the musculoskeletal system.

University of Arkansas Medical School Gross Anatomy for Medical Students

http://anatomy.uams.edu/anatomyhtml/grossresources.html

Dissections, anatomy tables, atlas images, links, etc.

Loyola University Medical Center: Structure of the Human Body

www.meddean.luc.edu/lumen/MedEd/GrossAnatomy/GA.html

An excellent site with many slides, dissections, tutorials, etc., for the study of human anatomy.

Wheeless' Textbook of Orthopaedics

www.wheelessonline.com

This site has an extensive index of links to the fractures, joints, muscles, nerves, trauma, medications, medical topics, lab tests, and links to orthopaedic journals and other orthopaedic and medical news.

Premiere Medical Search Engine

www.medsite.com/Default.asp?bhcp=1

This site allows the reader to enter any medical condition, and it will search the net to find relevant articles.

Arthroscopy.Com

www.arthroscopy.com/sports.htm

Patient information on various musculoskeletal problems of the upper and lower extremity.

Virtual Hospital

www.vh.org

Numerous slides, patient information, etc.

Medical Multimedia Group

www.healthpages.org/AHP/LIBRARY/HLTHTOP/CTD

A patient's guide to cumulative trauma disorder (CTD).

Baseball Almanac

www.baseball-almanac.com/chapters/cap-ch8.shtml

Coaching adult pitchers.

Lecture Topics in Kinesiology

http://moon.ouhsc.edu/dthompso/namics/shoulder.htm

Shoulder articulations, movements, and muscles that are within the shoulder girdle.

The Physician and Sportsmedicine

www.physsportsmed.com/issues/2003/0703/depalma.htm

Detecting and treating shoulder impingement syndrome: the role of scapulothoracic dyskinesis.

Southern California Orthopedic Institute

www.scoi.com/sholanat.htm

Anatomy of the shoulder.

FamilyDoctor.org

http://familydoctor.org/268.xml

Shoulder pain.

MedlinePlus

www.nlm.nih.gov/medlineplus/tutorials/shoulderarthroscopy/htm/index.htm

Shoulder arthroscopy interactive tutorial.

MedlinePlus

www.nlm.nih.gov/medlineplus/tutorials/rotatorcuffinjuries/htm/index.htm

Rotator cuff injuries interactive tutorial.

American Physical Therapy Association

www.apta.org/AM/Template.cfm?Section=Home&TEMPLATE=/CM/HTMLDisplay.cfm&CONTENTID=20448

Taking care of your shoulder.

American Academy of Orthapaedic Surgeons

http://orthoinfo.aaos.org/category.cfm?topcategory=Shoulder

Patient education library on the shoulder.

Orthopaedic Research Institute

www.ori.org.au/bonejoint/shoulder/contents.htm

Several Web pages, text, and graphics on glenohumeral instability.

American Sports Medicine Institute

www.asmi.org/asmiweb/mpresentations/mmp.htm

Biomechanics of the shoulder during throwing.

American Sports Medicine Institute

www.asmi.org.SportsMed/throwing/thrower10.html

Throwers' ten exercises.

Washington Musculoskeletal Tumor Center
www.sarcoma.org/main.php?page=shoulder
Shoulder girdle surgery.

Worksheet exercises

As an aid to learning, for in-class or out-of-class assignments, or for testing, tear-out worksheets are found at the end of the text (pp. 376 and 377).

Skeletal worksheet (no. 1)
Draw and label on the worksheet the following muscles:
a. Deltoid
b. Supraspinatus
c. Subscapularis
d. Teres major
e. Infraspinatus
f. Teres minor
g. Latissimus dorsi
h. Pectoralis major
i. Coracobrachialis

Human figure worksheet (no. 2)
Label and indicate with arrows the following movements of the shoulder joint:
a. Abduction
b. Adduction
c. Flexion
d. Extension
e. Horizontal adduction
f. Horizontal abduction
g. Internal rotation
h. External rotation

LABORATORY AND REVIEW EXERCISES

1. Locate the following parts of the humerus and scapula on a human skeleton and on a subject:
   a. Greater tubercle
   b. Lesser tubercle
   c. Neck
   d. Shaft
   e. Intertubercular groove
   f. Medial epicondyle
   g. Lateral epicondyle
   h. Trochlea
   i. Capitulum
   j. Supraspinatus fossa
   k. Infraspinatus fossa
   l. Spine of the scapula
2. How and where can the following muscles be palpated on a human subject?
   a. Deltoid
   b. Teres major
   c. Infraspinatus
   d. Teres minor
   e. Latissimus dorsi
   f. Pectoralis major (upper and lower)
      NOTE: Using the pectoralis major muscle, indicate how various actions allow muscle palpation.
3. Demonstrate and locate on a human subject the muscles that are primarily used in the following shoulder joint movements:
   a. Abduction
   b. Adduction
   c. Flexion
   d. Extension
   e. Horizontal adduction
   f. Horizontal abduction
   g. External rotation
   h. Internal rotation
4. List the planes in which each of the following glenohumeral joint movements occur. List the respective axis of rotation for each movement in each plane.
   a. Abduction
   b. Adduction
   c. Flexion
   d. Extension
   e. Horizontal adduction
   f. Horizontal abduction
   g. External rotation
   h. Internal rotation
5. Why is it essential that both anterior and posterior muscles of the shoulder joint be properly developed? What are some activities or sports

Chapter
5

that would cause unequal development? equal development?

6. Using an articulated skeleton, compare the relationship of the greater tubercle to the undersurface of the acromion in each of the following situations:
   a. Flexion with the humerus internally rotated versus externally rotated
   b. Abduction with the humerus internally versus externally rotated
   c. Horizontal adduction with the humerus internally versus externally rotated

7. What practical application do the activities or sports in question #5 support if
   a. The rotator cuff muscles are not functioning properly due to fatigue or lack of appropriate strength and endurance?
   b. The scapula stabilizers are not functioning properly due to fatigue or lack of strength and endurance?

8. Pair up with a partner with the back exposed. Use your hand to grasp your partner's right scapula along the lateral border to prevent scapula movement. Have your partner slowly abduct the glenohumeral joint as much as possible. Note the difference in total abduction possible normally versus when you restrict movement of the scapula. Repeat the same exercise, except hold the inferior angle of the scapula tightly against the chest wall while you have your partner internally rotate the humerus. Note the difference in total internal rotation possible normally versus when you restrict movement of the scapula.

9. Understanding the movements of the scapula in relation to the humerus can be explained by discussing the movement of the shoulder complex in its entirety. How does the position of the scapula affect shoulder joint abduction? How does the position of the scapula affect shoulder joint flexion?

10. Describe the bony articulations and movements specific to shoulder joint rotation during the acceleration phase of the throwing motion and how an athlete can work toward increasing the velocity of the throw. What factors affect the velocity of the throw?

11. Using the information from this chapter and other resources, how would you strengthen the four different rotator cuff muscles? Give several examples of how they are used in everyday activities.

12. Fill in the antagonistic muscle action chart below by listing the muscle(s) or parts of muscles that are antagonist in their actions to the muscles in the left column.

Antagonistic muscle action chart • Glenohumeral joint

Agonist	Antagonist
Deltoid (anterior fibers)	
Deltoid (middle fibers)	
Deltoid (posterior fibers)	
Supraspinatus	
Subscapularis	
Teres major	
Infraspinatus/teres minor	
Latissimus dorsi	
Pectoralis major (upper fibers)	
Pectoralis major (lower fibers)	
Coracobrachialis	

13. Fill in the muscle analysis chart below by listing the muscles primarily involved in each joint movement.
14. After analyzing each of the exercises in the shoulder joint movement analysis chart below, break each into two primary movement phases such as a lifting phase and lowering phase.

For each phase, determine the shoulder joint movements occurring, and then list the shoulder joint muscles primarily responsible for causing/controlling those movements. Beside each muscle in each movement, indicate the type of contraction as follows: I-isometric; C-concentric; E-eccentric.

Muscle analysis chart • Shoulder girdle and shoulder joint

Shoulder Girdle	Shoulder Joint
Adduction	Extension
Abduction	Flexion
Elevation	Horizontal adduction
Depression	Horizontal abduction
Upward rotation	Abduction
Downward rotation	Adduction
	External rotation
	Internal rotation
	Diagonal abduction (overhand activities)
	Diagonal adduction (overhand activities)

Shoulder joint movement analysis chart

Exercise	Initial movement phase		Secondary movement phase	
	Movement(s)	Agonist(s)–(contraction type)	Movement(s)	Agonist(s)–(contraction type)
Push-up				
Chin-up				
Bench press				
Dip				
Lat pull				
Overhead press				
Prone row				
Barbell shrugs				

15. Analyze each skill in the shoulder joint sport skill analysis chart below and list the movements of the right and left shoulder joint in each phase of the skill. You may prefer to list the initial position the shoulder joint is in for the stance phase. After each movement, list the shoulder joint muscle(s) primarily responsible for causing/controlling those movements. Beside each muscle in each movement, indicate the type of contraction as follows: I-isometric; C-concentric; E-eccentric. It may be desirable to review the concepts for analysis in Chapter 8 for the various phases.

Shoulder joint sport skill analysis chart

Exercise		Stance phase	Preparatory phase	Movement phase	Follow-through phase
Baseball pitch	(R)				
	(L)				
Volleyball serve	(R)				
	(L)				
Tennis serve	(R)				
	(L)				
Softball pitch	(R)				
	(L)				
Tennis backhand	(R)				
	(L)				
Batting	(R)				
	(L)				
Bowling	(R)				
	(L)				
Basketball free throw	(R)				
	(L)				

References

Andrews JR, Wilk KE: *The athlete's shoulder,* New York, 1994, Churchill Livingstone.

Andrews JR, Zarins B, Wilk KE: *Injuries in baseball,* Philadelphia, 1988, Lippincott-Raven.

Field D: *Anatomy: palpation and surface markings,* ed 3, Oxford, 2001, Butterworth-Heinemann.

Garth WP, et al: Occult anterior subluxations of the shoulder in noncontact sports, *American Journal of Sports Medicine* 15:579, November-December 1987.

Hislop HJ, Montgomery J: *Daniels and Worthingham's muscle testing: techniques of manual examination,* ed 7, Philadelphia, 2002, Saunders.

Muscolino JE: *The muscular system manual: the skeletal muscles of the human body,* ed 2, St. Louis, 2003, Elsevier Mosby.

Oatis CA: *Kinesiology: the mechanics and pathomechanics of human movement,* Philadelphia, 2004, Lippincott Williams & Wilkins.

Perry JF, Rohe DA, Garcia AO: *The kinesiology workbook,* Philadelphia, 1992, Davis.

Rasch PJ: *Kinesiology and applied anatomy,* ed 7, Philadelphia, 1989, Lea & Febiger.

Seeley RR, Stephens TD, Tate P: *Anatomy & physiology,* ed 7, New York, 2006, McGraw-Hill.

Sieg KW, Adams SP: *Illustrated essentials of musculoskeletal anatomy,* ed 2, Gainesville, FL, 1985, Megabooks.

Smith LK, Weiss EL, Lehmkuhl LD: *Brunnstrom's clinical kinesiology,* ed 5, Philadelphia, 1996, Davis.

Stacey E: Pitching injuries to the shoulder, *Athletic Journal* 65:44, January 1984.

The Elbow and Radioulnar Joints

Objectives

- To identify on a human skeleton selected bony features of the elbow and radioulnar joints

- To label selected bony features on a skeletal chart

- To draw and label the muscles on a skeletal chart

- To palpate the muscles on a human subject and list their antagonists

- To list the planes of motion and their respective axes of rotation

- To organize and list the muscles that produce the primary movements of the elbow joint and the radioulnar joint

Almost any movement of the upper extremity will involve the elbow and radioulnar joints. Quite often, these joints are grouped together because of their close anatomical relationship. The elbow joint is intimately associated with the radioulnar joint in that both bones of the radioulnar joint, the radius and ulna, share an articulation with the humerus to form the elbow joint. For this reason, some may confuse motions of the elbow with those of the radioulnar joint. In addition, radioulnar joint motion may be incorrectly attributed to the wrist joint because it appears to occur there. However, with close inspection, movements of the elbow joint can be clearly distinguished from those of the radioulnar joints, just as the radioulnar movements can be distinguished from those of the wrist. Even though the radius and ulna are both part of the articulation with the wrist, the relationship between the two is not nearly as intimate as that of the elbow and radioulnar joints.

Bones

The ulna is much larger proximally than the radius (Fig. 6.1), but distally the radius is much larger than the ulna (see Fig. 7.1 in Chapter 7). The scapula and humerus serve as the proximal attachments for the muscles that flex and extend the elbow. The ulna and radius serve as the distal attachments for the same muscles. The scapula, humerus, and ulna serve as proximal attachments for the muscles that pronate and supinate the radioulnar joints. The distal attachments of the radioulnar joint muscles are located on the radius.

The medial condyloid ridge, olecranon process, coranoid process, and radial tuberosity are important bony landmarks for these muscles. Additionally, the medial epicondyle, lateral epicondyle, and lateral supracondylar ridge are key bony landmarks for the muscles of the wrist and hand, discussed in Chapter 7.

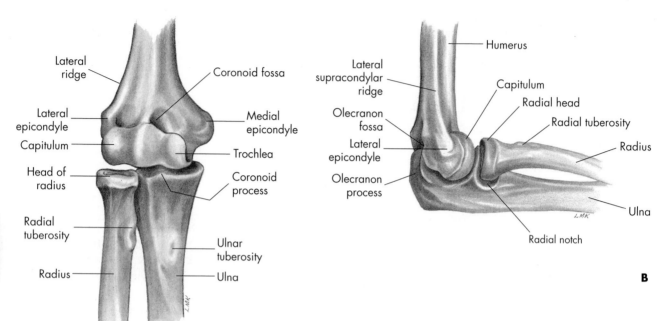

FIG. 6.1 ● Right elbow joint. **A,** Anterior view; **B,** Lateral view; **C,** Medial view.

Joints

The elbow joint is classified as a ginglymus or hinge-type joint that allows only flexion and extension (Fig. 6.1). The elbow may actually be thought of as two interrelated joints: the humeroulnar and the radiohumeral joints (Fig. 6.2). Elbow motions primarily involve movement between the articular surfaces of the humerus and ulna—specifically, the humeral trochlear fitting into the trochlear notch of the ulna. The head of the radius has a relatively small amount of contact with the capitulum of the humerus. As the elbow reaches full extension, the olecranon process of the ulna is received by the olecranon fossa of the humerus. This arrangement provides increased joint stability when the elbow is fully extended.

As the elbow flexes approximately 20 degrees or more, its bony stability is somewhat unlocked, allowing for more side-to-side laxity. The stability of the elbow in flexion is more dependent on the collateral ligaments, such as the lateral or radial col-

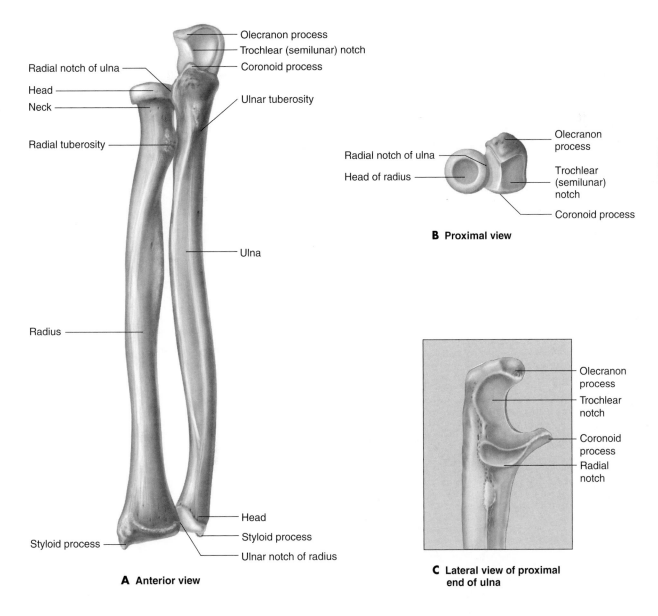

FIG. 6.2 ● Right radioulnar joint in supination. **A,** Anterior view; **B,** Proximal view of radioulnar joint; **C,** Lateral view of proximal end of ulna.

From Seeley RR, Stephens TD, Tate P: *Anantomy & physiology,* ed 7, New York, 2006, McGraw-Hill; Shier D, Butler J, Lewis R: *Hole's human anatomy & physiology,* ed 9, New York, 2002, McGraw-Hill.

lateral ligament and especially the medial or ulnar collateral ligament (Fig. 6.3). The ulnar collateral ligament is critical to providing medial support to prevent the elbow from abducting (not a normal movement of the elbow) when stressed in physical activity. Many contact sports, particularly sports with throwing activities, place stress on the medial aspect of the joint, resulting in injury. The radial collateral ligament on the opposite side provides lateral stability and is rarely injured. Additionally, the annular ligament is located laterally, providing a sling effect around the radial head to secure its stability.

The elbow is capable of moving from 0 degrees of extension to approximately 145 to 150 degrees of flexion, as detailed in Fig. 6.4.

The radioulnar joint is classified as a trochoid or pivot-type joint. The radial head rotates around in its location at the proximal ulna. This rotary movement is accompanied by the distal radius rotating around the distal ulna. The radial head is maintained in its joint by the annular ligament. The radioulnar joint can supinate approximately 80 to 90 degrees from the neutral position. Pronation varies from 70 to 90 degrees (Fig. 6.5).

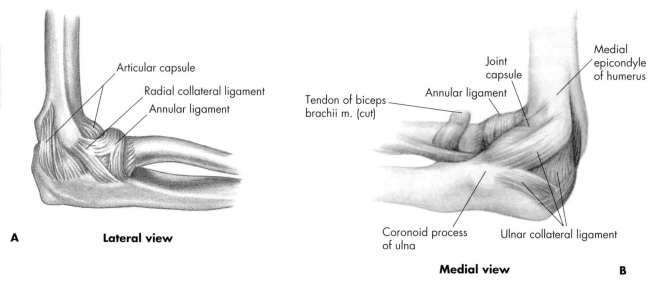

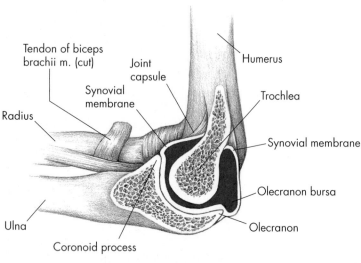

FIG. 6.3 ● Right elbow joint with ligaments detailed. **A,** Lateral view; **B,** Medial view; **C,** Sagittal cut view.

Due to the radius and ulna being held tightly together between the proximal and distal articulations by an interosseus membrane, the joint between the shafts of these bones is often referred to as syndesmosis type of joint. There is substantial rotary motion between the bones despite this classification.

Even though the elbow and radioulnar joint can and do function independently of each other, the muscles controlling each work together in synergy to perform actions at both to benefit the total function of the upper extremity. For this reason, dysfunction at one joint may affect normal function at the other.

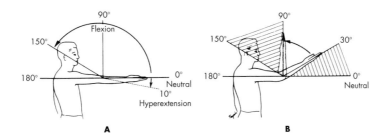

FIG. 6.4 ● ROM of the elbow: flexion, extension, and hyperextension. **A,** Flexion and hyperextension. *Flexion:* zero to 150°. *Extension:* 150° to zero. *Hyperextension:* measured in degrees beyond the zero starting point. This motion is not present in all persons. When it is present, it may vary from 5° to 15°; **B,** Measurement of limited motion. (The unshaded area indicates the limited ROM.) Limited motion may be expressed in the following ways: (1) the elbow flexes from 30° to 90° (30°–90°); (2) the elbow has a flexion deformity of 30° with further flexion to 90°.

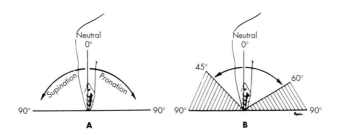

FIG. 6.5 ● ROM of the forearm: pronation and supination. **A,** Pronation and supination. *Pronation:* zero to 80° or 90°. *Supination:* zero to 80° or 90°. *Total forearm motion:* 160° to 180°. Persons may vary in the range of supination and pronation. Some may reach the 90° arc, and others may have only 70° plus; **B,** Limited motion. *Supination:* 45° (0–45°). *Total joint motion:* 105°.

Synergy between the glenohumeral, elbow, and radioulnar joint muscles

Just as there is synergy between the shoulder girdle and the shoulder joint in accomplishing upper extremity activities, there is also synergy between the glenohumeral joint and the elbow as well as the radioulnar joints.

As the radioulnar joint goes through its ranges of motion, the glenohumeral and elbow muscles contract to stabilize or assist in the effectiveness of movement at the radioulnar joints. For example,

when attempting to fully tighten (with the right hand) a screw with a screwdriver that involves radioulnar supination, we tend to externally rotate and flex the glenohumeral and elbow joints, respectively. Conversely, when attempting to loosen a tight screw with pronation, we tend to internally rotate and extend the elbow and glenohumeral joints, respectively. In either case, we depend on both the agonists and antagonists in the surrounding joints to assist in an appropriate amount of stabilization and assistance with the required task.

Movements FIGS. 6.4, 6.5, 6.6

Elbow movements

Flexion

Movement of the forearm to the shoulder by bending the elbow to decrease its angle.

Extension

Movement of the forearm away from the shoulder by straightening the elbow to increase its angle.

Radioulnar joint movements

Pronation

Internal rotary movement of the radius on the ulna that results in the hand moving from the palm-up to the palm-down position.

Supination

External rotary movement of the radius on the ulna that results in the hand moving from the palm-down to the palm-up position.

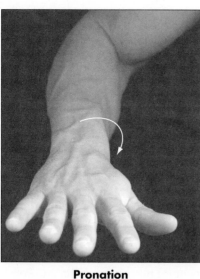

Flexion   **A**

Extension   **B**

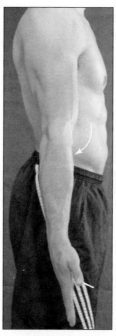

Pronation   **C**

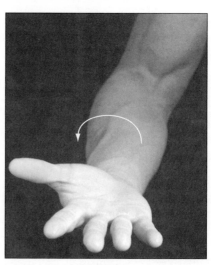

Supination   **D**

FIG. 6.6 ● Movements of the elbow and radioulnar joint. **A,** Elbow flexion; **B,** Elbow extension; **C,** Radioulnar pronation; **D,** Radioulnar supination.

Muscles

The muscles of the elbow and radioulnar joints may be more clearly understood when separated by function. The elbow flexors, located anteriorly, are the biceps brachii, the brachialis, and the brachioradialis, with some weak assistance from the pronator teres (Figs. 6.7, 6.8 and 6.11).

The triceps brachii, located posteriorly, is the primary elbow extensor, with assistance provided by the anconeus (Figs. 6.9, 6.10 and 6.11). The pronator group, located anteriorly, consists of the pronator teres, the pronator quadratus, and the brachioradialis. The brachioradialis also assists with supination, which is controlled mainly by the supinator muscle and the biceps brachii. The supinator muscle is located posteriorly. See Table 6.1.

A common problem associated with the muscles of the elbow is "tennis elbow," which usually involves the extensor digitorum muscle near its origin on the lateral epicondyle. This condition, known technically as **lateral epicondylitis**, is quite frequently associated with gripping and lifting activities. **Medial epicondylitis**, a somewhat less common problem frequently referred to as "golfer's elbow," is associated with the wrist flexor and pronator group near their origin on the medial epicondyle. Both of these conditions involve muscles that cross the elbow but act primarily on the wrist and hand. These muscles will be addressed in Chapter 7.

Elbow and radioulnar joint muscles—location

Anterior
    Primarily flexion and pronation
        Biceps brachii
        Brachialis
        Brachioradialis
        Pronator teres
        Pronator quadratus
Posterior
    Primarily extension and supination
        Triceps brachii
        Anconeus
        Supinator

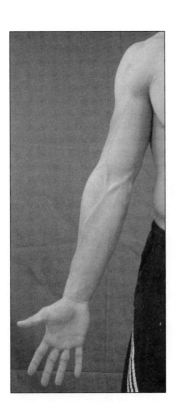

FIG. 6.7 ● Anterior upper extremity muscles.

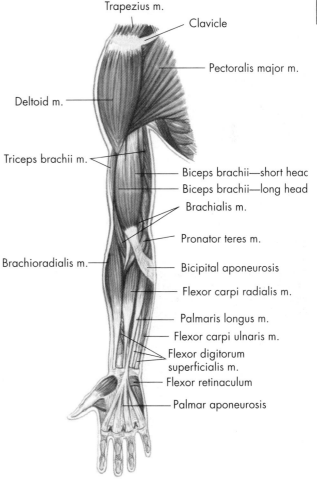

Trapezius m.
Clavicle
Pectoralis major m.
Deltoid m.
Triceps brachii m.
Biceps brachii—short heac
Biceps brachii—long head
Brachialis m.
Pronator teres m.
Brachioradialis m.
Bicipital aponeurosis
Flexor carpi radialis m.
Palmaris longus m.
Flexor carpi ulnaris m.
Flexor digitorum superficialis m.
Flexor retinaculum
Palmar aponeurosis

FIG. 6.8 ● Anterior upper extremity muscles.

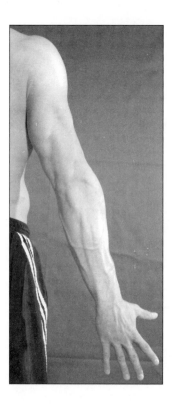

FIG. 6.9 ● Posterior upper extremity muscles.

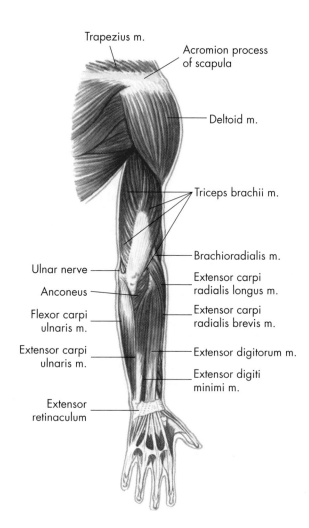

Trapezius m.

Acromion process of scapula

Deltoid m.

Triceps brachii m.

Ulnar nerve

Anconeus

Flexor carpi ulnaris m.

Extensor carpi ulnaris m.

Extensor retinaculum

Brachioradialis m.

Extensor carpi radialis longus m.

Extensor carpi radialis brevis m.

Extensor digitorum m.

Extensor digiti minimi m.

FIG. 6.10 ● Posterior upper extremity muscles.

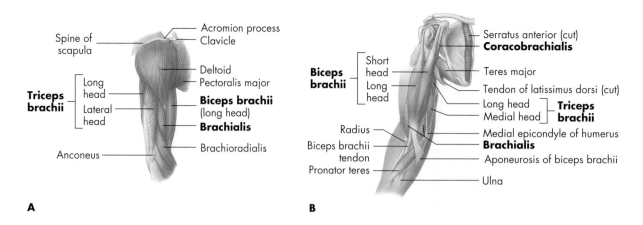

Spine of scapula

Acromion process

Clavicle

Deltoid

Pectoralis major

Biceps brachii (long head)

Brachialis

Brachioradialis

Triceps brachii

Long head

Lateral head

Anconeus

A

Biceps brachii

Short head

Long head

Radius

Biceps brachii tendon

Pronator teres

Serratus anterior (cut)

Coracobrachialis

Teres major

Tendon of latissimus dorsi (cut)

Long head

Medial head

Triceps brachii

Medial epicondyle of humerus

Brachialis

Aponeurosis of biceps brachii

Ulna

B

FIG. 6.11 ● Muscles of the arm. **A,** Lateral view of the right shoulder and arm; **B,** Anterior view of the right shoulder and arm (deep). Deltoid, pectoralis major, and pectoralis minor muscle removed to reveal deeper structures.

TABLE 6.1 • Agonist muscles of the elbow and radioulnar joints

	Muscle	Origin	Insertion	Action	Plane of motion	Palpation	Innervation
Anterior muscles (primarily flexors and pronators)	Biceps brachii long head	Supraglenoid tubercle above the superior lip of the glenoid fossa	Tuberosity of the radius and bicipital aponeurosis (lacertus fibrosis)	Supination of the forearm	Transverse	Easily palpated on the anterior humerus; the long head and short head tendons may be palpated in the intertubercular groove and just inferomedial to the coracoid process, respectively; distally, the biceps tendon is palpated just anteromedial to the elbow joint during supination and flexion	Musculocutaneous nerve (C5, C6)
				Flexion of the elbow	Sagittal		
				Weak flexion of the shoulder joint			
				Weak abduction of the shoulder joint	Frontal		
	Biceps brachii short head	Coracoid process of the scapula and upper lip of the glenoid fossa in conjunction with the proximal attachment of the coracobrachialis		Supination of the forearm	Transverse		
				Flexion of the elbow	Sagittal		
				Weak flexion of the shoulder joint			
				Weak abduction of the shoulder joint	Frontal		
	Brachialis	Distal half of the anterior shaft of the humerus	Coranoid process of the ulna	Flexion of the elbow	Sagittal	Deep on either side of the biceps tendon during flexion/extension with forearm in partial pronation; lateral margin may be palpated between biceps brachii and triceps brachii; belly may be palpated through biceps brachii when forearm in pronation during light flexion	Musculocutaneous nerve (C5, C6)
	Brachio-radialis	Distal 2/3 of the lateral condyloid (supracondylar) ridge of the humerus	Lateral surface of the distal end of the radius at the styloid process	Flexion of the elbow	Sagittal	Anterolaterally on the proximal forearm during resisted elbow flexion with the radioulnar joint positioned in neutral	Radial nerve (C5, C6)
				Pronation from supination to neutral	Transverse		
				Supination from pronation to neutral			
	Pronator teres	Distal part of the medial condyloid ridge of the humerus and medial side of the proximal ulna	Middle 1/3 of the lateral surface of the radius	Pronation of the forearm	Transverse	Anteromedial surface of the proximal forearm during resisted mid to full pronation	Median nerve (C6, C7)
				Weak flexion of the elbow	Sagittal		
	Pronator quadratus	Distal 1/4 of the anterior side of the ulna	Distal 1/4 of the anterior side of the radius	Pronation of the forearm	Transverse	Very deep and difficult to palpate, but with the forearm in supination, palpate immediately on either side of the radial pulse with resisted pronation	Median nerve (C6, C7)

Chapter 6

TABLE 6.1 (continued) • Agonist muscles of the elbow and radioulnar joints

	Muscle	Origin	Insertion	Action	Plane of motion	Palpation	Innervation
Posterior muscles (primarily extensors and supinators)	Triceps brachii long head	Infraglenoid tubercle below inferior lip of glenoid fossa of the scapula	Olecranon process of the ulna	Extension of the elbow joint	Sagittal	Proximally as a tendon on the posteromedial arm to underneath the posterior deltoid during resisted shoulder extension/ adduction	Radial nerve (C7, C8)
				Extension of the shoulder joint			
				Adduction of the shoulder joint	Frontal		
				Horizontal abduction of the shoulder joint	Transverse		
	Triceps brachii lateral head	Upper half of the posterior surface of the humerus		Extension of the elbow joint	Sagittal	Easily palpated on the proximal 2/3 of the posterior humerus during resisted extension	
	Triceps brachii medial head	Distal 2/3 of the posterior surface of the humerus		Extension of the elbow joint		Deep head: medially and laterally just proximal to the medial and lateral epicondyles	
	Supinator	Lateral epicondyle of the humerus and neighboring posterior part of the ulna	Lateral surface of the proximal radius just below the head	Supination of the forearm	Transverse	Position elbow and forearm in relaxed flexion and pronation, respectively; palpate deep to the brachioradialis, extensor carpi radialis longus, extensor carpi radialis brevis on the lateral aspect of the proximal radius with slight resistance to supination	Radial nerve (C6)
	Anconeus	Posterior surface of the lateral condyle of the humerus	Posterior surface of the lateral olecranon process and proximal 1/4 of the ulna	Extension of the elbow	Sagittal	Posterolateral aspect of the proximal ulna to the olecranon process during resisted extension of the elbow with the wrist in flexion	Radial nerve (C7, C8)

Note: The flexor carpi radialis, palmaris longus, flexor carpi ulnaris, and flexor digitorum superficialis assist in weak flexion of the elbow, while the extensor carpi ulnaris, extensor carpi radialis brevis, extensor carpi radialis longus, and extensor digitorum assist in weak extension of the elbow. Due to their coverage in Chapter 7, they are not listed above.

Chapter 6

Nerves   FIGS. 5.11, 6.12, 6.13

The muscles of the elbow and radioulnar joints are all innervated from the median, musculocutaneous, and radial nerves of the brachial plexus. The radial nerve, originating from C5, C6, C7, and C8, provides innervation for the triceps brachii, brachioradialis, supinator, and anconeus (Fig. 6.12). More specifically, the posterior interosseous nerve, derived from the radial nerve, supplies the supinator. The radial nerve also provides sensation to the posterolateral arm, forearm, and hand. The median nerve, illustrated in Fig. 6.13, innervates the pronator teres and further branches to become the anterior interosseus nerve, which supplies the pronator quadratus. The median nerve's most important related derivations are from C6 and C7. It provides sensation to the palmar aspect of the hand and first three phalanges. The palmar aspect of the radial side of the fourth finger is also provided sensation, along with the dorsal aspect of the index and long fingers. The musculocutaneous nerve, shown in Fig. 5.11, branches from C5 and C6 and supplies the biceps brachii and brachialis.

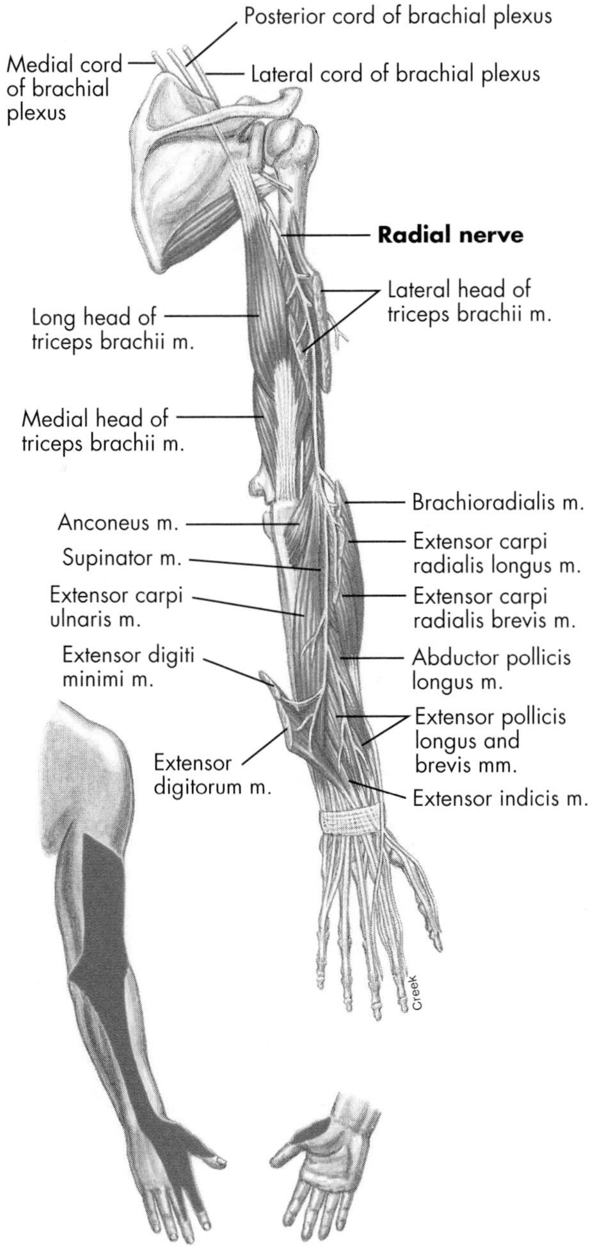

Posterior cord of brachial plexus

Medial cord of brachial plexus

Lateral cord of brachial plexus

Radial nerve

Long head of triceps brachii m.

Lateral head of triceps brachii m.

Medial head of triceps brachii m.

Anconeus m.

Supinator m.

Extensor carpi ulnaris m.

Extensor digiti minimi m.

Extensor digitorum m.

Brachioradialis m.

Extensor carpi radialis longus m.

Extensor carpi radialis brevis m.

Abductor pollicis longus m.

Extensor pollicis longus and brevis mm.

Extensor indicis m.

FIG. 6.12 ● Muscular and cutaneous distribution of the radial nerve.

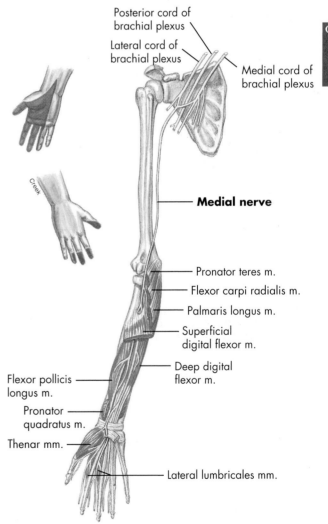

Posterior cord of brachial plexus

Lateral cord of brachial plexus

Medial cord of brachial plexus

Medial nerve

Pronator teres m.

Flexor carpi radialis m.

Palmaris longus m.

Superficial digital flexor m.

Deep digital flexor m.

Flexor pollicis longus m.

Pronator quadratus m.

Thenar mm.

Lateral lumbricales mm.

FIG. 6.13 ● Muscular and cutaneous distribution of the median nerve.

Biceps brachii muscle FIG. 6.14

(bi´seps bra´ki-i)

Origin

Long head: supraglenoid tubercle above the superior lip of the glenoid fossa

Short head: coracoid process of the scapula and upper lip of the glenoid fossa in conjunction with the proximal attachment of the coracobrachialis

Insertion

Tuberosity of the radius and bicipital aponeurosis (lacertus fibrosis)

Action

Flexion of the elbow
Supination of the forearm
Weak flexion of the shoulder joint
Weak abduction of the shoulder joint when the shoulder joint is in external rotation

Palpation

Easily palpated on the anterior humerus. The long head and short head tendons may be palpated in the intertubercular groove and just inferior to the coracoid process, respectively. Distally, the biceps tendon is palpated just anteromedial to the elbow joint during supination and flexion.

Innervation

Musculocutaneous nerve (C5, C6)

Application, strengthening, and flexibility

The biceps is commonly known as a two-joint (shoulder and elbow), or biarticular, muscle. However, technically it should be considered a three-joint (multiarticular) muscle—shoulder, elbow, and radio-ulnar. It is weak in actions of the shoulder joint, although it does assist in providing dynamic anterior stability to maintain the humeral head in the glenoid fossa. It is more powerful in flexing the elbow when the radioulnar joint is supinated. It is also a strong supinator, particularly if the elbow is flexed. Palms away from the face (pronation) decrease the effectiveness of the biceps, partly as a result of the disadvantageous pull of the muscle as the radius rotates. The same muscles are used in elbow joint flexion, regardless of forearm pronation or supination.

Flexion of the forearm with a barbell in the hands, known as "curling," is an excellent exercise to develop the biceps brachii. This movement can be performed one arm at a time with dumbbells or both arms simultaneously with a barbell. Other activities in which there is powerful flexion of the forearm are chinning and rope climbing.

Due to the multiarticular orientation of the biceps, all three joints must be positioned appropriately to achieve optimal stretching. The elbow must be extended maximally with the shoulder in full extension. The biceps may also be stretched by beginning with full elbow extension and progressing into full horizontal abduction at approximately 70 to 110 degrees of shoulder abduction. In all cases, the forearm should be fully pronated to achieve maximal lengthening of the biceps brachii.

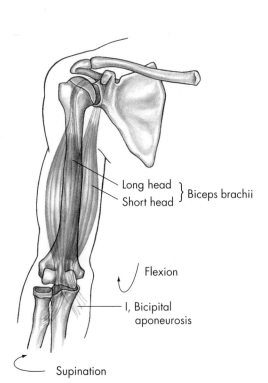

Long head / Short head } Biceps brachii

Flexion

I, Bicipital aponeurosis

Supination

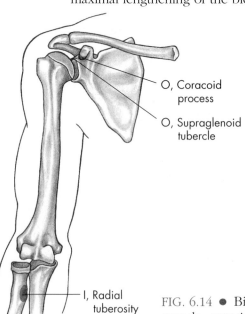

O, Coracoid process

O, Supraglenoid tubercle

I, Radial tuberosity

FIG. 6.14 ● Biceps brachii muscle, anterior view. *O,* Origin; *I,* Insertion.

Brachialis muscle   FIG. 6.15

(braʹki-aʹlis)

Origin
Distal half of the anterior shaft of the humerus

Insertion
Coronoid process of the ulna

Action
True flexion of the elbow

Palpation
Deep on either side of the biceps tendon during flexion/extension with forearm in partial pronation. The lateral margin may be palpated between the biceps brachii, and the triceps brachii and the belly may be palpated through the biceps brachii when the forearm is in pronation during light flexion.

Innervation
Musculocutaneous nerve and sometimes branches from radial and median nerves (C5, C6)

Application, strengthening, and flexibility

The brachialis muscle is used along with other flexor muscles, regardless of pronation or supination. It pulls on the ulna, which does not rotate, thus making this muscle the only pure flexor of this joint.

The brachialis muscle is called into action whenever the elbow flexes. It is exercised along with elbow curling exercises, as described for the biceps brachii, pronator teres, and brachioradialis muscles. Elbow flexion activities that occur with the forearm pronated isolate the brachialis to some extent by reducing the effectiveness of the biceps brachii. Since the brachialis is a pure flexor of the elbow, it can be stretched maximally only by extending the elbow with the shoulder relaxed and flexed. Forearm positioning should not affect the stretch on the brachialis unless the forearm musculature itself limits elbow extension, in which case the forearm is probably best positioned in neutral.

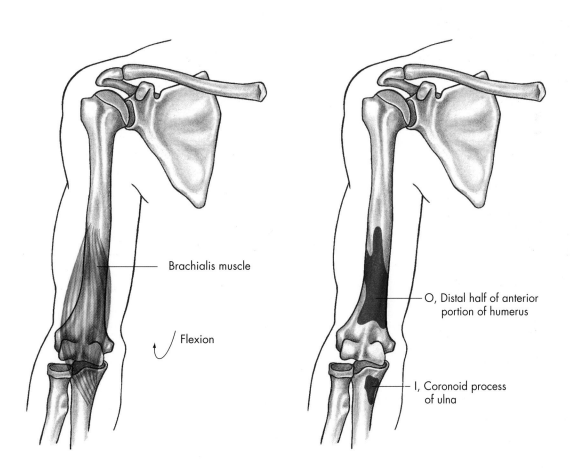

Brachialis muscle

Flexion

O, Distal half of anterior portion of humerus

I, Coronoid process of ulna

FIG. 6.15 ● Brachialis muscle, anterior view. *O*, Origin; *I*, Insertion.

Brachioradialis muscle   FIG. 6.16

(braˊki-o-raˊdi-aˊlis)

Origin

Distal two-thirds of the lateral condyloid (supracondylar) ridge of the humerus

Insertion

Lateral surface of the distal end of the radius at the styloid process

Action

Flexion of the elbow
Pronation from supinated position to neutral
Supination from pronated position to neutral

Palpation

Anterolaterally on the proximal forearm during resisted elbow flexion with the radioulnar joint positioned in neutral

Innervation

Radial nerve (C5, C6)

Application, strengthening, and flexibility

The brachioradialis is one of three muscles on the lateral forearm sometimes known as the mobile wad of three. The other two muscles are the extensor carpi radialis brevis and extensor carpi radialis longus to which it lies directly anterior. The bra-chioradialis muscle acts as a flexor best in a mid-position or neutral position between pronation and supination. In a supinated position of the forearm, it tends to pronate as it flexes. In a pronated position, it tends to supinate as it flexes. This muscle is favored in its action of flexion when the neutral position between pronation and supination is assumed, as previously suggested. Its insertion at the end of the radius makes it a strong elbow flexor. Its ability as a supinator decreases as the radioulnar joint moves toward neutral. Similarly, its ability to pronate decreases as the forearm reaches neutral. Because of its action of rotating the forearm to a neutral thumb-up position, it is referred to as the hitchhiker muscle although it has no action at the thumb. As you will see in Chapter 7, nearly all of the muscles originating off the lateral epicondyle have some action as weak elbow extensors. This is not the case with the brachioradialis, due to its line of pull being anterior to the elbow's axis of rotation.

The brachioradialis may be strengthened by performing elbow curls against resistance, particularly with the radioulnar joint in the neutral position. In addition, the brachioradialis may be developed by performing pronation and supination movements through the full range of motion against resistance.

The brachioradialis is stretched by maximally extending the elbow with the shoulder in flexion and the forearm in either maximal pronation or maximal supination.

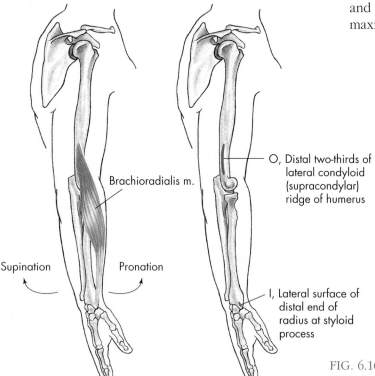

Brachioradialis m.

Supination   Pronation

O, Distal two-thirds of lateral condyloid (supracondylar) ridge of humerus

I, Lateral surface of distal end of radius at styloid process

FIG. 6.16 ● Brachioradialis muscle, lateral view. *O,* Origin; *I,* Insertion.

Triceps brachii muscle   FIG. 6.17

(tri´seps bra´ki-i)

Origin

Long head: infraglenoid tubercle below inferior lip of glenoid fossa of the scapula

Lateral head: upper half of the posterior surface of the humerus

Medial head: distal two-thirds of the posterior surface of the humerus

Insertion

Olecranon process of the ulna

Action

All heads: extension of the elbow

Long head: extension, adduction, and horizonal abduction of the shoulder joint

Palpation

Posterior arm during resisted extension from a flexed position and distally just proximal to its insertion on the olecranon process

Long head: proximally as a tendon on the posteromedial arm to underneath the posterior deltoid during resisted shoulder extension/adduction

Lateral head: easily palpated on the proximal two-thirds of the posterior humerus

Medial head (deep head): medially and laterally just proximal to the medial and lateral epicondyles

Innervation

Radial nerve (C7, C8)

Application, strengthening, and flexibility

Typical action of the triceps brachii is shown in push-ups when there is powerful extension of the elbow. It is used in hand balancing and in any pushing movement involving the upper extremity. The long head is an important extensor of the shoulder joint.

Two muscles extend the elbow—the triceps brachii and the anconeus. Push-ups demand strenuous contraction of these muscles. Dips on the parallel bars are more difficult to perform. Bench-pressing a barbell or a dumbbell is an excellent exercise. Overhead presses and triceps curls (elbow extensions from an overhead position) emphasize the triceps.

The triceps brachii should be stretched with both the shoulder and the elbow in maximal flexion.

Chapter 6

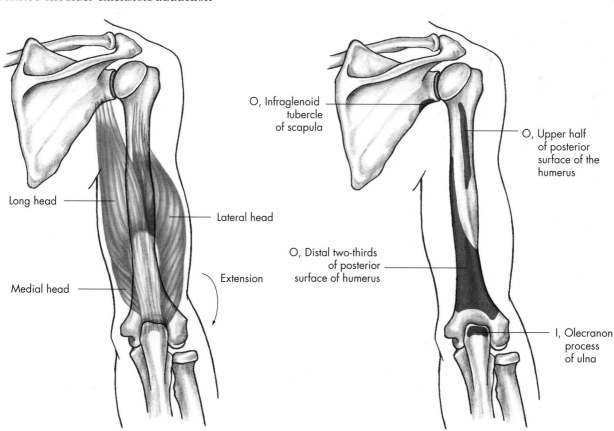

FIG. 6.17 ● Triceps brachii muscle, posterior view. *O,* Origin; *I,* Insertion.

Anconeus muscle   FIG. 6.18

(an-ko´ne-us)

Origin

Posterior surface of the lateral condyle of the humerus

Insertion

Posterior surface of the lateral olecranon process and proximal one-fourth of the ulna

Action

Extension of the elbow

Palpation

Posterolateral aspect of the proximal ulna to the olecranon process during resisted extension of the elbow with the wrist in flexion

Innervation

Radial nerve (C7, C8)

Application, strengthening, and flexibility

The chief function of the anconeus muscle is to pull the synovial membrane of the elbow joint out of the way of the advancing olecranon process during extension of the elbow. It contracts along with the triceps brachii. It is strengthened with any elbow extension exercise against resistance. Maximal elbow flexion stretches the anconeus.

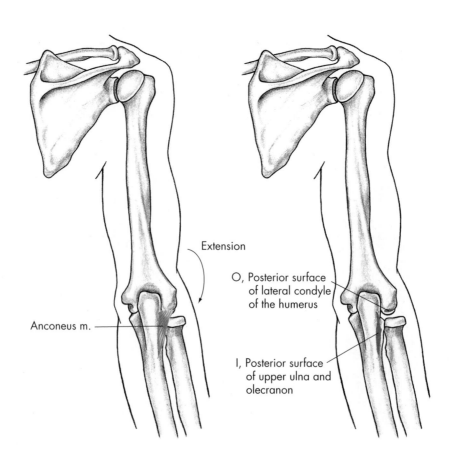

FIG. 6.18 ● Anconeus muscle, posterior view. *O,* Origin; *I,* Insertion.

Pronator teres muscle   FIG. 6.19

(pro-na´tor te´rez)

Origin

Distal part of the medial condyloid ridge of the humerus and medial side of the proximal ulna

Insertion

Middle third of the lateral surface of the radius

Action

Pronation of the forearm
Weak flexion of the elbow

Palpation

Anteromedial surface of the proximal forearm during resisted mid to full pronation

Innervation

Median nerve (C6, C7)

Application, strengthening, and flexibility

Typical movement of the pronator teres muscle is with the forearm pronating as the elbow flexes. Movement is weaker in flexion with supination. The use of the pronator teres alone in movement tends to bring the back of the hand to the face as it contracts. Pronation of the forearm with a dumbbell in the hand localizes action and develops the pronator teres muscle. Strengthening this muscle begins with holding a hammer in the hand with the hammer head suspended from the ulnar side of the hand while the forearm is supported on a desk or table. The hammer should be hanging toward the floor, with the forearm pronated to the palm-down position.

The elbow must be fully extended while taking the forearm into full supination to stretch the pronator teres.

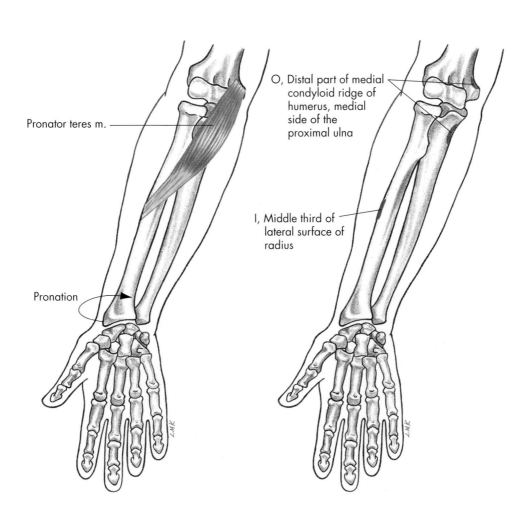

Pronator teres m.

O, Distal part of medial condyloid ridge of humerus, medial side of the proximal ulna

I, Middle third of lateral surface of radius

Pronation

FIG. 6.19 ● Pronator teres muscle, anterior view. *O,* Origin; *I,* Insertion.

Pronator quadratus muscle   FIG. 6.20

(pro-na´tor kwad-ra´tus)

Origin

Distal fourth of the anterior side of the ulna

Insertion

Distal fourth of the anterior side of the radius

Action

Pronation of the forearm

Palpation

Very deep and difficult to palpate, but with the forearm in supination, palpate immediately on either side of the radial pulse with resisted pronation

Innervation

Median nerve (palmar interosseous branch) (C6, C7)

Application, strengthening, and flexibility

The pronator quadratus muscle works in pronating the forearm in combination with the triceps in extending the elbow. It is commonly used in turning a screwdriver, as in taking out a screw, when extension and pronation are needed. It is used also in throwing a screwball, when extension and pronation are needed. It may be developed with similar pronation exercises against resistance, as described for the pronator teres. The pronator quadratus is best stretched by using a partner to grasp the wrist and passively take the forearm into extreme supination.

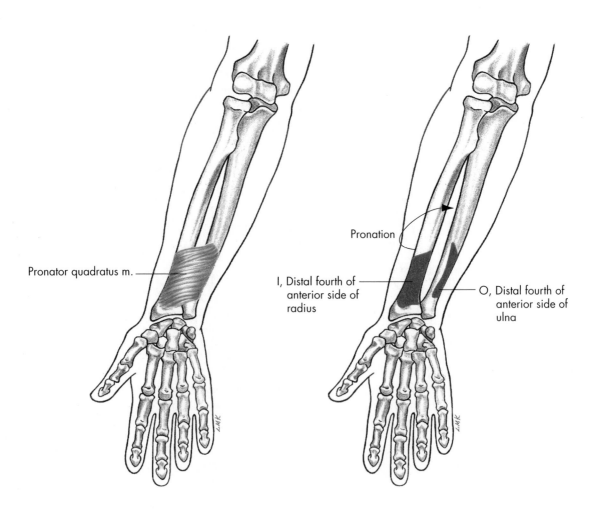

Pronator quadratus m.

Pronation

I, Distal fourth of anterior side of radius

O, Distal fourth of anterior side of ulna

FIG. 6.20 ● Pronator quadratus muscle, anterior view. *O,* Origin; *I,* Insertion.

Supinator muscle   FIG. 6.21

(su´pi-na´tor)

Origin

Lateral epicondyle of the humerus and neighboring posterior part of the ulna

Insertion

Lateral surface of the proximal radius just below the head

Action

Supination of the forearm

Palpation

Position the elbow and forearm in relaxed flexion and pronation, respectively, and palpate deep to the brachioradialis, extensor carpi radialis longus, extensor carpi radialis brevis on the lateral aspect of the proximal radius with slight resistance to supination

Innervation

Radial nerve (C6)

Application, strengthening, and flexibility

The supinator muscle is called into play when movements of extension and supination are required, as when turning a screwdriver. The curve in throwing a baseball calls this muscle into play as the elbow is extended just before ball release. It is most isolated in activities that require supination with elbow extension, because the biceps brachii assist with supination most when the elbow is flexed.

The hands should be grasped and the forearms extended, in an attempt to supinate the forearms against the grip of the hands. This localizes, to a degree, the action of the supinator.

The hammer exercise used for the pronator teres muscle may be modified to develop the supinator. In the beginning, the forearm is supported and the hand is free off the table edge. The hammer is again held suspended out of the ulnar side of the hand hanging toward the floor. The forearm is then supinated to the palm-up position to strengthen this muscle.

The supinator is stretched when the forearm is maximally pronated.

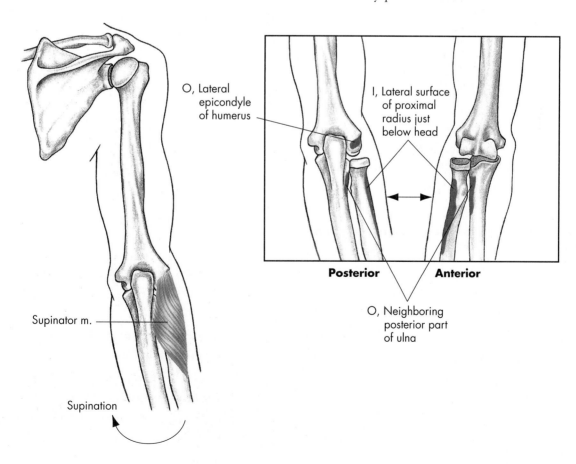

FIG. 6.21 ● Supinator muscle, posterior view. *O,* Origin; *I,* Insertion.

Web sites

American Family Physician

www.aafp.org/afp/200000201/691.html

Evaluation of overuse elbow injuries.

Medical Multimedia Group

www.healthpages.org/AHP/LIBRARY/HLTHTOP/CTD

A patient's guide to cumulative trauma disorder (CTD).

Lecture Topics in Kinesiology

http://moon.ouhsc.edu/dthompso/namics/elbow.htm

Describes motions caused by the muscles.

Huei Ming Chai

www.pt.ntu.edu.tw/hmchai/Kines04/KINupper/Elbow.htm

Functions, stability, and joint structure of elbow complex; kinematics, muscle action, and common injuries of the elbow.

Southern California Orthopedic Institute

www.scoi.com/teniselb.htm

Tennis elbow information.

MayoClinic.com

www.mayoclinic.com/invoke.cfm?id=AR00008

Elbow surgery for arthritis: relief when other therapies do not work.

National Aeronautics and Space Administration

http://rehabworks.ksc.nasa.gov/education/protocols/basicwristelbow.php

Basic wrist and elbow rehabilitation.

UpToDate

http://patients.uptodate.com/topic.asp?file=bone_joi/7086

Physical therapy for elbow tendinitis.

American Sports Medicine Institute

www.asmi.org/asmiweb/mpresentations/mmp.htm

Biomechanics of the elbow during throwing.

American Academy of Orthopaedic Surgeons

http://orthoinfo.aaos.org/category.cfm?topcategory=Hand

Patient education library on the elbow.

American Physical Therapy Association

www.apta.org/AM/Template.cfm?Section=Home&CONTENTID=20403&TEMPLATE=/CM/HTMLDisplay.cfm

Taking care of your hand, wrist, and elbow.

The Physician and Sportsmedicine

www.physsportsmed.com/issues/1996/05_96/nirschl.htm

Assessment and treatment guidelines for elbow injuries.

The Physician and Sportsmedicine

www.physsportsmed.com/issues/1999/06_99/whiteside.htm

Elbow injuries in young baseball players.

Radiologic Anatomy Browser

radlinux1.usuf1.usuhs.mil/rad/iong/index.html

Numerous radiological views of the musculoskeletal system.

University of Arkansas Medical School Gross Anatomy for Medical Students

http://anatomy.uams.edu/anatomyhtml/grossresources.html

Dissections, anatomy tables, atlas images, links, etc.

Loyola University Medical Center: Structure of the Human Body

www.meddean.luc.edu/lumen/MedEd/GrossAnatomy/GA.html

An excellent site with many slides, dissections, tutorials, etc., for the study of human anatomy.

Wheeless' Textbook of Orthopaedics

www.wheelessonline.com

An extensive index of links to the fractures, joints, muscles, nerves, trauma, medications, medical topics, lab tests, and links to orthopaedic journals and other orthopaedic and medical news.

Arthroscopy.com

www.arthroscopy.com/sports.htm

Patient information on various musculoskeletal problems of the upper and lower extremity.

Premiere Medical Search Engine

www.medsite.com/Default.asp?bhcp=1

Allows the reader to enter any medical condition and it will search the net to find relevant articles.

Chapter
6

Virtual Hospital

www.vh.org

Numerous slides, patient information, etc.

Worksheet exercises

As an aid to learning, for in-class and out-of-class assignments, or for testing, tear-out worksheets are found at the end of the text (pp. 378 and 379).

Skeletal worksheet (no. 1)

Draw and label on the worksheet the following muscles:
a. Biceps brachii
b. Brachioradialis
c. Brachialis
d. Pronator teres
e. Pronator quadratus
f. Supinator
g. Triceps brachii
h. Anconeus

Human figure worksheet (no. 2)

Label and indicate with arrows the following movements of the elbow and radioulnar joints:
1. Elbow joint
   a. Flexion
   b. Extension
2. Radioulnar joint
   a. Pronation
   b. Supination

LABORATORY AND REVIEW EXERCISES

1. Locate the following parts of the humerus, radius, and ulna on a human skeleton and on a subject.
   a. Skeleton
      1. Medial epicondyle
      2. Lateral epicondyle
      3. Lateral supracondylar ridge
      4. Trochlea
      5. Capitulum
      6. Olecranon fossa
      7. Olecranon process
      8. Coronoid process
      9. Coronoid fossa
      10. Tuberosity of the radius
      11. Ulnar tuberosity
   b. Subject
      1. Medial epicondyle
      2. Lateral epicondyle
      3. Lateral supracondylar ridge
         4. Proximal radioulnar joint
         5. Olecranon process
         6. Olecranon fossa

2. How and where can the following muscles be palpated on a human subject?
   a. Biceps brachii
   b. Brachioradialis
   c. Brachialis
   d. Pronator teres
   e. Supinator
   f. Triceps brachii
   g. Anconeus

3. Palpate and list the muscles primarily responsible for the following movements as you demonstrate each:
   a. Flexion
   b. Extension
   c. Pronation
   d. Supination

4. List the planes in which each of the following elbow and radioulnar joint movements occurs. List the respective axis of rotation for each movement in each plane.
   a. Flexion
   b. Extension
   c. Pronation
   d. Supination

5. Discuss the difference between chinning with the palms toward the face and chinning with the palms away from the face. Consider this muscularly and anatomically.

6. Analyze and list the differences in elbow and radioulnar joint muscle activity between turning a door knob clockwise and pushing the door open, versus turning the knob counterclockwise, pulling the door open.

7. Fill in the antagonistic muscle action chart (p. 156) by listing the muscle(s) or parts of muscles that are antagonist in their actions to the muscles in the left column.

8. Fill in the muscle analysis chart (p. 156) by listing the muscles primarily involved in each movement.

9. The elbow joint is the insertion of what biarticular muscle? Describe what motions it is involved in at the elbow and at the superior joint of its muscular origin.

10. Lifting a television set as you help your roommate move in requires appropriate lifting techniques and an effective angle of pull. Describe the angle of pull chosen at the elbow joint and why it is chosen as opposed to other angles.

11. List which muscles are involved with tennis elbow and describe specifically how you

Antagonistic Muscle Action Chart • Elbow and radioulnar joints

Agonist	Antagonist
Biceps brachii	
Brachioradialis	
Brachialis	
Pronator teres	
Supinator	
Triceps brachii	
Anconeus	

would encourage someone to work on both strength and flexibility of these muscles.

12. After analyzing each of the exercises in the elbow and radioulnar joint movement analysis chart below, break each into two primary movement phases such as a lifting phase and lowering phase. For each phase, determine the elbow and radioulnar joint movements occurring, and then list the elbow and radioulnar joint muscles primarily responsible for causing/controlling those movements. Beside each muscle in each movement indicate the type of contraction as follows: I-isometric; C-concentric; E-eccentric.

13. Analyze each skill in the elbow and radioulnar joint sport skill analysis chart (p. 157) and list the movements of the right and left elbow and radioulnar joints in each phase of the skill. You may prefer to list the initial positions that the elbow and radioulnar joints are in for the stance phase. After each movement, list the elbow and radioulnar joint muscle(s) primarily responsible for causing/controlling those movements. Beside each muscle in each movement indicate the type of contraction as follows: I-isometric; C-concentric; E-eccentric. It may be desirable to review the concepts for analysis in Chapter 8 for the various phases.

Muscle Analysis Chart • Elbow and radioulnar joints

Elbow and radioulnar joints	
Flexion	Extension
Pronation	Supination

Elbow and Radioulnar Joint Movement Analysis Chart

Exercise	Initial movement phase		Secondary movement phase	
	Movement(s)	Agonist(s)–(contraction type)	Movement(s)	Agonist(s)–(contraction type)
Push-up				
Chin-up				
Bench press				
Dip				
Lat pull				
Overhead press				
Prone row				

Elbow and Radioulnar Joint Sport Skill Analysis Chart

Exercise		Stance phase	Preparatory phase	Movement phase	Follow-through phase
Baseball pitch	(R)				
	(L)				
Volleyball serve	(R)				
	(L)				
Tennis serve	(R)				
	(L)				
Softball pitch	(R)				
	(L)				
Tennis backhand	(R)				
	(L)				
Batting	(R)				
	(L)				
Bowling	(R)				
	(L)				
Basketball free throw	(R)				
	(L)				

Chapter
6

References

Andrews JR, Wilk KE: *The athlete's shoulder,* New York, 1994, Churchill Livingstone.

Andrews JR, Zarins B, Wilk KE: *Injuries in baseball,* Philadelphia, 1998, Lippincott-Raven.

Back BR Jr, et al: Triceps rupture: a case report and literature review, *American Journal of Sports Medicine* 15:285, May–June 1987.

Gabbard CP, et al: Effects of grip and forearm position on flex arm hang performance, *Research Quarterly for Exercise and Sport,* July 1983.

Guarantors of Brain: *Aids to the examination of the peripheral nervous system,* ed 4, London, 2000, Saunders.

Herrick RT, Herrick S: Ruptured triceps in powerlifter presenting as cubital tunnel syndrome—a case report, *American Journal of Sports Medicine* 15:514, September–October 1987.

Hislop HJ, Montgomery J: *Daniels and Worthingham's muscle testing: techniques of manual examination,* ed 7, Philadelphia, 2002, Saunders.

Magee DJ: *Orthopedic physical assessment,* ed 4, Philadelphia, 2002, Saunders.

Muscolino JE: *The muscular system manual: the skeletal muscles of the human body,* ed 2, St. Louis, 2003, Elsevier Mosby.

Oatis CA: *Kinesiology: the mechanics and pathomechanics of human movement,* Philadelphia, 2004, Lippincott Williams & Wilkins.

Rasch PJ: *Kinesiology and applied anatomy,* ed 7, Philadelphia, 1989, Lea & Febiger.

Shier D, Butler J, Lewis R: *Hole's human anatomy & physiology,* ed 9, New York, 2002, McGraw-Hill.

Sieg KW, Adams SP: *Illustrated essentials of musculoskeletal anatomy,* ed 2, Gainesville, FL, 1985, Megabooks.

Sisto DJ, et al: An electromyographic analysis of the elbow in pitching, *American Journal of Sports Medicine* 15:260, May–June 1987.

Smith LK, Weiss EL, Lehmkuhl LD: *Brunnstrom's clinical kinesiology,* ed 5, Philadelphia, 1996, Davis.

Springer SI: Racquetball and elbow injuries, *National Racquetball* 16:7, March 1987.

Van De Graaff KM: *Human anatomy,* ed 6, Dubuque, IA. 2002, McGraw-Hill.

Chapter 7

The Wrist and Hand Joints

Objectives

- To identify on a human skeleton selected bony features of the wrist, hand, and fingers

- To label selected bony features on a skeletal chart

- To draw and label the muscles on a skeletal chart

- To palpate the muscles on a human subject while demonstrating their actions

- To list the planes of motion and their respective axes of rotation

- To organize and list the muscles that produce the primary movements of the wrist, hand, and fingers

---

Online Learning Center Resources

Visit *Manual of Structural Kinesiology's* Online Learning Center at **www.mhhe.com/floyd16e** for additional information and study material for this chapter, including:

- *Self-grading quizzes*
- *Anatomy flashcards*
- *Animations*

The joints of the wrist, hand, and fingers are often overlooked in their importance to us in comparison with the larger joints needed for ambulation. This should not be the case, because even though the fine motor skills characteristic of this area are not essential in some sports, many sports with skilled activities require precise functioning of the wrist and hand. Several sports, such as archery, bowling, golf, baseball, and tennis, require the combined use of all of these joints. Beyond this, appropriate function in the joints and muscles of our hands is critical for activities of daily living throughout our life.

Because of numerous muscles, bones, and ligaments, combined with relatively small joint size, the functional anatomy of the wrist and hand is complex and overwhelming to some. This complexity may be simplified by relating the functional anatomy to the major actions of the joints: flexion, extension, abduction, and adduction of the wrist and hand.

A large number of muscles are used in these movements. Anatomically and structurally, the human wrist and hand have highly developed, complex mechanisms capable of a variety of movements, which is a result of the arrangement of the 29 bones, more than 25 joints, and more than 30 muscles, of which 18 are intrinsic (both origin and insertion found in the hand) muscles.

For most who use this text, an extensive knowledge of these intrinsic muscles is not necessary. However, athletic trainers, physical therapists, occupational therapists, chiropractors, anatomists, physicians, and nurses require a more extensive knowledge. The intrinsic muscles are listed, illustrated, and discussed to a limited degree at the end of this chapter. References at the end of this chapter provide additional sources from which to gain further information.

Our discussion is limited to a review of the muscles, joints, and movements involved in gross motor activities. The muscles included are those of the forearm and the extrinsic muscles of the wrist, hand, and fingers. The larger, more important extrinsic muscles of each joint are included, providing a limited knowledge of this area. The prescription of exercises for strengthening these muscles will be somewhat redundant, since there are primarily only four movements accomplished by their combined actions. An exercise that will strengthen many of these muscles is fingertip push-ups.

Bones

The wrist and hand contain 29 bones, including the radius and ulna (Fig. 7.1). Eight carpal bones in two rows of four bones form the wrist. The proximal row consists, from the radial (thumb) side to the ulnar (little finger) side, of the scaphoid (boat-shaped) or navicular as it is commonly known, the lunate (moon-shaped), the triquetrum (three-cornered), and the pisiform (pea-shaped) bones. The distal row, from the radial to ulnar side, consists of the trapezium (greater multangular), trapezoid (lesser multangular), capitate (head-shaped), and hamate (hooked) bones. These bones form a three-sided arch that is concave on the palmar side. This bony arch is spanned by the transverse carpal and volar carpal ligaments creating the **carpal tunnel**, which

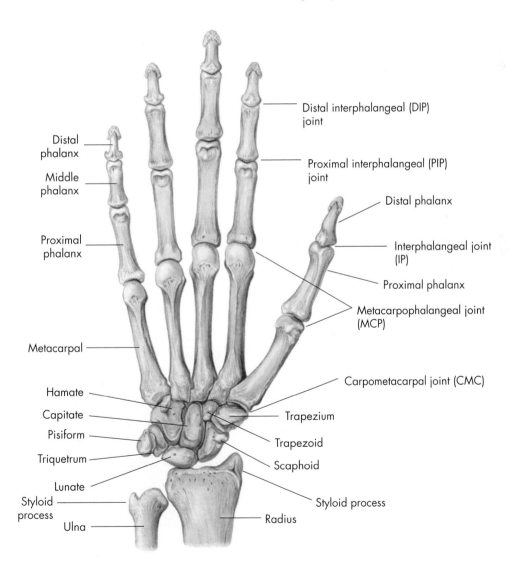

FIG. 7.1 ● Right wrist and hand, palmar surface.

From Anthony CP, Kolthoff NJ: *Textbook of anatomy and physiology,* ed 9, St. Louis, 1975, Mosby.

is frequently a source of problems known as carpal tunnel syndrome (see Fig. 7.8). Of these carpal bones, the scaphoid is by far the most commonly fractured and is usually due to severe wrist hyperextension from falling on the outstretched hand. Unfortunately, this particular fracture is often dismissed as a sprain after initial injury only to cause significant problems in the long term if not properly treated. Treatment often requires precise immobilization for periods longer than that of many fractures and/or surgery. Five metacarpal bones, numbered one to five from the thumb to the little finger, join the wrist bones. There are 14 phalanxes (digits), three for each phalange except the thumb, which has only two. They are indicated as proximal, middle, and distal from the metacarpals. Additionally, the thumb has a sesamoid bone within its flexor tendon, and other sesamoids may occur in the fingers.

The medial epicondyle, medial condyloid ridge, and coranoid process serve as a point of origin for many of the wrist and finger flexors, whereas the lateral epicondyle and lateral supracondylar ridge serve as the point of origin for many extensors of the wrist and fingers (Figs. 6.1 and 6.3). Distally, the key bony landmarks for the muscles involved in wrist motion are the base of the second, third, and fifth metacarpals and the pisiform and hamate. The muscles of the fingers, which are also involved in wrist motion, insert on the base of the proximal, middle, and distal phalanxes (Figs. 7.1 and 7.2). The base of the first metacarpal, proximal and distal phalanxes of the thumb serve as key insertion points for the muscles involved in thumb motion (Fig. 7.1).

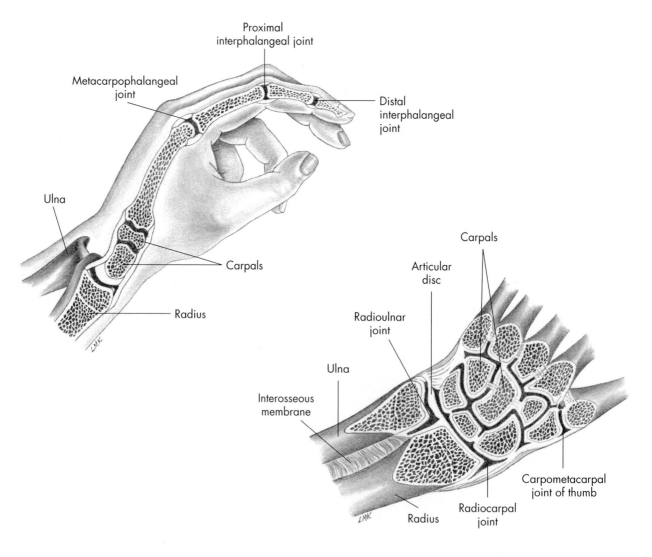

FIG. 7.2 ● Wrist and hand joint structures.

Joints

The wrist joint is classified as a condyloid-type joint, allowing flexion, extension, abduction (radial deviation), and adduction (ulnar deviation) (Fig. 7.2). Wrist motion occurs primarily between the distal radius and the proximal carpal row, consisting of the scaphoid, lunate, and triquetrum. The joint allows 70 to 90 degrees of flexion and 65 to 85 degrees of extension. The wrist can abduct 15 to 25 degrees and adduct 25 to 40 degrees (Fig. 7.3).

Each finger has three joints. The metacarpophalangeal (MCP) joints are classified as condyloid. In these joints, 0 to 40 degrees of extension and 85 to 100 degrees of flexion are possible. The proximal interphalangeal (PIP) joints, classified as ginglymus, can move from full extension to approximately 90 to 120 degrees of flexion.

The distal interphalangeal (DIP) joints, also classified as ginglymus, can flex 80 to 90 degrees from full extension (Fig. 7.5).

The thumb has only two joints, both of which are classified as ginglymus. The metacarpophalangeal (MCP) joint moves from full extension into 40 to 90 degrees of flexion. The interphalangeal (IP) joint can flex 80 to 90 degrees. The carpometacarpal (CMC) joint of the thumb is a unique saddle-type joint having 50 to 70 degrees of abduction. It can flex approximately 15 to 45 degrees and extend 0 to 20 degrees (Fig. 7.6).

Ligaments, too numerous to mention in this discussion, support and provide static stability to the many joints of the wrist and hand. Some of the finger ligaments are detailed in Fig. 7.4.

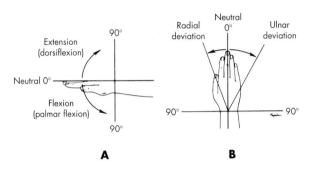

FIG. 7.3 ● ROM of the wrist. **A,** Flexion and extension. *Flexion* (palmar flexion): zero to ±80°. *Extension* (dorsiflexion): zero to ±70°; **B,** Radial and ulnar deviation. *Radial deviation* zero to 20°. *Ulnar deviation:* zero to 30°.

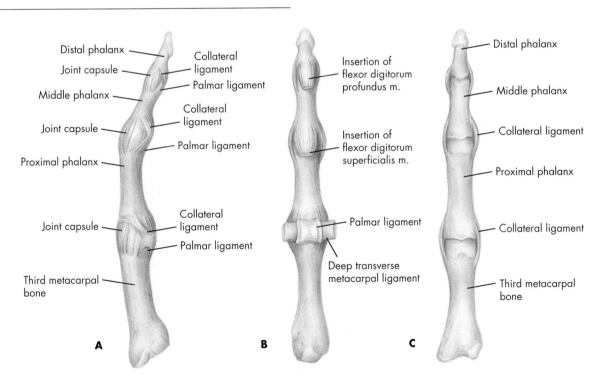

FIG. 7.4 ● Metacarpophalangeal and interphalangeal joints of ring finger. **A,** A lateral view; **B,** An anterior (palmar) view; and **C,** A posterior view.

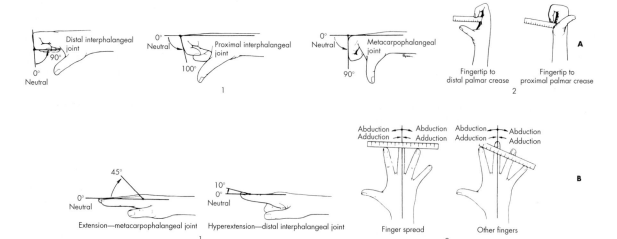

FIG. 7.5 ● ROM of the fingers. **A,** Flexion. *1,* Motion can be estimated or measured in degrees. *2,* Motion can be estimated by a ruler as the distance from the tip of the finger to the distal palmar crease (*left*) (measures flexion of the middle and distal joints) and the proximal palmar crease *(right)* (measures the distal, middle, and proximal joints of the fingers); **B,** Extension, abduction, and adduction. *1,* Extension and hyperextension. *2,* Abduction and adduction. These motions take place in the plane of the palm away from and to the long, or middle, finger of the hand. The spread of fingers can be measured from the tip of the index finger to the tip of the little finger (*right*). Individual fingers spread from tip to tip of indicated fingers (*left*).

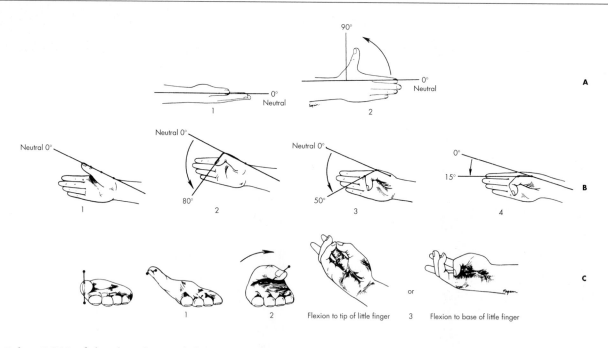

FIG. 7.6 ● ROM of the thumb. **A,** Abduction. *1,* Zero starting position: the extended thumb alongside the index finger, which is in line with the radius. *Abduction* is the angle created between the metacarpal bones of the thumb and index finger. This motion may take place in two planes. *2,* Radial abduction or *extension* takes place parallel to the plane of the palm; **B,** Flexion. *1,* Zero starting position: the extended thumb. *2,* Flexion of the interphalangeal joint: zero to ±80°. *3,* Flexion of the metacarpophalangeal joint: zero to ±50°. *4,* Flexion of the carpometacarpal joint: zero to ±15°; **C,** Opposition. Zero starting position (*far left*): the thumb in line with the index fingers. *Opposition* is a composite motion consisting of three elements: (1) abduction, (2) rotation, and (3) flexion. Motion is usually considered complete when the tip of the thumb touches the tip of the fifth finger. Some consider the arc of opposition complete when the tip of the thumb touches the base of the fifth finger. Both methods are illustrated.

Movements

The common actions of the wrist are flexion, extension, abduction, and adduction (Fig. 7.7, *A–D*). The fingers can only flex and extend (Fig. 7.7, *E–F*), except at the metacarpophalangeal joints, where abduction and adduction (Fig. 7.7, *G–H*) are controlled by the intrinsic hand muscles. In the hand, the middle phalange is regarded as the reference point by which to differentiate abduction and adduction. Abduction of the index and middle fingers occurs when they move laterally toward the radial side of the forearm. Abduction of the ring and little fingers occurs when they move medially toward the ulnar side of the hand. Movement medially of the index and middle fingers toward the ulnar side of the forearm is adduction. Ring and little finger adduction occurs when these fingers move laterally toward the radial side of the hand. The thumb is abducted when it moves away from the palm and is adducted when it moves toward the palmar aspect of the second metacarpal. These movements, together with pronation and supination of the forearm, make possible the many fine, coordinated movements of the forearm, wrist, and hand.

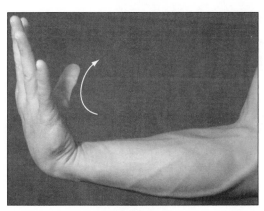

Wrist flexion

A

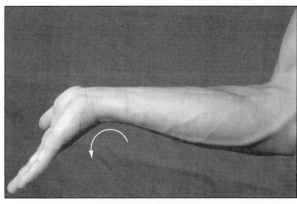

Wrist extension

B

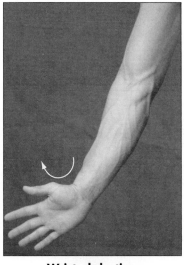

**Wrist abduction
(radial deviation)**

C

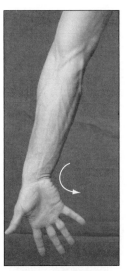

**Wrist adduction
(ulnar deviation)**

D

FIG. 7.7 ● Wrist and hand movements. **A,** Wrist flexion; **B,** Wrist extension; **C,** Wrist abduction; **D,** Wrist adduction.

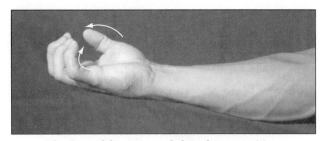

Flexion of fingers and thumb, opposition

E

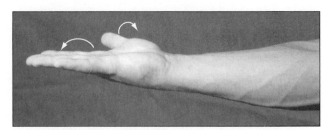

Extension of fingers and thumb, reposition

F

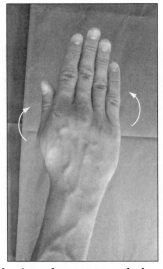

**Adduction of metacarpophalangeal
joints and the thumb**

G

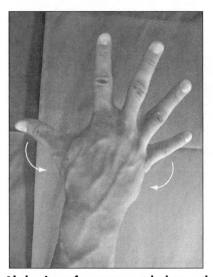

**Abduction of metacarpophalangeal
joints and the thumb**

H

FIG. 7.7 continued ● Wrist and hand movements. **E,** Flexion of the fingers and thumb, opposition;
F, Extension of the fingers and thumb, reposition; **G,** Adduction of metacarpophalangeal joints and the
thumb; **H,** Abduction of metacarpophalangeal joints and the thumb.

Flexion (palmar flexion)

Movement of the palm of the hand and/or the
phalanges toward the anterior or volar aspect of
the forearm

Extension (dorsiflexion)

Movement of the back of the hand and/or the phalanges toward the posterior or dorsal aspect of the
forearm; sometimes referred to as hyperextension

Abduction (radial deviation, radial flexion)

Movement of the thumb side of the hand toward the
lateral aspect or radial side of the forearm; also, movement of the fingers away from the middle finger

Adduction (ulnar deviation, ulnar flexion)

Movement of the little finger side of the hand toward the medial aspect or ulnar side of the forearm; also, movement of the fingers back together
toward the middle finger

Opposition

Movement of the thumb across the palmar aspect
to oppose any or all of the phalanges

Reposition

Movement of the thumb as it returns to the anatomical position from opposition with the hand
and/or fingers

Muscles TABLE 7.1

The extrinsic muscles of the wrist and hand can be grouped according to function and location (Table 7.1). There are six muscles that move the wrist but do not cross the hand to move the fingers and thumb. The three wrist flexors in this group are the flexor carpi radialis, flexor carpi ulnaris, and palmaris longus—all of which have their origin on the medial epicondyle of the humerus. The extensors of the wrist have their origins on the lateral epicondyle and include the extensor carpi radialis longus, extensor carpi radialis brevis, and extensor carpi ulnaris (Figs. 6.8 and 6.10).

Another nine muscles function primarily to move the phalanges but are also involved in wrist joint actions because they originate on the forearm and cross the wrist. These muscles generally are weaker in their actions on the wrist. The flexor digitorum superficialis and the flexor digitorum profundus are finger flexors; however, they also assist in wrist flexion along with the flexor pollicis longus, which is a thumb flexor. The extensor digitorum, the extensor indicis, and the extensor digiti minimi are finger extensors but also assist in wrist extension, along with the extensor pollicis longus and the extensor pollicis brevis, which extend the thumb. The abductor pollicis longus abducts the thumb and assists in wrist abduction.

TABLE 7.1 • Agonist muscles of the wrist and hand joints

	Muscle	Origin	Insertion	Action	Plane of motion	Palpation	Innervation
Anterior muscles (wrist flexors)	Flexor carpi radialis	Medial epicondyle of humerus	Base of 2nd and 3rd metacarpals on palmar surface	Flexion of the wrist	Sagittal	Anterior distal forearm and wrist surface, slightly lateral, in line with the 2nd and 3rd metacarpals with resisted flexion and abduction	Median nerve (C6, C7)
				Abduction of the wrist	Frontal		
				Weak flexion of the elbow	Sagittal		
				Weak pronation of the forearm	Transverse		
	Palmaris longus	Medial epicondyle of humerus	Palmar aponeurosis of the 2nd, 3rd, 4th, and 5th metacarpals	Flexion of the wrist	Sagittal	Anteromedial and central aspect of the anterior forearm just proximal to the wrist, particularly with slight wrist flexion and opposition of thumb to the 5th finger	Median nerve (C6, C7)
				Weak flexion of the elbow			
	Flexor carpi ulnaris	Medial epicondyle of humerus and posterior aspect of proximal ulna	Base of 5th metacarpal (palmar surface), pisiform, and hamate	Flexion of the wrist	Sagittal	Anteromedial surface of the forearm, a few inches below the medial epicondyle of the humerus to just proximal to the wrist, with resisted flexion/adduction	Ulnar nerve (C8, T1)
				Adduction of the wrist	Frontal		
				Weak flexion of the elbow	Sagittal		

TABLE 7.1 (continued) • **Agonist muscles of the wrist and hand joints**

	Muscle	Origin	Insertion	Action	Plane of motion	Palpation	Innervation
Anterior muscles (wrist and phalangeal flexors)	Flexor digitorum superficialis	Medial epicondyle of humerus Ulnar head: medial coronoid process Radial head: upper 2/3 of anterior border of the radius just distal to the radial tuberosity	Each tendon splits and attaches to the sides of middle phalanx of four fingers on palmar surface	Flexion of the fingers at the metacarpo-phalangeal and proximal inter-phalangeal joints	Sagittal	In depressed area between palmaris longus and flexor carpi ulnaris tendons, particularly when making a fist but keeping the distal interphalangeals extended and slightly resisted wrist flexion; also on anterior mid forearm during same activity	Median nerve (C7, C8, T1)
				Flexion of the wrist			
				Weak flexion of the elbow			
	Flexor digitorum profundus	Proximal 3/4 of anterior and medial ulna	Base of distal phalanges of four fingers	Flexion of the four fingers at the metacar-pophalangeal, proximal inter-phalangeal, and distal inter-phalangeal joints	Sagittal	Deep to the flexor digitorum superficialis, but on anterior mid forearm while flexing the distal interphalangeal joints and keeping the proximal interphalangeal joints in extension; over the palmar surface of the 2nd, 3rd, 4th, and 5th meta-carpophalangeal joints during finger flexion against resistance	Median nerve (C8, T1) to 2nd and 3rd fingers; ulnar nerve (C8, T1) to 4th and 5th fingers
				Flexion of the wrist			
	Flexor pollicis longus	Middle anterior surface of the radius and the anterior medial border of the ulna just distal to the coronoid process; occasionally a small head is present attaching on the medial epicondyle of the humerus	Base of distal phalanx of thumb on palmar surface	Flexion of the thumb carpo-metacarpal, metacarpo-phalangeal, and inter-phalangeal joints	Sagittal	Anterior surface of the thumb on the proximal phalanx, and just lateral to the palmaris longus and medial to the flexor carpi radialis on the anterior distal forearm especially during active flexion of the thumb inter-phalangeal joint	Median nerve palmar interosseous branch (C8, T1)
				Flexion of the wrist			
				Abduction of the wrist	Frontal		

Chapter 7

TABLE 7.1 (continued) • **Agonist muscles of the wrist and hand joints**

	Muscle	Origin	Insertion	Action	Plane of motion	Palpation	Innervation
Posterior muscles (wrist extensors)	Extensor carpi ulnaris	Lateral epicondyle of humerus and middle 1/2 of the posterior border of the ulna	Base of 5th metacarpal on dorsal surface	Extension of the wrist	Sagittal	Just lateral to the ulnar styloid process and crossing the posteromedial wrist, particularly with wrist extension/adduction	Radial nerve (C6, C7, C8)
				Adduction of the wrist	Frontal		
				Weak extension of the elbow	Sagittal		
	Extensor carpi radialis brevis	Lateral epicondyle of humerus	Base of 3rd metacarpal on dorsal surface	Extension of the wrist	Sagittal	Just proximal to the dorsal aspect of the wrist and approximately 1 cm medial to the radial styloid process, the tendon may be felt during extension and traced to base of 3rd metacarpal, particularly when making a fist; proximally and posteriorly, just medial to the bulk of the brachioradialis	Radial nerve (C6, C7)
				Abduction of the wrist	Frontal		
				Weak flexion of the elbow	Sagittal		
	Extensor carpi radialis longus	Lower third of lateral supracondylar ridge of humerus and lateral epicondyle of the humerus	Base of 2nd metacarpal on dorsal surface	Extension of the wrist	Sagittal	Just proximal to the dorsal aspect of the wrist and approximately 1 cm medial to the radial styloid process, the tendon may be felt during extension and traced to base of 2nd metacarpal, particularly when making a fist; proximally and posteriorly, just medial to the bulk of the brachioradialis	Radial nerve (C6, C7)
				Abduction of the wrist	Frontal		
				Weak extension of the elbow	Sagittal		
				Weak pronation	Transverse		

Chapter

7

TABLE 7.1 (continued) • Agonist muscles of the wrist and hand joints

	Muscle	Origin	Insertion	Action	Plane of motion	Palpation	Innervation
Posterior muscles (wrist and phalangeal extensors)	Extensor digitorum	Lateral epicondyle of humerus	Four tendons to bases of middle and distal phalanxes of four fingers on dorsal surface	Extension of the 2nd, 3rd, 4th, and 5th phalanges at the metacarpophalangeal joints	Sagittal	With all four fingers extended, on the posterior surface of the distal forearm immediately medial to extensor pollicis longus tendon and lateral to the extensor carpi ulnaris and extensor digiti minimi, then dividing into four separate tendons which are over the dorsal aspect of the hand and metacarpophalangeal joints	Radial nerve (C6, C7, C8)
				Extension of the wrist			
				Weak extension of the elbow			
	Extensor indicis	Middle to distal 1/3 of posterior ulna	Base of the middle and distal phalanxes of the 2nd phalange, dorsal surface	Extension of the index finger at the metacarpophalangeal joint	Sagittal	With forearm pronated on the posterior aspect of the distal forearm and dorsal surface of the hand just medial to the extensor digitorum tendon of the index finger with extension of the index fingers and flexion of the 3rd, 4th, and 5th fingers	Radial nerve (C6, C7, C8)
				Weak wrist extension	Sagittal		
				Weak supination	Transverse		
	Extensor digiti minimi	Lateral epicondyle of humerus	Base of the middle and distal phalanxes of the 5th phalange (dorsal surface)	Extension of the little finger at the metacarpophalangeal joint	Sagittal	Passing over the dorsal aspect of the distal radioulnar joint, particularly with relaxed flexion of other fingers and alternating 5th finger extension and relaxation; dorsal surface of forearm immediately medial to the extensor digitorum and lateral to the extensor carpi ulnaris	Radial nerve (C6, C7, C8)
				Weak wrist extension			
				Weak elbow extension			
	Extensor pollicis longus	Posterior lateral surface of the lower middle ulna	Base of distal phalanx of thumb on dorsal surface	Extension of the thumb at the carpometacarpal, metacarpophalangeal, and interphalangeal joints	Sagittal	Dorsal aspect of the hand to its insertion on the base of the distal phalanx; also on posterior surface of lower forearm between radius and ulnar just proximal to the extensor indicis and medial to the extensor pollicis brevis and abductor pollicis longus with forearm pronated and fingers in relaxed flexion while actively extending the thumb	Radial nerve (C6, C7, C8)
				Extension of the wrist			
				Abduction of the wrist	Frontal		
				Weak supination	Transverse		

Chapter 7

TABLE 7.1 (continued) • **Agonist muscles of the wrist and hand joints**

	Muscle	Origin	Insertion	Action	Plane of motion	Palpation	Innervation
Posterior muscles (wrist and phalangeal extensors)	Extensor pollicis brevis	Posterior surface of lower middle radius	Base of proximal phalanx of thumb on dorsal surface	Extension of the thumb at the carpometacarpal and metacarpo-phalangeal joint	Sagital	Just lateral to the extensor pollicis longus tendon on the dorsal side of the hand to its insertion on the proximal phalanx with extension of the thumb carpometacarpal and metacarpophalangeal and flexion of the interphalangeal joints	Radial nerve (C6, C7)
				Weak wrist extension			
				Abduction of the wrist	Frontal		
Posterior muscles	Abductor pollicis longus	Posterior aspect of radius and midshaft of the ulna	Base of 1st metacarpal on dorsal lateral surface	Abduction of the thumb at the carpometacarpal joint	Frontal	Lateral aspect of the wrist joint just proximal to the 1st metacarpal	Radial nerve (C6, C7)
				Abduction of the wrist			
				Extension of the thumb at the carpometacarpal joint	Sagital		
				Weak wrist flexion			
				Weak supination	Transverse		

Chapter

7

All of the wrist flexors generally have their origins on the anteromedial aspect of the proximal forearm and medial epicondyle of the humerus, whereas their insertions are on the anterior aspect of the wrist and hand. All of the flexor tendons except for the flexor carpi ulnaris and palmaris longus pass through the carpal tunnel, along with the median nerve. Conditions leading to swelling and inflammation in this area can result in increased pressure in the carpal tunnel, which interferes with normal function of the median nerve, leading to reduced motor and sensory function of its distribution (Fig. 7.8). Known as **carpal tunnel syndrome**, this condition is particularly common with repetitive use of the hand and wrist in manual labor and clerical work such as typing and keyboarding. Often slight modifications in work habits and the positions of the hand and wrist during these activities can be preventative. Additionally, flexibility exercises for the wrist and finger flexors may be helpful.

The wrist extensors generally have their origins on the posterolateral aspect of the proximal forearm and lateral humeral epicondyle, whereas their insertions are located on the posterior aspect of the wrist and hand. The flexor and extensor tendons at the distal forearm immediately proximal to the wrist are held in place on the palmar and dorsal aspects by transverse bands of tissue. These bands, respectively known as the flexor and extensor retinaculum, prevent these tendons from bowstringing during flexion and extension.

The wrist abductors are the flexor carpi radialis, extensor carpi radialis longus, extensor carpi radialis brevis, abductor pollicis longus, extensor pollicis longus, and extensor pollicis brevis. These muscles generally cross the wrist joint anterolaterally and posterolaterally to insert on the radial side of the hand. The flexor carpi ulnaris and extensor carpi ulnaris adduct the wrist and cross the wrist joint anteromedially and posteromedially to insert on the ulnar side of the hand.

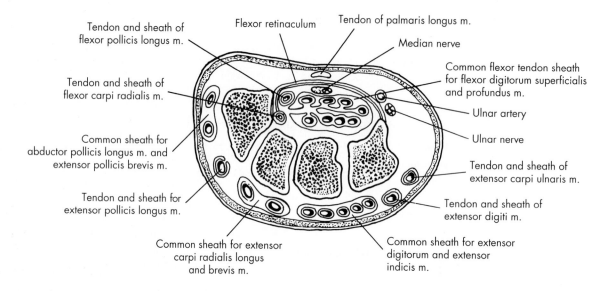

Tendon and sheath of flexor pollicis longus m.

Tendon and sheath of flexor carpi radialis m.

Common sheath for abductor pollicis longus m. and extensor pollicis brevis m.

Tendon and sheath for extensor pollicis longus m.

Common sheath for extensor carpi radialis longus and brevis m.

Flexor retinaculum

Tendon of palmaris longus m.

Median nerve

Common flexor tendon sheath for flexor digitorum superficialis and profundus m.

Ulnar artery

Ulnar nerve

Tendon and sheath of extensor carpi ulnaris m.

Tendon and sheath of extensor digiti m.

Common sheath for extensor digitorum and extensor indicis m.

FIG. 7.8 ● Cross-section of the right wrist. Note the carpal tunnel and the arrangement of the tendon sheaths.

The intrinsic muscles of the hand (see Table 7.2 and Fig. 7.26) have their origins and insertions on the bones of the hand. Grouping of the intrinsic muscles into three groups according to location is helpful in understanding and learning these muscles. On the radial side there are four muscles of the thumb—the opponens pollicis, the abductor pollicis brevis, the flexor pollicis brevis, and the adductor pollicis. On the ulnar side, there are three muscles of the little finger—the opponens digiti minimi, the abductor digiti minimi, and the flexor digiti minimi brevis. In the remainder of the hand, there are 11 muscles, which can be further grouped as the 4 lumbricals, the 3 palmar interossei, and the 4 dorsal interossei.

Wrist and hand muscles—location

Anteromedial at the elbow and forearm and anterior at the hand (Fig. 7.9 A–C)

Primarily wrist flexion

Flexor carpi radialis

Flexor carpi ulnaris

Palmaris longus

Primarily wrist and phalangeal flexion

Flexor digitorum superficialis

Flexor digitorum profundus

Flexor pollicis longus

Posterolateral at the elbow and forearm and posterior at the hand (Fig. 7.9 D)

Primarily wrist extension

Extensor carpi radialis longus

Extensor carpi radialis brevis

Extensor carpi ulnaris

Primarily wrist and phalangeal extension

Extensor digitorum

Extensor indicis

Extensor digiti minimi

Extensor pollicis longus

Extensor pollicis brevis

Abductor pollicis longus

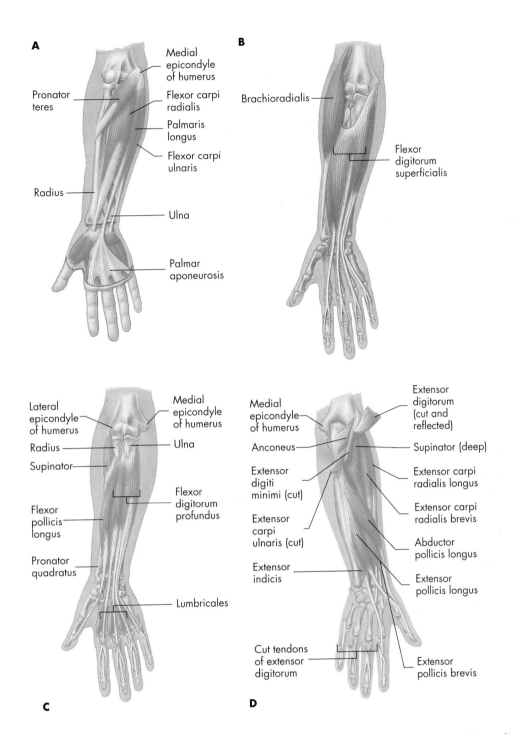

A, Anterior view of the right forearm (superficial). Brachioradialis muscle

- Medial epicondyle of humerus
- Pronator teres
- Flexor carpi radialis
- Palmaris longus
- Flexor carpi ulnaris
- Radius
- Ulna
- Palmar aponeurosis

B
- Brachioradialis
- Flexor digitorum superficialis

C
- Lateral epicondyle of humerus
- Radius
- Supinator
- Flexor pollicis longus
- Pronator quadratus
- Medial epicondyle of humerus
- Ulna
- Flexor digitorum profundus
- Lumbricales

D
- Medial epicondyle of humerus
- Anconeus
- Extensor digiti minimi (cut)
- Extensor carpi ulnaris (cut)
- Extensor indicis
- Cut tendons of extensor digitorum
- Extensor digitorum (cut and reflected)
- Supinator (deep)
- Extensor carpi radialis longus
- Extensor carpi radialis brevis
- Abductor pollicis longus
- Extensor pollicis longus
- Extensor pollicis brevis

Chapter 7

FIG. 7.9 ● Muscles of the forearm. **A,** Anterior view of the right forearm (superficial). Brachioradialis muscle is removed; **B,** Anterior view of the right forearm (deeper than **A**). Pronator teres, flexor carpi radialis and ulnaris, and palmaris longus muscles are removed; **C,** Anterior view of the right forearm (deeper than **A** or **B**). Brachioradialis, pronator teres, flexor carpi radialis and ulnaris, palmaris longus, and flexor digitorum superficialis muscles are removed; **D,** Deep muscles of the right posterior forearm, Extensor digitorum, extensor digiti minimi, and extensor carpi ulnaris muscles are cut to reveal deeper muscles.

Nerves

The muscles of the wrist and hand are all innervated from the radial, median, and ulnar nerves of the brachial plexus as illustrated in Figs. 6.12, 6.13, and 7.10. The radial nerve, originating from C6, C7, and C8, provides innervation for the extensor carpi radialis brevis and extensor carpi radialis longus. It then branches to become the posterior interosseous nerve, which supplies the extensor carpi ulnaris, extensor digitorum, extensor digiti minimi, abductor pollicis longus, extensor pollicis longus, extensor pollicis brevis, and extensor indicis. The median nerve, arising from C6, C7, C8, and T1, innervates the flexor carpi radialis, palmaris longus, and flexor digitorum superficialis. It then branches to become the anterior interosseous nerve, which innervates the flexor digitorum profundus for the index and long finger as well as the flexor pollicis longus. Regarding the intrinsic muscles of the hand, the median nerve innervates the abductor pollicis brevis, flexor pollicis brevis (superficial head), opponens pollicis, and the first and second lumbrical. The ulnar nerve, branching from C8 and T1, supplies the flexor digitorum profundus for the fourth and fifth fingers and the flexor carpi ulnaris. Additionally, it innervates the remaining intrinsic muscles of the hand (the deep head of the flexor pollicis brevis, adductor pollicis, palmar interossei, dorsal interossei, third and fourth lumbrical, opponens digiti minimi, abductor digiti minimi, and flexor digiti minimi brevis). Sensation to the ulnar side of the hand, ulnar one-half of the ring finger, and the entire little finger are provided by the ulnar nerve.

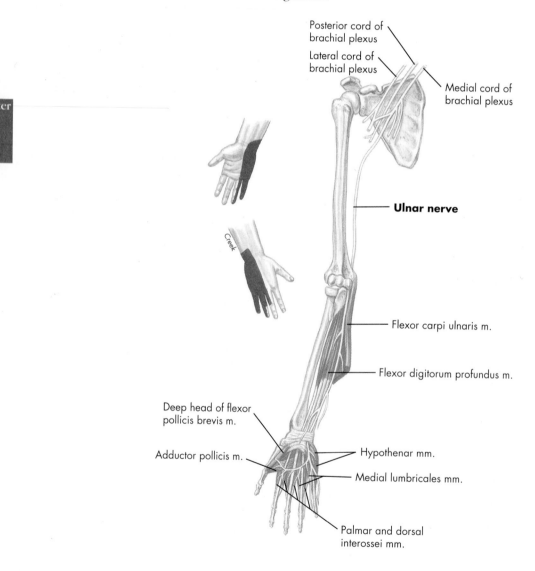

FIG. 7.10 ● Muscular and cutaneous distribution of the ulnar nerve.

Flexor carpi radialis muscle FIG. 7.11

(fleks´or kar´pi ra´di-a´lis)

Origin

Medial epicondyle of the humerus

Insertion

Base of the second and third metacarpals, anterior (palmar surface)

Action

Flexion of the wrist
Abduction of the wrist
Weak flexion of the elbow
Weak pronation of the forearm

Palpation

Anterior surface of the wrist, slightly lateral, in line with the second and third metacarpals with resisted flexion and abduction

Innervation

Median nerve (C6, C7)

Application, strengthening, and flexibility

The flexor carpi radialis, flexor carpi ulnaris, and palmaris longus are the most powerful of the wrist flexors. They are brought into play during any activity that requires wrist curling or stabilization of the wrist against resistance, particularly if the forearm is supinated.

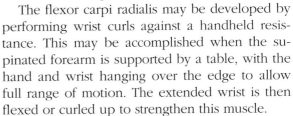

The flexor carpi radialis may be developed by performing wrist curls against a handheld resistance. This may be accomplished when the supinated forearm is supported by a table, with the hand and wrist hanging over the edge to allow full range of motion. The extended wrist is then flexed or curled up to strengthen this muscle.

To stretch the flexor carpi radialis, the elbow must be fully extended with the forearm supinated while a partner passively extends and adducts the wrist.

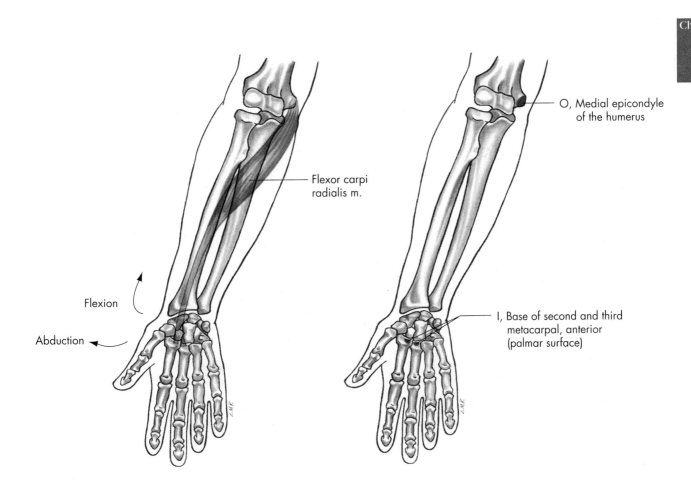

Flexor carpi radialis m.

Flexion

Abduction

O, Medial epicondyle of the humerus

I, Base of second and third metacarpal, anterior (palmar surface)

FIG. 7.11 ● Flexor carpi radialis muscle, anterior view. *O*, Origin; *I*, Insertion.

Palmaris longus muscle FIG. 7.12

(pal-ma´ris long´gus)

Origin

Medial epicondyle of the humerus

Insertion

Palmar aponeurosis of the second, third, fourth, and fifth metacarpals

Action

Flexion of the wrist
Weak flexion of the elbow

Palpation

The palmaris longus is absent in either one or both forearms in some people. Anteromedial and central aspect of the anterior forearm just proximal to the wrist, particularly with slight wrist flexion and opposition of thumb to the 5th finger

Innervation

Median nerve (C6, C7)

Application, strengthening, and flexibility

Unlike the flexor carpi radialis and flexor carpi ulnaris, which are not only wrist flexors but also abductors and adductors, respectively, the palmaris longus is involved only in wrist flexion from the anatomical position because of its central location on the anterior forearm and wrist. It can, however, assist in abducting the wrist from an extremely adducted position back to neutral and assist in adducting the wrist from an extremely abducted position back to neutral. It may also assist slightly in forearm pronation because of its slightly lateral insertion in relation to its origin on the medial epicondyle. It may also be strengthened with any type of wrist-curling activity, such as the ones described for the flexor carpi radialis muscle.

Maximal elbow and wrist extension stretches the palmaris longus.

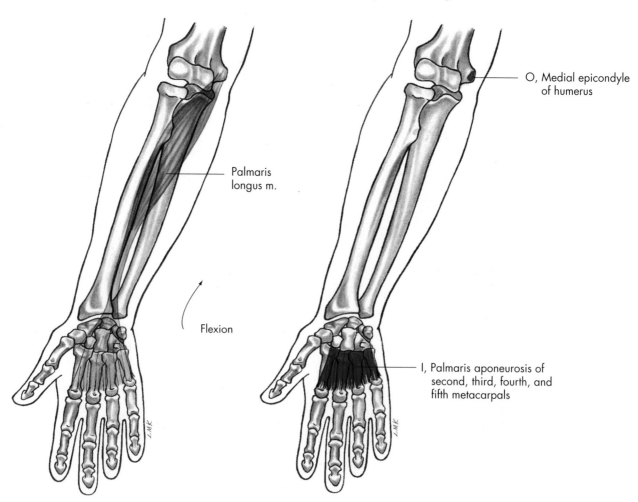

Palmaris longus m.

Flexion

O, Medial epicondyle of humerus

I, Palmaris aponeurosis of second, third, fourth, and fifth metacarpals

FIG. 7.12 ● Palmaris longus muscle, anterior view. *O,* Origin; *I,* Insertion.

Flexor carpi ulnaris muscle FIG. 7.13

(fleks´or kar´pi ul-na´ris)

Origin

Medial epicondyle of the humerus
Posterior aspect of the proximal ulna

Insertion

Pisiform, hamate, and base of the fifth metacarpal
(palmar surface)

Action

Flexion of the wrist
Adduction of the wrist, together with the extensor
carpi ulnaris muscle
Weak flexion of the elbow

Palpation

Anteromedial surface of the forearm, a few inches
below the medial epicondyle of the humerus to just
proximal to the wrist, with resisted flexion/adduction

Innervation

Ulnar nerve (C8, T1)

Application, strengthening, and flexibility

The flexor carpi ulnaris is very important in wrist
flexion or curling activities. In addition, it is one of
only two muscles involved in wrist adduction or
ulnar flexion. It may be strengthened with any type
of wrist-curling activity against resistance, such as
those described for the flexor carpi radialis muscle.

To stretch the flexor carpi ulnaris, the elbow
must be fully extended with the forearm supinated while a partner passively extends and abducts the wrist.

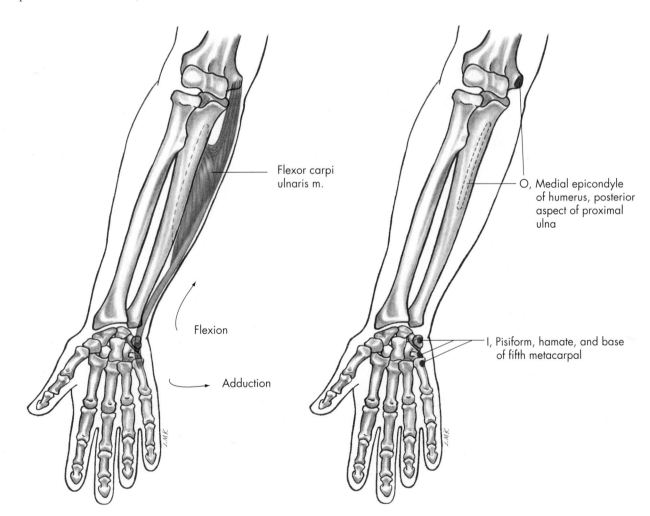

FIG. 7.13 ● Flexor carpi ulnaris muscle, anterior view. *O,* Origin; *I,* Insertion.

Extensor carpi ulnaris muscle FIG. 7.14

(eks-ten´sor kar´pi ul-na´ris)

Origin

Lateral epicondyle of the humerus
Middle two-fourths of the posterior border of the ulna

Insertion

Base of the fifth metacarpal (dorsal surface)

Action

Extension of the wrist
Adduction of the wrist together with the flexor carpi
 ulnaris muscle
Weak extension of the elbow

Palpation

Just lateral to the ulnar styloid process and crossing the posteromedial wrist, particularly with wrist
extension/adduction

Innervation

Radial nerve (C6–C8)

Application, strengthening, and flexibility

Besides being a powerful wrist extensor, the extensor carpi ulnaris muscle is the only muscle other than
the flexor carpi ulnaris involved in wrist adduction or
ulnar deviation. The extensor carpi ulnaris, the extensor carpi radialis brevis, and the extensor carpi radialis
longus are the most powerful of the wrist extensors.
Any activity requiring wrist extension or stabilization
of the wrist against resistance, particularly if the forearm is pronated, depends greatly on the strength of
these muscles. They are often brought into play with
the backhand in racquet sports.

The extensor carpi ulnaris may be developed by
performing wrist extension against a handheld resistance. This may be accomplished with the pronated
forearm being supported by a table with the hand
hanging over the edge to allow full range of motion.
The wrist is then moved from the fully flexed position
to the fully extended position against the resistance.

Stretching the extensor carpi ulnaris requires the
elbow to be extended with the forearm pronated while
the wrist is passively flexed and slightly abducted.

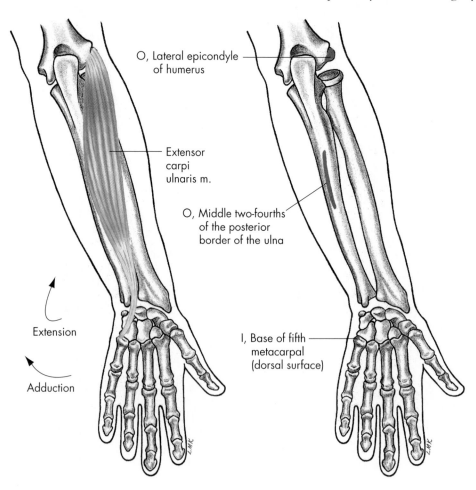

FIG. 7.14 ● Extensor carpi ulnaris muscle, posterior view. *O,* Origin; *I,* Insertion.

Extensor carpi radialis brevis muscle FIG. 7.15

(eks-ten´sor kar´pi ra´di-a´lis bre´vis)

Origin
Lateral epicondyle of the humerus

Insertion
Base of the third metacarpal (dorsal surface)

Action
Extension of the wrist
Abduction of the wrist
Weak flexion of the elbow

Palpation
Dorsal side of the forearm, and difficult to distinguish between the extensor carpi radialis longus and the extensor digitorum

Just proximal to the dorsal aspect of the wrist and approximately 1 cm medial to the radial styloid process, the tendon may be felt during extension and traced to base of third metacarpal particularly when making a fist. Proximally and posteriorly, just medial to the bulk of the brachioradialis

Innervation
Radial nerve (C6, C7)

Application, strengthening, and flexibility
The extensor carpi radialis brevis is important in any sports activity that requires powerful wrist extension, such as golf or tennis. Wrist extension exercises, such as those described for the extensor carpi ulnaris, are appropriate for development of the muscle.

Stretching the extensor carpi radialis brevis and longus requires the elbow to be extended with the forearm pronated while the wrist is passively flexed and slightly adducted.

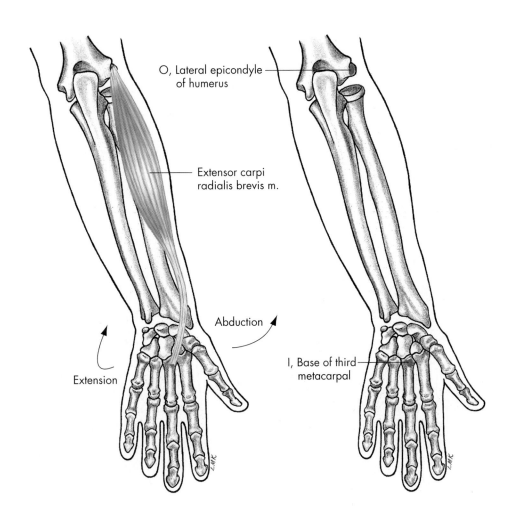

FIG. 7.15 ● Extensor carpi radialis brevis muscle, posterior view. *O,* Origin; *I,* Insertion.

Extensor carpi radialis longus muscle FIG. 7.16

(eks-ten´sor kar´pi ra´di-a´lis long´gus)

Origin

Lower third of lateral supracondylar ridge of humerus and lateral epicondyle of the humerus

Insertion

Base of the second metacarpal (dorsal surface)

Action

Extension of the wrist
Abduction of the wrist
Weak flexion of the elbow
Weak pronation to neutral from a fully supinated position

Palpation

Just proximal to the dorsal aspect of the wrist and approximately 1 cm medial to the radial styloid pro-

cess, the tendon may be felt during extension and traced to base of second metacarpal particularly when making a fist. Proximally and posteriorly, just medial to the bulk of the brachioradialis

Innervation

Radial nerve (C6, C7)

Application, strengthening, and flexibility

The extensor carpi radialis longus, like the extensor carpi radialis brevis, is also important in any sports activity that requires powerful wrist extension. In addition, both muscles are involved in abduction of the wrist. The extensor carpi radialis longus may be developed with the same wrist extension exercises as described for the extensor carpi ulnaris muscle.

The extensor carpi radialis longus is stretched in the same manner as the extensor carpi radialis brevis.

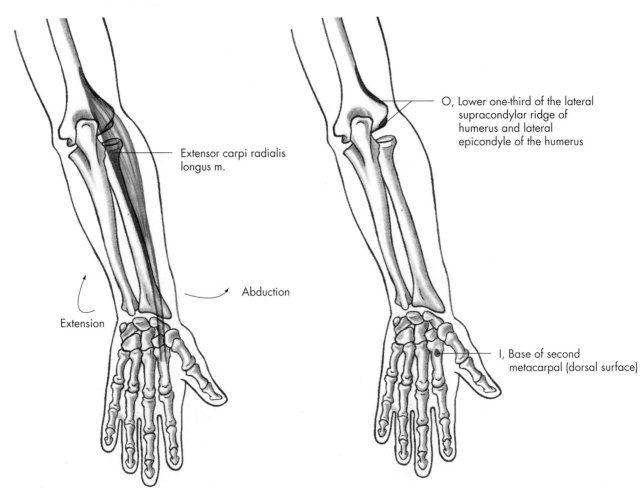

Extensor carpi radialis longus m.

Abduction

Extension

O, Lower one-third of the lateral supracondylar ridge of humerus and lateral epicondyle of the humerus

I, Base of second metacarpal (dorsal surface)

FIG. 7.16 ● Extensor carpi radialis longus muscle, posterior view. *O,* Origin; *I,* Insertion.

Flexor digitorum superficialis muscle FIG. 7.17

(fleks´or dij-i-to´rum su´per-fish-e-al´is)

Origin

Medial epicondyle of the humerus
Ulnar head: medial coronoid process
Radial head: upper two-thirds of anterior border of radius just distal to the radial tuberosity

Insertion

Each tendon splits and attaches to the sides of the middle phalanx of four fingers (palmar surface)

Action

Flexion of the fingers at the metacarpophalangeal and proximal interphalangeal joints
Flexion of the wrist
Weak flexion of the elbow

Palpation

In depressed area between palmaris longus and flexor carpi ulnaris tendons, particularly when making a fist but keeping the distal interphalangeal extended and slightly resisted wrist flexion; also on anterior mid forearm during same activity

Innervation

Median nerve (C7, C8, and T1)

Application, strengthening, and flexibility

The flexor digitorum superficialis muscle, also known as the flexor digitorum sublimis, divides into four tendons on the palmar aspect of the wrist and hand to insert on each of the four fingers. The flexor digitorum superficialis and the flexor digitorum profundus are the only muscles involved in flexion of all four fingers. Both of these muscles are vital in any type of gripping activity.

Squeezing a sponge rubber ball in the palm of the hand, along with other gripping and squeezing activities, can be used to develop these muscles.

The flexor digitorum superficialis is stretched by passively extending the elbow, wrist, metacarpophalangeal, and proximal interphalangeal joints while maintaining the forearm in full supination.

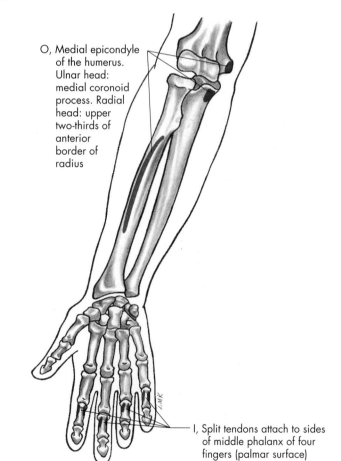

Flexor digitorum superficialis m.

Flexion of wrist

Flexion of fingers

O, Medial epicondyle of the humerus. Ulnar head: medial coronoid process. Radial head: upper two-thirds of anterior border of radius

I, Split tendons attach to sides of middle phalanx of four fingers (palmar surface)

FIG. 7.17 ● Flexor digitorum superficialis muscle, anterior view. *O,* Origin; *I,* Insertion.

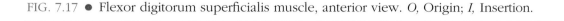

Flexor digitorum profundus muscle

FIG. 7.18

(fleks´or dij-i-to´rum pro-fun´dus)

Origin

Proximal three-fourths of the anterior and medial ulna

Insertion

Base of the distal phalanxes of the four fingers

Action

Flexion of the four fingers at the metacarpophalangeal, proximal interphalangeal, and distal interphalangeal joints

Flexion of the wrist

Palpation

Difficult to distinguish deep to the flexor digitorum superficialis, but on anterior mid forearm while flexing the distal interphalangeal joints and keeping the proximal interphalangeal joints in extension; over the palmar surface of the second, third, fourth, and fifth metacarpophalangeal joints during finger flexion against resistance

Innervation

Median nerve (C8, T1) to the second and third fingers
Ulnar nerve (C8, T1) to the fourth and fifth fingers

Application, strengthening, and flexibility

Both the flexor digitorum profundus muscle and the flexor digitorum superficialis muscle assist in wrist flexion because of their palmar relationship to the wrist. The flexor digitorum profundus is used in any type of gripping, squeezing, or hand-clenching activity, such as gripping a racket or climbing a rope.

The flexor digitorum profundus muscle may be developed through these activities, in addition to the strengthening exercises described for the flexor digitorum superficialis muscle.

The flexor digitorum profundus is stretched similarly to the flexor digitorum superficialis, except that the distal interphalangeal joints must be passively extended in addition to the elbow, wrist, metacarpophalangeal, and proximal interphalangeal joints while maintaining the forearm in full supination.

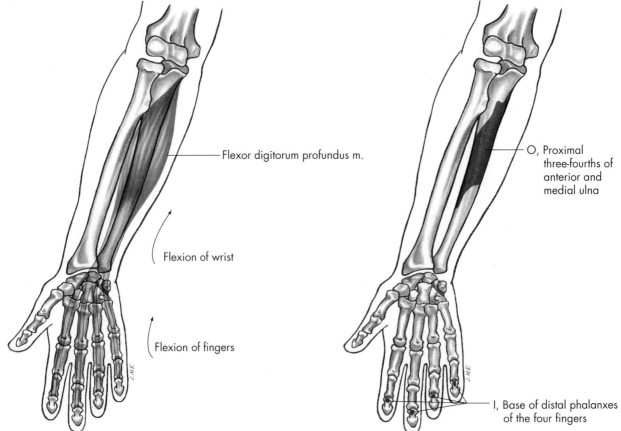

FIG. 7.18 ● Flexor digitorum profundus muscle, anterior view. *O*, Origin; *I*, Insertion.

Flexor pollicis longus muscle FIG. 7.19

(fleks´or pol´i-sis long´gus)

Origin

Middle anterior surface of the radius and the anterior medial border of the ulna just distal to the coronoid process; occasionally a small head is present attaching on the medial epicondyle of the humerus

Insertion

Base of the distal phalanx of the thumb (palmar surface)

Action

Flexion of the thumb carpometacarpal, metacarpophalangeal, and interphalangeal joints
Flexion of the wrist
Abduction of the wrist

Palpation

Anterior surface of the thumb on the proximal phalanx, and just lateral to the palmaris longus and medial to the flexor carpi radialis on the anterior distal forearm especially during active flexion of the thumb interphalangeal joint

Innervation

Median nerve, palmar interosseous branch (C8, T1)

Application, strengthening, and flexibility

The primary function of the flexor pollicis longus muscle is flexion of the thumb, which is vital in gripping and grasping activities of the hand. Because of its palmar relationship to the wrist, it provides some assistance in wrist flexion.

It may be strengthened by pressing a sponge rubber ball into the hand with the thumb and by many other gripping or squeezing activities.

The flexor pollicis longus is stretched by passively extending the entire thumb while simultaneously maintaining maximal wrist extension.

Chapter 7

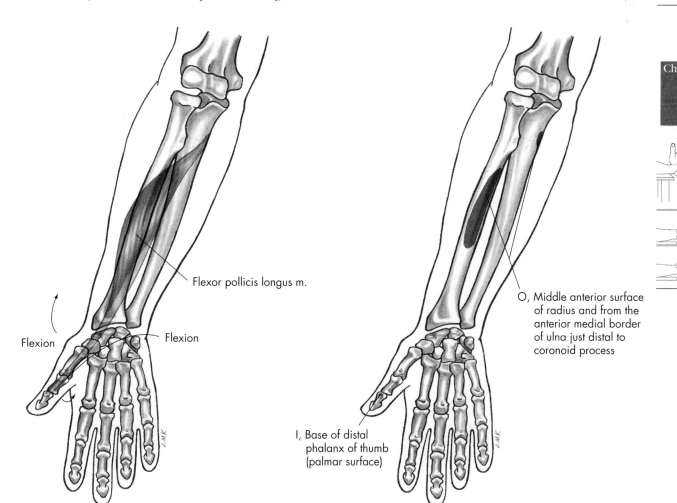

Flexor pollicis longus m.

Flexion

Flexion

Flexion

O, Middle anterior surface of radius and from the anterior medial border of ulna just distal to coronoid process

I, Base of distal phalanx of thumb (palmar surface)

FIG. 7.19 ● Flexor pollicis longus muscle, anterior view. *O,* Origin; *I,* Insertion.

Extensor digitorum muscle FIG. 7.20

(eks-ten´sor dij-i-to´rum)

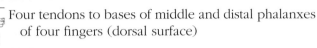

Origin

Lateral epicondyle of the humerus

Insertion

Four tendons to bases of middle and distal phalanxes of four fingers (dorsal surface)

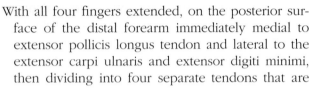

Action

Extension of the second, third, fourth, and fifth phalanges at the metacarpophalangeal joints
Extension of the wrist
Weak extension of the elbow

Palpation

With all four fingers extended, on the posterior surface of the distal forearm immediately medial to extensor pollicis longus tendon and lateral to the extensor carpi ulnaris and extensor digiti minimi, then dividing into four separate tendons that are

over the dorsal aspect of the hand and metacarpophalangeal joints

Innervation

Radial nerve (C6–C8)

Application, strengthening, and flexibility

The extensor digitorum, also known as the extensor digitorum communis, is the only muscle involved in extension of all four fingers. This muscle divides into four tendons on the dorsum of the wrist to insert on each of the fingers. It also assists with wrist extension movements. It may be developed by applying manual resistance to the dorsal aspect of the flexed fingers and then extending the fingers fully. When performed with the wrist in flexion, this exercise increases the workload on the extensor digitorum.

To stretch the extensor digitorum, the fingers must be maximally flexed at the metacarpophalangeal, proximal interphalangeal, and distal interphalangeal joints while the wrist is fully flexed.

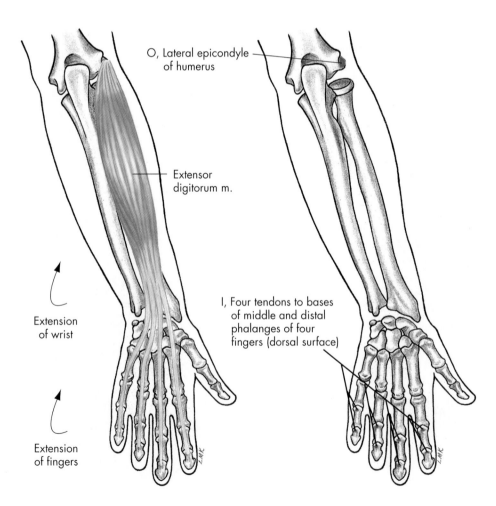

O, Lateral epicondyle of humerus

Extensor digitorum m.

Extension of wrist

Extension of fingers

I, Four tendons to bases of middle and distal phalanges of four fingers (dorsal surface)

FIG. 7.20 ● Extensor digitorum muscle, posterior view. *O,* Origin; *I,* Insertion.

Extensor indicis muscle FIG. 7.21

(eks-ten´sor in´di-sis)

Origin
Middle to distal one-third of the posterior ulna

Insertion
Base of the middle and distal phalanxes of second phalange (dorsal surface)

Action
Extension of the index finger at the metacarpophalangeal joint

Weak wrist extension

Weak supination of the forearm from a pronated position

Palpation
With forearm pronated on the posterior aspect of the distal forearm and dorsal surface of the hand just medial to the extensor digitorum tendon of the index finger with extension of the index fingers and flexion of the third, fourth, and fifth fingers

Innervation
Radial nerve (C6–C8)

Application, strengthening, and flexibility
The extensor indicis muscle is the pointing muscle. That is, it is responsible for extending the index finger, particularly when the other fingers are flexed. It also provides weak assistance to wrist extension and may be developed through exercises similar to those described for the extensor digitorum.

The extensor indicis is stretched by passively taking the index finger into maximal flexion at its metacarpophalangeal, proximal interphalangeal, and distal interphalangeal joints while fully flexing the wrist.

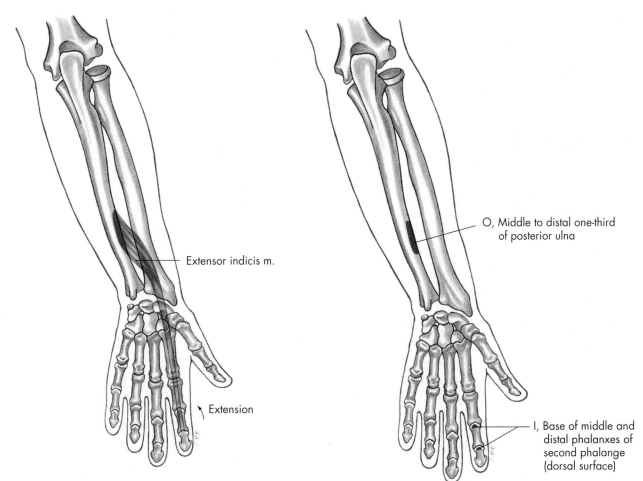

Extensor indicis m.

Extension

O, Middle to distal one-third of posterior ulna

I, Base of middle and distal phalanxes of second phalange (dorsal surface)

FIG. 7.21 ● Extensor indicis muscle, posterior view. *O,* Origin; *I,* Insertion.

Extensor digiti minimi muscle FIG. 7.22

(eks-ten´sor dij´i-ti min´im-i)

Origin

Lateral epicondyle of the humerus

Insertion

Base of the middle and distal phalanxes of the fifth phalange (dorsal surface)

Action

Extension of the little finger at the metacarpophalangeal joint

Weak wrist extension

Weak elbow extension

Palpation

Passing over the dorsal aspect of the distal radioulnar joint, particularly with relaxed flexion of other fingers and alternating fifth finger extension and relaxation; dorsal surface of forearm immediately medial to the extensor digitorum and lateral to the extensor carpi ulnaris

Innervation

Radial nerve (C6–C8)

Application, strengthening, and flexibility

The primary function of the extensor digiti minimi muscle is to assist the extensor digitorum in extending the little finger. Because of its dorsal relationship to the wrist, it also provides weak assistance in wrist extension. It is strengthened with the same exercises described for the extensor digitorum.

The extensor digiti minimi is stretched by passively taking the little finger into maximal flexion at its metacarpophalangeal, proximal interphalangeal, and distal interphalangeal joints while fully flexing the wrist.

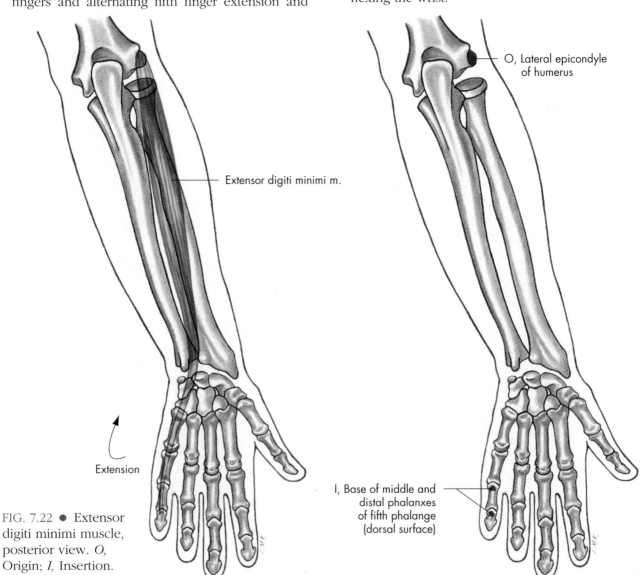

Extensor digiti minimi m.

Extension

O, Lateral epicondyle of humerus

I, Base of middle and distal phalanxes of fifth phalange (dorsal surface)

FIG. 7.22 ● Extensor digiti minimi muscle, posterior view. *O,* Origin; *I,* Insertion.

Chapter 7

Extensor pollicis longus muscle FIG. 7.23

(eks-ten´sor pol´i-sis long´gus)

Origin
Posterior lateral surface of the lower middle ulna

Insertion
Base of the distal phalanx of the thumb (dorsal surface)

Action
Extension of the thumb at the carpometacarpal, meta-carpophalangeal, and interphalangeal joints
Extension of the wrist
Abduction of the wrist
Weak supination of the forearm from a pronated position

Palpation
Dorsal aspect of the hand to its insertion on the base of the distal phalanx; also on posterior surface of lower forearm between radius and ulnar just proximal to the extensor indicis and medial to the extensor pollicis brevis and abductor pollicis longus with forearm pronated and fingers in relaxed flexion while actively extending the thumb

Innervation
Radial nerve (C6–C8)

Application, strengthening, and flexibility

The primary function of the extensor pollicis longus muscle is extension of the thumb, although it does provide weak assistance in wrist extension.

It may be strengthened by extending the flexed thumb against manual resistance. It is stretched by passively taking the entire thumb into maximal flexion at its carpometacarpal, metacarpophalangeal, and interphalangeal joints while fully flexing the wrist.

The tendons of the extensor pollicis longus and extensor pollicis brevis, along with the tendon of the abductor pollicis brevis, forms the anatomical snuffbox, which is the small depression that develops between these two tendons when they contract. The name "anatomical snuffbox" originates from tobacco users placing their snuff in this depression. Deep in the snuffbox the scaphoid bone can be palpated and is often a site of point tenderness when it is fractured.

Chapter 7

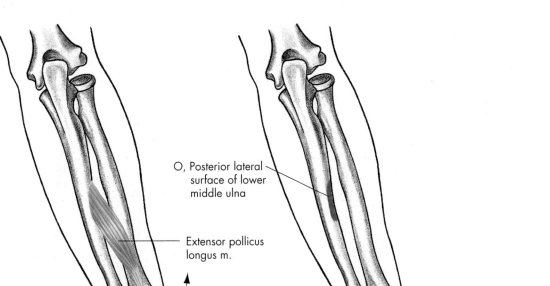

O, Posterior lateral surface of lower middle ulna

Extensor pollicus longus m.

Extension

Extension

I, Base of distal phalanx of thumb (dorsal surface)

FIG. 7.23 ● Extensor pollicis longus muscle, posterior view. *O,* Origin; *I,* Insertion.

Extensor pollicis brevis muscle FIG. 7.24

(eks-ten´sor pol´i-sis bre´vis)

Origin

Posterior surface of the lower middle radius

Insertion

Base of the proximal phalanx of the thumb (dorsal surface)

Action

Extension of the thumb at the carpometacarpal and metacarpophalangeal joint

Weak wrist extension

Wrist abduction

Palpation

Just lateral to the extensor pollicis longus tendon on the dorsal side of the hand to its insertion on the proximal phalanx with extension of the thumb carpometacarpal and metacarpophalangeal and flexion of the interphalangeal joints

Innervation

Radial nerve (C6, C7)

Application, strengthening, and flexibility

The extensor pollicis brevis assists the extensor pollicis longus in extending the thumb. Because of its dorsal relationship to the wrist, it, too, provides weak assistance in wrist extension.

It may be strengthened through the same exercises described for the extensor pollicis longus muscle. It is stretched by passively taking the first carpometacarpal joint and the metacarpophalangeal joint of the thumb into maximal flexion while fully flexing the wrist.

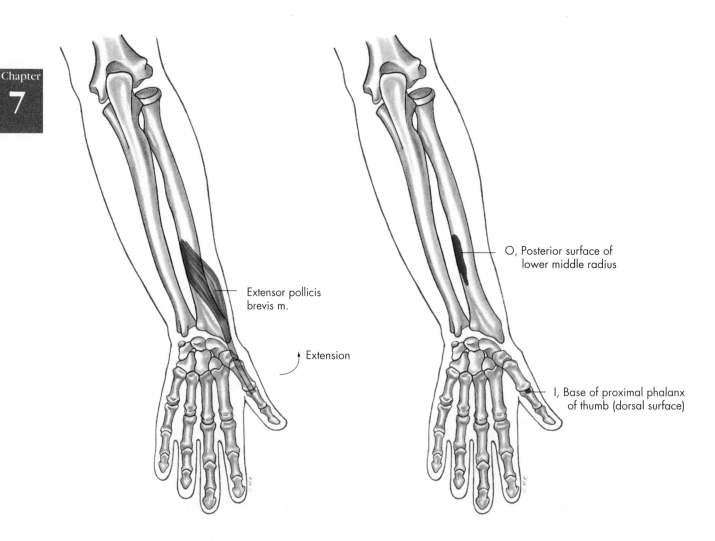

Extensor pollicis brevis m.

Extension

O, Posterior surface of lower middle radius

I, Base of proximal phalanx of thumb (dorsal surface)

FIG. 7.24 ● Extensor pollicis brevis muscle, posterior view. *O*, Origin; *I*, Insertion.

Abductor pollicis longus muscle FIG. 7.25

(ab-duk´tor pol´i-sis lon´gus)

Origin

Posterior aspect of the radius and midshaft of the ulna

Insertion

Base of the first metacarpal (dorsal lateral surface)

Action

Abduction of the thumb at the carpometacarpal joint
Abduction of the wrist
Extension of the thumb at the carpometacarpal joint
Weak supination of the forearm from a pronated position
Weak flexion of the wrist joint

Palpation

With forearm in neutral pronation/supination on the lateral aspect of the wrist joint just proximal to the first metacarpal during active thumb and wrist abduction

Innervation

Radial nerve (C6, C7)

Application, strengthening, and flexibility

The primary function of the abductor pollicis longus muscle is abduction of the thumb, although it does provide some assistance in abduction of the wrist. It may be developed by abducting the thumb from the adducted position against a manually applied resistance. Stretching of the abductor pollicis longus is accomplished by fully flexing and adducting the entire thumb across the palm with the wrist fully adducted.

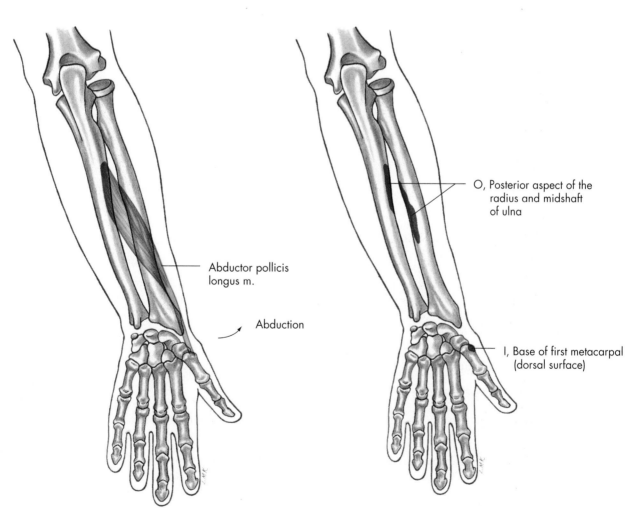

Abductor pollicis longus m.

Abduction

O, Posterior aspect of the radius and midshaft of ulna

I, Base of first metacarpal (dorsal surface)

FIG. 7.25 ● Abductor pollicis longus muscle, posterior view. *O,* Origin; *I,* Insertion.

Intrinsic muscles of the hand

The intrinsic hand muscles may be grouped according to location as well as according to the parts of the hand they control (Fig. 7.26). The abductor pollicis brevis, opponens pollicis, flexor pollicis brevis, and adductor pollicis make up the thenar eminence—the muscular pad on the palmar surface of the first metacarpal. The hypothenar eminence is the muscular pad that forms the ulnar border on the palmar surface of the hand and is made up of the abductor digiti minimi, flexor digiti minimi brevis, and opponens digiti minimi. The intermediate muscles of the hand consist of three palmar interossei, four dorsal interossei, and four lumbrical muscles.

Four intrinsic muscles act on the carpometacarpal joint of the thumb. The opponens pollicis is the muscle that causes opposition in the thumb metacarpal. The abductor pollicis brevis abducts the thumb metacarpal and is assisted in this action by the flexor pollicis brevis, which also flexes the thumb metacarpal. The metacarpal of the thumb is adducted by the adductor pollicis. Both the flexor pollicis brevis and the adductor pollicis flex the proximal phalanx of the thumb.

The three palmar interossei are adductors of the second, fourth, and fifth phalanges. The four dorsal interossei both flex and abduct the index, middle, and ring proximal phalanxes, in addition to assisting with extension of the middle and distal phalanxes of these fingers. The third dorsal

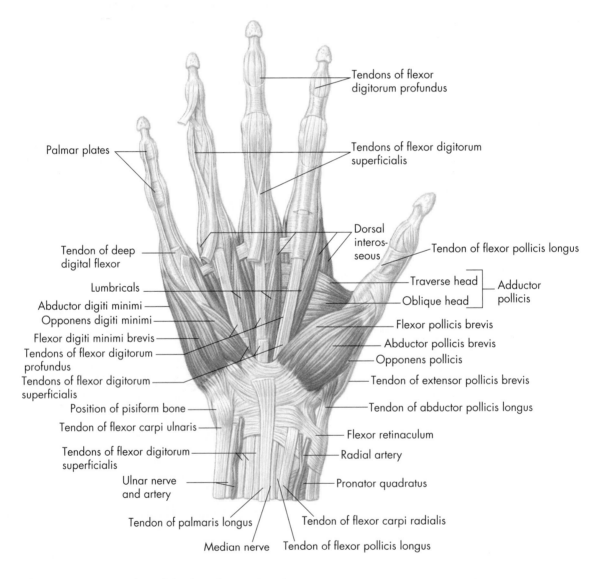

FIG. 7.26 ● Intrinsic muscles of the hand, anterior view.

Modified from Van De Graaff KM: *Human anatomy,* ed 4, New York, 1995, McGraw-Hill.

interossei also adducts the middle finger. The four lumbricals flex the index, middle, ring, and little proximal phalanxes and extend the middle and distal phalanxes of these fingers.

Three muscles act on the little finger. The opponens digiti minimi causes opposition of the little finger metacarpal. The abductor digiti minimi abducts the little finger metacarpal, and the flexor digiti minimi brevis flexes this metacarpal.

Refer to Table 7.2 for further details regarding the intrinsic muscles of the hand.

TABLE 7.2 • Intrinsic muscles of the hand

	Muscle	Origin	Insertion	Action	Palpation	Innervation
Thenar muscles	Opponens pollicis	Anterior surface of transverse carpal ligament, trapezium	Lateral border of 1st metacarpal	CMC opposition of thumb	Palmar aspect of 1st metacarpal with opposition of fingertips to thumb	Median nerve (C6, C7)
	Abductor pollicis brevis	Anterior surface of transverse carpal ligament, trapezium, scaphoid	Base of 1st proximal phalanx	CMC abduction of thumb	Radial aspect of palmar surface of 1st metacarpal with 1st CMC abduction	Median nerve (C6, C7)
	Flexor pollicis brevis	Superficial head: trapezium and transverse carpal ligament Deep head: ulnar aspect of 1st metacarpal	Base of proximal phalanx of 1st metacarpal	CMC flexion and abduction; MCP flexion of thumb	Medial aspect of thenar eminence just proximal to 1st MCP joint with 1st MCP flexion against resistance	Superficial head: median nerve (C6, C7) Deep head: ulnar nerve (C8, T1)
Intermediate muscles	Adductor pollicis	Transverse head: anterior shaft of 3rd metacarpal Oblique head: base of 2nd and 3rd metacarpals, capitate, trapezoid	Ulnar aspect of base of proximal phalanx of 1st metacarpal	CMC adduction; MCP flexion of thumb	Palmar surface between 1st and 2nd metacarpal with 1st CMC adduction	Ulnar nerve (C8, T1)
	Palmar interossei	Shaft of 2nd, 4th, and 5th metacarpals and extensor expansions	Bases of 2nd, 4th, and 5th proximal phalanxes and extensor expansions	MCP adduction of 2nd, 4th, and 5th phalanges	Cannot be palpated	Ulnar nerve (C8, T1)
	Dorsal interossei	Two heads on shafts on adjacent metacarpals	Bases of 2nd, 3rd, and 4th proximal phalanxes and extensor expansions	MCP flexion and abduction; PIP/DIP extension of 2nd, 3rd, and 4th phalanges; MCP adduction of 3rd phalange	Dorsal surface between 1st and 2nd metacarpals, between shafts of 2nd through 5th metacarpals with active abduction/adduction of 2nd, 3rd, and 4th MCP joints	Ulnar nerve, palmar branch (C8, T1)
	Lumbricals	Flexor digitorum profundus tendon in center of palm	Extensor expansions on radial side of 2nd, 3rd, 4th, and 5th proximal phalanxes	MCP flexion and PIP/DIP extension of 2nd, 3rd, 4th, and 5th phalanges	Cannot be palpated	1st and 2nd: median nerve (C6, C7) 3rd and 4th: ulnar nerve (C8, T1)

TABLE 7.2 (continued) • Intrinsic muscles of the hand

	Muscle	Origin	Insertion	Action	Palpation	Innervation
Hypothenar muscles	Opponens digiti minimi	Hook of hamate and adjacent transverse carpal ligament	Medial border of 5th metacarpal	MCP opposition of 5th phalange	On radial aspect of hypothenar eminence with opposition of 5th phalange to thumb	Ulnar nerve (C8, T1)
	Abductor digiti minimi	Pisiform and flexor carpi ulnaris tendon	Ulnar aspect of base of 5th proximal phalanx	MCP abduction of 5th phalange	Ulnar aspect of hypothenar eminence with 5th MCP abduction	Ulnar nerve (C8, T1)
	Flexor digiti minimi brevis	Hook of hamate and adjacent transverse carpal ligament	Ulnar aspect of base of 5th proximal phalanx	MCP flexion of 5th phalange	Palmar surface of 5th metacarpal, lateral to opponens digiti minimi with 5th MCP flexion against resistance	Ulnar nerve (C8, T1)

Web sites

Radiologic Anatomy Browser

http://radlinux1.usuf1.usuhs.mil/rad/iong/index.html

Numerous radiological views of the musculoskeletal system.

University of Arkansas Medical School Gross Anatomy for Medical Students

http://anatomy.uams.edu/anatomyhtml/grossresources.html

Dissections, anatomy tables, atlas images, links, etc.

Loyola University Medical Center: Structure of the Human Body

www.meddean.luc.edu/lumen/MedEd/GrossAnatomy/GA.html

An excellent site with many slides, dissections, tutorials, etc., for the study of human anatomy.

Wheeless' Textbook of Orthopaedics

www.wheelessonline.com

An extensive index of links to the fractures, joints, muscles, nerves, trauma, medications, medical topics, lab tests, and links to orthopaedic journals and other orthopaedic and medical news.

Arthroscopy.com

www.arthroscopy.com/sports.htm

Patient information on various musculoskeletal problems of the upper and lower extremity.

Premiere Medical Search Engine

www.medsite.com/Default.asp?bhcp=1

Allows the reader to enter any medical condition, and it will search the net to find relevant articles.

Virtual Hospital

www.vh.org

Numerous slides, patient information, etc.

Medical Multimedia Group

www.healthpages.org/AHP/LIBRARY/HLTHTOP/CTD

A patient's guide to cumulative trauma disorder (CTD).

Medical Multimedia Group

www.healthpages.org/AHP/LIBRARY/HLTHTOP/CTS/ctsndx.htm

A patient's guide to carpal tunnel syndrome.

Physioroom.com

www.physioroom.com/injuries/arm/index.shtml

Articles on hand and wrist injuries.

Dartmouth Medical School

www.dartmouth.edu/~anatomy/wrist-hand/muscles

Muscles of the wrist and hand.

American Academy of Orthopaedic Surgeons

http://orthoinfo.aaos.org/category.cfm?topcategory=Arm

Patient education library on the hand.

MayoClinic.com

www.mayoclinic.com/invoke.cfm?id=AR00030

Hand exercises for people with arthritis.

American Society for Surgery of the Hand

**www.assh.org/Content/NavigationMenu/
Patients_and_Public/Extensor_Tendon_Injuries/
Extensor_Tendon_Injuries.htm**

Extensor tendon injuries.

Worksheet exercises

As an aid to learning, for in-class or out-of-class assignments, or for testing, tear-out worksheets are found at the end of the text (pp. 380 and 381).

Skeletal worksheet (no. 1)

Draw and label on the worksheet the following muscles:
  a. Flexor pollicis longus
  b. Flexor carpi radialis
  c. Flexor carpi ulnaris
  d. Extensor digitorum
  e. Extensor pollicis longus
  f. Extensor carpi ulnaris

Human figure worksheet (no. 2)

Label and indicate with arrows the following movements of the wrist and hand:
  a. Flexion
  b. Extension
  c. Abduction (ulnar flexion)
  d. Adduction (radial flexion)

LABORATORY AND REVIEW EXERCISES

1. Locate the following parts of the humerus, radius, ulna, carpals, and metacarpals on a human skeleton and on a subject:
  a. Skeleton
      1. Medial epicondyle
      2. Lateral epicondyle
      3. Lateral supracondylar ridge
      4. Trochlea
      5. Capitulum
      6. Coronoid process
      7. Tuberosity of the radius
      8. Styloid process—radius
      9. Styloid process—ulna
      10. First and third metacarpals
      11. Wrist bones
      12. First phalanx of third metacarpal

  b. Subject
      1. Medial epicondyle
      2. Lateral epicondyle
      3. Lateral supracondylar ridge
      4. Pisiform
      5. Scaphoid (navicular)
2. How and where can the following muscles be palpated on a human subject?
  a. Flexor pollicis longus
  b. Flexor carpi radialis
  c. Flexor carpi ulnaris
  d. Extensor digitorum communis
  e. Extensor pollicis longus
  f. Extensor carpi ulnaris
3. Demonstrate the action and list the muscles primarily responsible for these movements at the wrist joint.
  a. Flexion
  b. Extension
  c. Abduction
  d. Adduction
4. List the planes in which each of the following wrist, hand, and finger joint movements occur. List the respective axis of rotation for each movement in each plane.
  a. Abduction
  b. Adduction
  c. Flexion
  d. Extension
5. Discuss why the thumb is the most important part of the hand.
6. How should boys and girls be taught to do push-ups? Justify your answer.
  a. Hands flat on the floor
  b. Fingertips
7. With a laboratory partner, determine how and why maintaining full flexion of all the fingers is impossible when passively moving the wrist into maximal flexion. Is it also difficult to maintain maximal extension of all the finger joints while passively taking the wrist into full extension?

Chapter

7

Muscle analysis chart • Wrist, hand, and fingers

Wrist and hand	
Flexion	Extension
Adduction	Abduction

Fingers—metacarpophalangeal joints	
Flexion	Extension

Fingers—proximal interphalangeal joints	
Flexion	Extension

Fingers—distal interphalangeal joints	
Flexion	Extension

Thumb	
Flexion	Extension

8. Fill in the muscle analysis chart above by listing the muscles primarily involved in each movement.

9. List the muscles involved in the little finger as you would type on a computer keyboard and reach for the left tab key as the wrists are properly stabilized in an ergonomic position.

10. Describe the importance of the intrinsic muscles in the hand as you reach to turn a door knob.

11. Determine the kind of flexibility exercises that would be indicated for a patient with carpal tunnel syndrome and explain in detail how they should be performed.

12. After analyzing each of the exercises in the wrist and hand joint movement analysis chart (p. 193), break each into two primary movement phases such as a lifting phase and lowering phase. For each phase, determine the wrist and hand joint movements occurring, and then list the wrist and hand joint muscles primarily responsible for causing/controlling those movements. Beside each muscle in each movement indicate the type of contraction as follows: I-isometric; C-concentric; E-eccentric.

13. Analyze each skill in the wrist and hand joint sport skill analysis chart (p. 193) and list the movements of the right and left wrist and hand joints in each phase of the skill. You may prefer to list the initial positions that the wrist and hand joints are in for the stance phase. After each movement, list the wrist and hand joint muscle(s) primarily responsible for causing/controlling those movements. Beside each muscle in each movement indicate the type of contraction as follows: I-isometric; C-concentric; E-eccentric. It may be desirable to review the concepts for analysis in Chapter 8 for the various phases.

Wrist and hand joint movement analysis chart

Exercise	Initial movement phase		Secondary movement phase	
	Movement(s)	Agonist(s)–(contraction type)	Movement(s)	Agonist(s)–(contraction type)
Push-up				
Chin-up				
Bench press				
Dip				
Lat pull				
Ball squeeze				
Frisbee throw				

Wrist and hand joint sport skill analysis chart

Exercise		Stance phase	Prepatory phase	Movement phase	Follow-through phase
Baseball pitch	(R)				
	(L)				
Volleyball serve	(R)				
	(L)				
Tennis serve	(R)				
	(L)				
Softball pitch	(R)				
	(L)				
Tennis backhand	(R)				
	(L)				
Batting	(R)				
	(L)				
Bowling	(R)				
	(L)				
Basketball free throw	(R)				
	(L)				

References

Gabbard CP, et al: Effects of grip and forearm position on flex arm hang performance, *Research Quarterly for Exercise and Sport,* July 1983.

Gench BE, Hinson MM, Harvey PT: *Anatomical kinesiology,* Dubuque, IA, 1995, Eddie Bowers.

Guarantors of Brain: *Aids to the examination of the peripheral nervous system,* ed 4, London, 2000, Saunders.

Hislop HJ, Montgomery J: *Daniels and Worthingham's muscle testing: techniques of manual examination,* ed 7, Philadelphia, 2002, Saunders.

Lindsay DT: *Functional human anatomy,* St. Louis, 1996, Mosby.

Luttgens K, Hamilton N: *Kinesiology: scientific basis of human motion,* ed 10, New York, 2002, McGraw-Hill.

Magee DJ: *Orthopedic physical assessment,* ed 4, Philadelphia, 2002, Saunders.

Muscolino JE: *The muscular system manual: the skeletal muscles of the human body,* ed 2, St. Louis, 2003, Elsevier Mosby.

Norkin CC, Levangie PK: *Joint structure and function—a comprehensive analysis,* Philadelphia, 1983, Davis.

Norkin CC, White DJ: *Measurement of joint motion: a guide to goniometry,* Philadelphia, 1985, Davis.

Oatis CA: *Kinesiology: the mechanics and pathomechanics of human movement,* Philadelphia, 2004, Lippincott Williams & Wilkins.

Rasch PJ: *Kinesiology and applied anatomy,* ed 7, Philadelphia, 1989, Lea & Febiger.

Seeley RR, Stephens TD, Tate P: *Anatomy & physiology,* ed 7, New York, 2006, McGraw-Hill.

Sieg KW, Adams SP: *Illustrated essentials of musculoskeletal anatomy,* ed 2, Gainesville, FL, 1985, Megabooks.

Sisto DJ, et al: An electromyographic analysis of the elbow in pitching, *American Journal of Sports Medicine* 15:260, May–June 1987.

Smith LK, Weiss EL, Lehmkuhl LD: *Brunnstrom's clinical kinesiology,* ed 5, Philadelphia, 1996, Davis.

Springer SI: Racquetball and elbow injuries, *National Racquetball* 16:7, March 1987.

Stone RJ, Stone JA: *Atlas of the skeletal muscles,* New York, 1990, McGraw-Hill.

Van De Graaff KM: *Human anatomy,* ed 6, Dubuque, IA, 2002, McGraw-Hill.

Chapter 7

Muscular Analysis of Upper Extremity Exercises

Objectives

- To begin analyzing sports skills in terms of phases and the various joint movements occurring in those phases

- To understand various conditioning principles and how to apply them to strengthening major muscle groups

- To analyze an exercise to determine the joint movements and the types of contractions occurring in the specific muscles involved in those movements

- To learn and understand the concept of open versus closed kinetic chain

- To learn to group individual muscles into units that produce certain joint movements

- To begin to think of exercises that increase the strength and endurance of individual muscle groups

- To learn to analyze and prescribe exercises to strengthen major muscle groups.

Proper functioning of the upper extremity is critical for most sports activities as well as many activities of daily living. Strength and endurance in this part of the human body are essential for improved appearance and posture, as well as for more efficient skill performance. Unfortunately, it is often one of the body's weaker areas when considering the number of muscles involved. Specific exercises and activities to condition this area should be intelligently selected by becoming thoroughly familiar with the muscles involved.

At this stage, simple exercises are used to begin teaching individuals how to group muscles to produce joint movement. Some of these simple introductory exercises are included in this chapter.

The early analysis of exercise makes the study of structural kinesiology more meaningful as students come to better understand the importance of individual muscles and groups of muscles in bringing about joint movements in various exercises. Chapter 13 contains analysis of exercises for the entire body, with emphasis on the trunk and lower extremities. Contrary to what most beginning students in structural kinesiology believe, muscular analysis of activities is not difficult once the basic concepts are understood.

Upper extremity activities

Children seem to have an innate desire to climb, swing, and hang. Such movements use the muscles of the hands, wrists, elbows, and shoulder joints. But the opportunity to perform these types of activities is limited in our modern culture. Unless emphasis is placed on the development of this area of our bodies by physical education teachers in elementary schools, for both boys and girls, it will continue to be muscularly the weakest area of our bodies. Weakness in the upper extremities can impair skill development and performance in many common enjoyable recreational activities, such as golf, tennis, softball, and racquetball. People enjoy what they can do well, and they can be taught to enjoy activities that will increase the strength of this part of the body. These and other such activities are ones that can be enjoyed throughout the entire adult life; therefore, adequate skill development built on an appropriate base of muscular strength and endurance is essential for enjoyment and prevention of injury.

Often, we perform typical strengthening exercises in the weight room, such as the bench press, overhead press, and biceps curl. While being good exercises, these all concentrate primarily on the muscles of the anterior upper extremity. This can lead to an overdevelopment of these muscles with respect to the posterior muscles. As a result, individuals may become strong and tight anteriorly and weak and flexible posteriorly. It is for these reasons that one must be able to analyze specific strengthening exercises and determine the muscles involved so that overall muscular balance is addressed through appropriate exercise prescription.

Concepts for analysis

In analyzing activities, it is important to understand that muscles are usually grouped according to their concentric function and work in paired opposition to an antagonistic group. An example of this **aggregate** muscle grouping to perform a given joint action is seen with the elbow flexors all working together. In this example, the elbow flexors (biceps, brachii, brachialis, and brachioradialis) are concentrically contracting as an agonist group to achieve flexion. As they flex the elbow, each muscle contributes significantly to the task. They are working in opposition to their antagonists, the triceps brachii and anconeus. The triceps brachii and anconeus both work together as an aggregate muscle group to cause elbow extension, but in this example they are cooperating in their lengthening to allow the flexors to perform their task. In this cooperative lengthening, the triceps and anconeus may or may not be under tension. If there is no tension, then the lengthening is passive, caused totally by the elbow flexors. If there is tension, then the elbow extensors are contracting eccentrically to control the amount and speed of lengthening.

An often confusing aspect is that, depending on the activity, these muscle groups can function to control the exact opposite actions by contracting eccentrically. That is, through eccentric contractions, the elbow flexors may control elbow extension, and the triceps brachii and anconeus may control elbow flexion. Exercise professionals should be able to view an activity and not only determine which muscles are performing the movement but also know what type of contraction is occurring and what kind of exercises are appropriate for developing the muscles. Chapter 2 provides a review of how muscles contract to work in groups to function in joint movement.

Analysis of movement

When analyzing various exercises and sport skills, it is essential to break down all of the movements into phases. The number of phases, usually three to five, will vary, depending on the skill. Practically all sport skills will have at least a preparatory phase, a movement phase, and a follow-through phase. Many will also begin with a stance phase and end with a recovery phase. The names of the phases will vary from skill to skill to fit in with the terminology used in various sports, and may also vary depending upon the body part involved. In some cases, these major phases may also be divided even further, as with baseball, where the preparatory phase for the pitching arm is broken into early cocking and late cocking.

The **stance phase** allows the athlete to assume a comfortable and appropriately balanced body position from which to initiate the sport skill. The emphasis is on setting the various joint angles in the correct positions with respect to one another and to the sport surface. Generally, with respect to the subsequent phases, the stance phase is a relatively static phase with fairly short ranges of motion involved. Due to the minimal amount of movement in this phase, the majority of the joint position maintenance throughout the body will be accomplished through isometric contractions.

Chapter
8

196   www.mhhe.com/floyd16e

The **preparatory phase**, often referred to as the cocking or wind-up phase, is used to lengthen the appropriate muscles so that they will be in position to generate more force and momentum as they concentrically contract in the next phase. It is the most critical phase in leading toward the desired result of the activity and becomes more dynamic as the need for explosiveness increases. Generally, to lengthen the muscles needed in the next phase, concentric contractions occur in their antagonist muscles in this phase.

The **movement phase**, sometimes known as the acceleration, action, motion, or contact phase, is the action part of the skill. It is the phase in which the summation of force is generated directly to the ball, sport object, or opponent, and is usually characterized by near-maximal concentric activity in the involved muscles.

The **follow-through phase** begins immediately after the climax of the movement phase, in order to bring about negative acceleration of the involved limb or body segment. In this phase, often referred to as the deceleration phase, the velocity of the body segment progressively decreases, usually over a wide range of motion. This velocity decrease is usually attributable to high eccentric activity in the muscles that were antagonist to the muscles used in the movement phase. Generally, the greater the acceleration in the movement phase, the greater the length and the importance of the follow-though phase. Occasionally, some athletes may begin the follow-through phase too soon, thereby cutting short the movement phase and having a less-than-desirable result in the activity.

The **recovery phase** is used after follow-through to regain balance and positioning to be ready for the next sport demand. To a degree, the muscles used eccentrically in the follow-through phase to decelerate the body or body segment will be used concentrically in recovery to bring about the initial return to a functional position.

Skill analysis can be seen with the example of a baseball pitch in Fig. 8.1. The stance phase begins when the player assumes a position with the ball in the glove before receiving the signal from the catcher. The pitcher begins the preparatory phase by extending the throwing arm posteriorly and rotating the trunk to the right in conjunction with left hip flexion. The right shoulder girdle is fully retracted in combination with abduction and

FIG. 8.1 ● Skill analysis phases—baseball pitch.

maximum external rotation of the glenohumeral joint to complete this phase. Immediately following, the movement phase begins with forward movement of the arm and continues until ball release. At ball release, the follow-through phase begins as the arm continues moving in the same direction established by the movement phase until the velocity decreases to the point that the arm can safely change movement direction. This deceleration of the body, and especially the arm, is accomplished by high amounts of eccentric activity. At this point, the recovery phase begins, enabling the player to reposition to field the batted ball. In this example, reference has been made briefly only to the throwing arm, but there are many similarities in other overhand sports skills, such as the tennis serve, javelin throw, and volleyball serve. In actual practice, the movements of each joint in the body should be analyzed into the various phases.

The kinetic chain concept

As you have learned, our extremities consist of several bony segments linked by a series of joints. These bony segments and their linkage system of joints may be likened to a chain. Just as with a chain, any one link in the extremity may be moved individually without significantly affecting the other links if the chain is open or not attached at one end. However, if the chain is securely attached or closed, substantial movement of any one link cannot occur without substantial and subsequent movement of the other links.

In the body, an extremity may be seen as representing an **open kinetic chain** if the distal end of the extremity is not fixed to any surface. This arrangement allows any one joint in the extremity to move or function separately without necessitating movement of other joints in the extremity. Examples in the upper extremity of these single

joint exercises would be a shoulder shrug, deltoid raise (shoulder abduction), or biceps curl. Lower extremity examples include seated hip flexion, knee extension, and ankle dorsiflexion exercises. In all of these examples, the core of the body and the proximal segment is stabilized while the distal segment is free to move in space through a single plane. These types of exercises are known as joint-isolation exercises and are beneficial in isolating a particular joint to concentrate on specific muscle groups; however, they are not very functional in that most physical activity, particularly for the lower extremity, requires multiple-joint activity involving numerous muscle groups simultaneously. Furthermore, since the joint is stable proximally and loaded distally, shear forces are acting on the joint with potential negative consequences.

If the distal end of the extremity is fixed, as in a push-up, dip, squat, or dead lift, the extremity represents a **closed kinetic chain**. In this closed

TABLE 8.1 • **Differences between open- and closed-chain exercises**

Variable	Open-chain exercise	Closed-chain exercise
Distal end of extremity	Free in space and not fixed	Fixed to something
Movement pattern	Characterized by rotary stress in the joint (often nonfunctional)	Characterized by linear stress in the joint (functional)
Joint movements	Occur in isolation	Multiple occur simultaneously
Muscle recruitment	Isolated (minimal muscular co-contraction)	Multiple (significant muscular co-contraction)
Joint axis	Stable during movement patterns	Primarily transverse
Movement plane	Usually single	Multiple (triplanar)
Proximal segment of joint	Stable	Mobile
Distal segment of joint	Mobile	Mobile, except for most distal aspect
Motion occurs	Distal to instantaneous axis of rotation	Proximal and distal to instantaneous axis of rotation
Functionality	Often nonfunctional, especially lower extremity	Functional
Joint forces	Shear	Compressive
Joint stability	Decreased due to shear and distractive forces	Increased due to compressive forces
Stabilization	Artificial	Not artificial, realistic and functional
Loading	Artificial	Physiological, provides for normal proprioceptive and kinesthetic feedback

Adapted from Ellenbecker TS, Davies GJ: *Closed kinetic chain exercise: a comprehensive guide to multiple-joint exercise,* Champaign, IL, 2001, Human Kinetics.

system, movement of one joint cannot occur without causing predictable movements of the other joints in the extremity. Closed-chain activities are very functional and involve the body moving in relation to the relatively fixed distal segment. The advantage of these multiple-joint exercises is that several joints are involved, and numerous muscle groups must participate in causing and controlling the multiple-plane movements, which strongly correlate to most physical activities. Additionally, the joint is more stable due to the joint compressive forces from weight bearing.

To state the differences another way, open-chain exercises involve the extremity being moved to or from the stabilized body whereas closed-chain exercises involve the body being moved to or from the stabilized extremity. Table 8.1 provides a comparison of variables that differ with open- versus closed-chain exercises, and Fig. 8.2 provides examples of each.

FIG. 8.2 ● Open- versus closed-chain kinetic activities. **A,** Open-chain activity for the upper extremity; **B,** Closed-chain activity for the upper extremity; **C,** Open-chain activity for the lower extremity; **D,** Closed-chain activity for the lower extremity.

Not every exercise or activity can be classified totally as either an open- or closed-chain exercise. For example, walking and running are both open and closed due to their swing and stance phases, respectively. Another case is bicycle riding, which is mixed in that the pelvis on the seat is the most stable segment, but the feet are attached to movable pedals.

Consideration of the open versus closed kinetic chain is important in determining both the muscles and their types of contractions when analyzing sport activities. Realizing the relative differences in demands on the musculoskeletal system through detailed analysis of skilled movements is critical for determining the most appropriate conditioning exercises to improve performance. Generally, closed kinetic chain exercises are more functional and applicable to the demands of sports and physical activity. Most sports involve closed-chain activities in the lower extremities and open-chain activities in the upper extremities. However, there are many exceptions, and closed-chain conditioning exercises may be beneficial for extremities primarily involved in open-chain sporting activities. Open-chain exercises are useful in concentrating on the development of a specific muscle group at a single joint.

Conditioning considerations

Overload principle

A basic physiological principle of exercise is the overload principle. It states that, within appropriate parameters, a muscle or muscle group increases in strength in direct proportion to the overload placed on it. While it is beyond the scope of this text to fully explain specific applications of the overload principle for each component of physical fitness, some general concepts will be mentioned. To improve the strength and functioning of major muscles, this principle should be applied to every large muscle group in the body, progressively throughout each year, at all age levels. In actual practice, the amount of overload applied varies significantly based on several factors. For example, an untrained person beginning a strength-training program will usually make significant gains in the amount of weight he or she is able to lift in the first few weeks of the exercise program. Most of this increased ability is due to a refinement of neuromuscular function, rather than an actual increase in muscle tissue strength. Similarly, a well-trained person will see

relatively minor improvements in the amount of weight that can be lifted over a much longer period of time. Therefore, the amount and rate of progressive overload are extremely variable and must be adjusted to match the specific needs of the individual's exercise objectives.

Overload may be modified by changing any one or a combination of three exercise variables—**frequency**, **intensity**, and **duration**. Frequency usually refers to the number of times per week. Intensity is usually a certain percentage of the absolute maximum, and duration usually refers to the number of minutes per exercise bout. Increasing the speed of doing the exercise, the number of repetitions, the weight, and more bouts of exercise are all ways to modify these variables and apply this principle. All of these factors are important in determining the total exercise volume.

Overload is not always progressively increased. In certain periods of conditioning, the overload should actually be prescriptively reduced or increased to improve the total results of the entire program. This intentional variance in a training program at regular intervals is known as **periodization** and is done to bring about optimal gains in physical performance. A part of the basis for periodization is so that the athlete will be at his or her peak level during the most competitive part of the season. To achieve this a number of variables may be manipulated including the number of sets per exercise, repetitions per set, types of exercises, number of exercises per training session, rest periods between sets and exercises, resistance used for a set, type of muscle contraction, and the number of training sessions per day and per week.

SAID principle

The SAID (**S**pecific **A**daptations to **I**mposed **D**emands) principle should be considered in all aspects of physiological conditioning and training. This principle, which states that the body will gradually, over time, adapt very specifically to the various stresses and overloads to which it is subjected, is applicable in every form of muscle training, as well as to the other systems of the body. For example, if an individual were to undergo several weeks of strength-training exercises for a particular joint through a limited range of motion, the specific muscles involved in performing the strengthening exercises would improve primarily in the ability to move against increased resistance through the specific range of motion used. There would be, in most cases, minimal strength gains significantly beyond the range of motion used in the training. Additionally, other components of physical fitness—such as flexibility, cardiorespiratory endurance, and muscular endurance—would be enhanced minimally, if at all. In other words, to achieve specific benefits, exercise programs must be specifically designed for the adaptation desired.

It should be recognized that this adaptation may be positive or negative, depending on whether or not the correct techniques are used and stressed in the design and administration of the conditioning program. Inappropriate or excessive demands placed on the body in too short a time span can result in injury. If the demands are too minimal or administered too infrequently over too long a time period, less than desired improvement will occur. Conditioning programs and the exercises included in them should be analyzed to determine if they are using the specific muscles for which they were intended in the correct manner.

Specificity

Specificity of exercise strongly relates to the discussion of the SAID principle. The components of physical fitness—such as muscular strength, muscular endurance, and flexibility—are not general body characteristics. They are specific to each body area and muscle group. Therefore, the specific needs of the individual must be addressed when designing an exercise program. Quite often, it will be necessary to analyze an individual's exercise and skill technique to design an exercise program to meet his or her specific needs. Potential exercises to be used in the conditioning program must be analyzed to determine their appropriateness for the individual's specific needs. The goals of the exercise program should be determined regarding specific areas of the body, preferred time to physically peak, and physical fitness needs such as strength, muscular endurance, flexibility, cardiorespiratory endurance, and body composition. After establishing goals, a regimen incorporating the overload variables of frequency, intensity, and duration may be prescribed to include the entire body or specific areas in a manner to address the improvement of the preferred physical fitness components. Regular observation and follow-up exercise analysis is necessary to ensure proper adherence to correct technique.

Muscular development

For years it was thought that a person developed adequate muscular strength, endurance, and flexibility through participation in sport activities. Now the philosophy is that one needs to develop muscular strength, endurance, and flexibility in order to be able to participate safely and effectively in sport activities.

Adequate muscular strength, endurance, and flexibility of the entire body from head to toe should be developed through correct use of the appropriate exercise principles. Individuals responsible for this development need to prescribe exercises that will meet these objectives.

In schools this development should start at an early age and continue throughout the school years. Results of fitness tests reveal the need for considerable improvement in this area.

Sit-ups, the standing long jump, the mile run, and other tests all indicated fitness deficiencies in the children of the United States. Adequate muscular strength and endurance are important in the adult years for the activities of daily living, as well as job-related requirements and recreational needs. Many back problems and other physical ailments could be avoided through proper maintenance of the musculoskeletal system.

Analysis of upper body exercises

Presented over the next several pages are brief analyses of several common upper body exercises. Following and perhaps expanding on the approach used is encouraged in analyzing other upper body activities. All muscles listed in the analysis are contracting concentrically unless specifically noted to be contracting eccentrically or isometrically.

Valsalva maneuver

Many people bear down by holding the breath without thinking when attempting to lift something heavy. This bearing down, known as the Valsalva maneuver, is accomplished by trying to exhale against a closed epiglottis (the flap of cartilage behind the tongue that shuts the air passage when swallowing) and is thought by many to enhance one's lifting ability. It is mentioned here to caution against its use due to the dramatic increase in blood pressure followed by an equally dramatic drop in blood pressure it causes. Using the Valsalva maneuver can cause lightheadedness and fainting and lead to complications in people with heart disease. Instead of using the Valsalva maneuver, one should always be sure to use rhythmic and consistent breathing. It is usually advisable to exhale during the lifting or contracting phase and inhale during the lowering or recovery phase.

Shoulder pull

Description

In a standing or sitting position, the subject interlocks the fingers in front of the chest and then attempts to pull them apart (Fig. 8.3). This contraction is maintained for 5 to 20 seconds.

Analysis

In this type of exercise, there is little or no movement of the contracting muscles. In certain isometric exercises, contraction of the antagonistic muscles is as strong as contraction of the muscles attempting to produce the force for movement. The muscle groups contracting to produce a movement are designated the **agonists**. In the exercise just described, the agonists in the right upper extremity are antagonistic to the agonists in the left upper extremity, and vice versa (Table 8.2). This exercise results in isometric contractions of the wrist and hand, elbow, shoulder joint, and shoulder girdle muscles. The strength of the contraction depends on the angle of pull and the leverage of the joint involved. Thus, it is not the same at each point.

FIG. 8.3 ● Shoulder pull

Isometric exercises vary in the number of muscles contracting, depending on the type of exercise and the joints at which movement is attempted. The shoulder pull exercise produces some contraction of antagonistic muscles at four sets of joints.

TABLE 8.2 ● **Shoulder pull**

Chapter 8

| Joint | This entire exercise is isometric in design, so that all contractions presented are isometric | | | | |
	Action	Agonists	Action	Agonists
Wrist and hand	Extension	Resisted by flexors of wrists and hand Agonists—wrist and hand extensors Antagonists—wrist and hand flexors	Flexion	Resisted by extensors of wrist and hand Agonists—wrist and hand flexors Antagonists—wrist and hand extensors
Elbow joint	Extension	Resisted by flexors of wrist, elbow, and hand Agonists—triceps brachii, anconeus Antagonists—biceps brachii, brachialis, brachioradialis	Flexion	Resisted by extensors of wrist, elbow, and hand Agonists—biceps brachii, brachialis, brachioradialis Antagonists—triceps brachii, anconeus
Shoulder joint	Abduction	Resisted by adductors of shoulder joint Agonists—deltoid and supraspinatus Antagonists—teres major, latissimus dorsi, pectoralis major	Adduction	Resisted by adductors of shoulder joint Agonists—teres major, latissimus dorsi, pectoralis major Antagonists—deltoid and supraspinatus
Shoulder girdle	Adduction and depression	Resisted by abductors of shoulder girdle Agonists—rhomboid and trapezius (lower) Antagonists—serratus anterior, pectoralis minor, trapezius (upper and middle)	Abduction and elevation	Resisted by adductors of shoulder girdle Agonists—serratus anterior, pectoralis minor, trapezius (upper and middle) Antagonists—rhomboid and trapezius (lower)

Arm curl

Description

With the subject in a standing position, the dumbbell is held in the hands with the palms to the front. The dumbbell is lifted until the elbow is completely flexed (Fig. 8.4). Then it is returned to the starting position.

Analysis

This exercise is divided into two phases for analysis: (1) lifting phase to flexed position and (2) lowering phase to extended position (Table 8.3). Note: An assumption is made that no movement occurs in the shoulder joint and shoulder girdle.

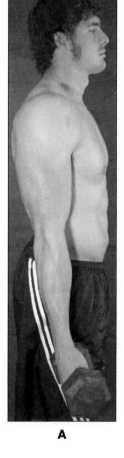

A **B**

FIG. 8.4 ● Arm curl. **A,** Beginning position in extension; **B,** Flexed position.

TABLE 8.3 • Arm curl

Joint	Lifting phase to flexed position		Lowering phase to extended position	
	Action	Agonists	Action	Agonists

Wrist and hand	Flexion*	Wrist and hand flexors (isometric contraction)
		   Flexor carpi radialis
		   Flexor carpi ulnaris
		   Palmaris longus
		   Flexor digitorum profundus
		   Flexor digitorum superficialis
		   Flexor pollicis longus
			Flexion*	Wrist and hand flexors (isometric contraction)
				   Flexor carpi radialis
				   Flexor carpi ulnaris
				   Palmaris longus
				   Flexor digitorum profundus
				   Flexor digitorum superficialis
				   Flexor pollicis longus

Elbow	Flexion	Elbow flexors
		   Biceps brachii
		   Brachialis
		   Brachioradialis
			Extension	Elbow flexors (eccentric contraction)
				   Biceps brachii
				   Brachialis
				   Brachioradialis

*Note: The wrist is in a position of slight extension to facilitate greater active finger flexion in gripping the dumbbell. (The flexors remain isometrically contracted throughout the entire exercise, to hold the dumbbell.)

Triceps extension

Description

The subject may use the opposite hand to assist in maintaining the arm in a shoulder-flexed position. The subject then, grasping the dumbbell and beginning in full elbow flexion, extends the elbow until the arm and forearm are straight. The shoulder joint and shoulder girdle are stabilized by the opposite hand. Consequently, no movement is assumed to occur in these areas (Fig. 8.5).

Analysis

This exercise is divided into two phases for analysis: (1) lifting phase to extended position and (2) lowering phase to flexed position (Table 8.4). Note: An assumption is made that no movement occurs in the shoulder joint and shoulder girdle.

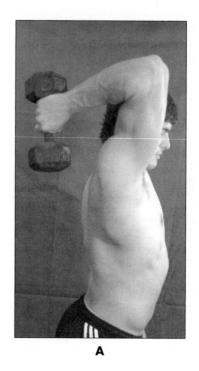
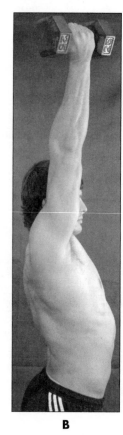

A   B

FIG. 8.5 ● Triceps extension. **A,** Beginning position in flexion; **B,** Extended position.

TABLE 8.4 • Triceps extension

Joint	Lifting phase to extended position		Lowering phase to flexed position	
	Action	Agonists	Action	Agonists
Wrist and hand	Flexion*	Wrist and hand flexors (isometric contraction) Flexor carpi radialis Flexor carpi ulnaris Palmaris longus Flexor digitorum profundus Flexor digitorum superficialis Flexor pollicis longus	Flexion*	Wrist and hand flexors (isometric contraction) Flexor carpi radialis Flexor carpi ulnaris Palmaris longus Flexor digitorum profundus Flexor digitorum superficialis Flexor pollicis longus
Elbow	Extension	Elbow extensors Triceps brachii Anconeus	Flexion	Elbow extensors (eccentric contraction) Triceps brachii Anconeus

***Note:** The wrist is in a position of slight extension to facilitate greater active finger flexion in gripping the dumbbell. (The flexors remain isometrically contracted throughout the entire exercise, to hold the dumbbell.)

Barbell press

Description

This exercise is sometimes referred to as the *overhead* or *military press*. The barbell is held in a position high in front of the chest, with palms facing forward, feet comfortably spread, and back and legs straight (Fig. 8.6, *A*). From this position, the barbell is pushed upward until fully overhead (Fig. 8.6, *B*), and then it is returned to the starting position.

Analysis

This exercise is separated into two phases for analysis: (1) lifting phase to full overhead position and (2) lowering phase to starting position (Table 8.5).

A

B

FIG. 8.6 ● Barbell press. **A,** Starting position; **B,** Full overhead position.

TABLE 8.5 • **Barbell press**

| Joint | Lifting phase to full overhead position | | Lowering phase to starting position | |
	Action	Agonists	Action	Agonists

Wrist and hand	Flexion*	Wrist and hand flexors (isometric contraction)
		    Flexor carpi radialis
		    Flexor carpi ulnaris
		    Palmaris longus
		    Flexor digitorum profundus
		    Flexor digitorum superficialis
		    Flexor pollicis longus
			Flexion*	Wrist and hand flexors (isometric contraction)
				    Flexor carpi radialis
				    Flexor carpi ulnaris
				    Palmaris longus
				    Flexor digitorum profundus
				    Flexor digitorum superficialis
				    Flexor pollicis longus

Elbow joint	Extension	Elbow extensors
		    Triceps brachii
		    Anconeus
			Flexion	Elbow extensors (eccentric contraction)
				    Triceps brachii
				    Anconeus

Shoulder	Flexion	Shoulder joint flexors
		    Pectoralis major (clavicular head or upper fibers)
		    Anterior deltoid
		    Coracobrachialis
		    Biceps brachii
			Extension	Shoulder joint flexors (eccentric contraction)
				    Pectoralis major (clavicular head or upper fibers)
				    Anterior deltoid
				    Coracobrachialis
				    Biceps brachii

Shoulder girdle	Upward rotation and elevation	Shoulder girdle upward rotators and elevators
		    Trapezius
		    Levator scapulae
		    Serratus anterior
			Downward rotation and depression	Shoulder girdle upward rotators and elevators (eccentric contraction)
				    Trapezius
				    Levator scapulae
				    Serratus anterior

***Note:** The wrist is in a position of slight extension to facilitate greater active finger flexion in gripping the bar.

Chest press (bench press)

Description

The subject lies on the exercise bench in the supine position, grasps the barbell, and presses the weight upward through the full range of arm and shoulder movement (Fig. 8.7). Then the weight is lowered to the starting position.

Analysis

This exercise can be divided into two phases for analysis: (1) lifting phase to upward position and (2) lowering phase to starting position (Table 8.6).

A

B

FIG. 8.7 ● Chest press (bench press). **A,** Starting position; **B,** Up position.

TABLE 8.6 ● **Chest press (bench press)**

Joint	Lifting phase to upward position		Lowering phase to starting position	
	Action	Agonists	Action	Agonists
Wrist and hand	Flexion*	Wrist and hand flexors (isometric contraction) Flexor carpi radialis Flexor carpi ulnaris Palmaris longus Flexor digitorum profundus Flexor digitorum superficialis Flexor pollicis longus	Flexion*	Wrist and hand flexors (isometric contraction) Flexor carpi radialis Flexor carpi ulnaris Palmaris longus Flexor digitorum profundus Flexor digitorum superficialis Flexor pollicis longus
Elbow	Extension	Elbow extensors Triceps brachii Anconeus	Flexion	Elbow extensors (eccentric contraction) Triceps brachii Anconeus
Shoulder	Flexion and horizontal adduction	Shoulder flexors and horizontal adductors Pectoralis major Anterior deltoid Coracobrachialis Biceps brachii	Extension and horizontal abduction	Shoulder joint flexors and horizontal adductors (eccentric contraction) Pectoralis major Anterior deltoid Coracobrachialis Biceps brachii
Shoulder girdle	Abduction	Shoulder girdle abductors Serratus anterior Pectoralis minor	Adduction	Shoulder girdle abductors (eccentric contraction) Serratus anterior Pectoralis minor

*__Note:__ The wrist is in a position of slight extension to facilitate greater active finger flexion in gripping the bar.

Chin-up (pull-up)

Description

The subject grasps a horizontal bar or ladder with the palms away from the face (Fig. 8.8, *A*). From a hanging position on the bar, the subject pulls up until the chin is over the bar (Fig. 8.8, *B*) and returns to the starting position (Fig. 8.8, *C*).

Analysis

This exercise is separated into two phases for analysis: (1) pulling up phase to chinning position and (2) lowering phase to starting position (Table 8.7).

| A | B | C |

FIG. 8.8 ● Pull-up. **A,** Straight-arm hang; **B,** Chin over bar; **C,** Bent-arm hang on way up or down.

TABLE 8.7 • Chin-up (pull-up)

Joint	Pulling up phase to chinning position		Lowering phase to starting position	
	Action	Agonists	Action	Agonists

Wrist and hand	Flexion	Wrist and hand flexors (isometric contraction)
		  Flexor carpi radialis
		  Flexor carpi ulnaris
		  Palmaris longus
		  Flexor digitorum profundus
		  Flexor digitorum superficialis
		  Flexor pollicis longus
			Flexion	Wrist and hand flexors (isometric contraction)
				  Flexor carpi radialis
				  Flexor carpi ulnaris
				  Palmaris longus
				  Flexor digitorum profundus
				  Flexor digitorum superficialis
				  Flexor pollicis longus

Elbow	Flexion	Elbow flexors
		  Biceps brachii
		  Brachialis
		  Brachioradialis
			Extension	Elbow flexors (eccentric contraction)
				  Biceps brachii
				  Brachialis
				  Brachioradialis

Shoulder	Extension	Shoulder joint extensors
		  Latissimus dorsi
		  Teres major
		  Posterior deltoid
		  Pectoralis major
		  Triceps brachii (long head)
			Flexion	Shoulder joint extensors (eccentric contraction)
				  Latissimus dorsi
				  Teres major
				  Posterior deltoid
				  Pectoralis major
				  Triceps brachii (long head)

Shoulder girdle	Adduction, depression, and downward rotation	Shoulder girdle adductors, depressors, and downward rotators
		  Trapezius (lower and middle)
		  Pectoralis minor
		  Rhomboids
			Elevation, abduction, and upward rotation	Shoulder girdle adductors, depressors, and downward rotators (eccentric contraction)
				  Trapezius (lower and middle)
				  Pectoralis minor
				  Rhomboids

Latissimus pull (lat pull)

Description

From a sitting position, the subject reaches up and grasps a horizontal bar (Fig. 8.9, *A*). The bar is pulled down to a position below the chin (Fig. 8.9, *B*). Then it is returned slowly to the starting position.

Analysis

This exercise is separated into two phases for analysis: (1) pull-down phase to below the chin position, and (2) return phase to starting position (Table 8.8).

A   **B**

FIG. 8.9 ● Latissimus pull (lat pull). **A,** Starting position; **B,** Downward position.

TABLE 8.8 • Latissimus pull (lat pull)

Joint	Pull-down phase to below the chin position		Return phase to starting position	
	Action	Agonists	Action	Agonists
Wrist and hand	Flexion	Wrist and hand flexors (isometric contraction) Flexor carpi radialis Flexor carpi ulnaris Palmaris longus Flexor digitorum profundus Flexor digitorum superficialis Flexor pollicis longus	Flexion	Wrist and hand flexors (isometric contraction) Flexor carpi radialis Flexor carpi ulnaris Palmaris longus Flexor digitorum profundus Flexor digitorum superficialis Flexor pollicis longus
Elbow	Flexion	Elbow flexors Biceps brachii Brachialis Brachioradialis	Extension	Elbow flexors (eccentric contraction) Biceps brachii Brachialis Brachioradialis
Shoulder	Adduction	Shoulder joint adductors Pectoralis major Latissimus dorsi Teres major Subscapularis	Abduction	Shoulder joint adductors (eccentric contraction) Pectoralis major Latissimus dorsi Teres major Subscapularis
Shoulder girdle	Adduction, depression, and downward rotation	Shoulder girdle adductors, depressors, and downward rotators Trapezius (lower and middle) Pectoralis minor Rhomboids	Abduction, elevation, and upward rotation	Shoulder girdle adductors, depressors, and downward rotators (eccentric contraction) Trapezius (lower and middle) Pectoralis minor Rhomboids

Chapter

8

Push-up

Description

The subject lies on the floor in a prone position with the legs together, the palms touching the floor, and the hands pointed forward and approximately under the shoulders (Fig. 8.10, *A*). Keeping the back and legs straight, the subject pushes up to the up position and returns to the starting position (Fig. 8.10, *B*).

Analysis

This exercise is separated into two phases for analysis: (1) pushing phase to up position and (2) lowering phase to starting position (Table 8.9).

Chin-ups and push-ups are excellent exercises for the shoulder area, shoulder girdle, shoulder joint, elbow joint, and wrist and hand (see Figs. 8.8 and 8.10). The use of free weights, machines, and other conditioning exercises helps develop strength and endurance for this part of the body.

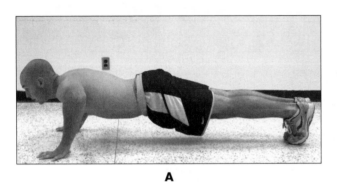
A

B

FIG. 8.10 ● Push-up. **A,** Starting position; **B,** Up position.

TABLE 8.9 • **Push-up**

Joint	Pushing phase to up position		Lowering phase to starting position	
	Action	Agonists	Action	Agonists
Wrist and hand	Flexion	Wrist and hand flexors (isometric contraction) Flexor carpi radialis Flexor carpi ulnaris Palmaris longus Flexor digitorum profundus Flexor digitorum superficialis Flexor pollicis longus	Flexion	Wrist and hand flexors (isometric contraction) Flexor carpi radialis Flexor carpi ulnaris Palmaris longus Flexor digitorum profundus Flexor digitorum superficialis Flexor pollicis longus
Elbow	Extension	Elbow extensors Triceps brachii Anconeus	Flexion	Elbow extensors (eccentric contraction) Triceps brachii Anconeus
Shoulder	Horizontal adduction	Shoulder joint horizontal adductors Pectoralis major Anterior deltoid Biceps brachii Coracobrachialis	Horizontal abduction	Shoulder joint horizontal adductors (eccentric contraction) Pectoralis major Anterior deltoid Biceps brachii Coracobrachialis
Shoulder girdle	Abduction	Shoulder girdle abductors Serratus anterior Pectoralis minor	Adduction	Shoulder girdle abductors (eccentric contraction) Serratus anterior Pectoralis minor

Prone row

Description

This exercise is sometimes referred to as a bent-over row. The subject is lying prone on a table or kneeling on a bench so that the involved arm is free from contact with the floor (Fig. 8.11, *A*). When kneeling, the subject uses the contralateral arm to support the body. The dumbbell is held in the hand with arm and the shoulder hanging straight to the floor. From this position, the subject adducts the shoulder girdle and horizontally abducts the shoulder joint (Fig. 8.11, *B*). Then the dumbbell is lowered slowly to the starting position.

Analysis

This exercise is separated into two phases for analysis: (1) pull-up phase to horizontal abducted position, and (2) lowering phase to starting position (Table 8.10).

FIG. 8.11 ● Prone row.

A **B**

TABLE 8.10 • **Prone row**

| Joint | Pull-up phase to horizontal abducted position | | Lowering phase to starting position | |
	Action	Agonists	Action	Agonists
Hand	Flexion	Hand flexors (isometric contraction) Flexor digitorum profundus Flexor digitorum superficialis Flexor pollicis longus	Flexion	Hand flexors (isometric contraction) Flexor digitorum profundus Flexor digitorum superficialis Flexor pollicis longus
Elbow	Flexion	Passive flexion occurs as the arm becomes parallel to the floor due to gravity	Extension	Passive extension occurs as the arm becomes perpendicular to the floor due to gravity
Shoulder	Horizontal abduction	Shoulder joint horizontal abductors Posterior deltoid Infraspinatus Teres minor Latissimus dorsi	Horizontal adduction	Shoulder joint horizontal abductors (eccentric contraction) Posterior deltoid Infraspinatus Teres minor Latissimus dorsi
Shoulder girdle	Adduction	Shoulder girdle adductors Trapezius (lower and middle) Rhomboids	Abduction	Shoulder girdle adductors (eccentric contraction) Trapezius (lower and middle) Rhomboids

Web sites

American College of Sports Medicine

www.acsm.org

Scientific research, education, and practical applications of sports medicine and exercise science to maintain and enhance physical performance, fitness, health, and quality of life.

Concept II

www.concept2.com

Information on the technique of rowing and the muscles used.

Fitness World

www.fitnessworld.com

Fitness in general; includes access to *Fitness Management* magazine.

National Council of Strength & Fitness

www.ncsf.org

Personal training certification and continuing education for the fitness professional.

National Strength and Conditioning Association

www.nsca-lift.org

Information on the profession of strength and conditioning specialists and personal trainers.

NSCA Certification Commission

www.nsca-cc.org

The certifying body for the National Strength and Conditioning Association.

Presidents Council on Physical Fitness and Sports

www.fitness.gov

Information and links from the U.S. government on fitness.

ExRx.net

www.exrx.net/Lists/Directory.html

A resource for the exercise professional, coach, or fitness enthusiast consisting of over 1500 pages of exercises and anatomy illustrations.

National Academy of Sports Medicine

www.nasm.org

Offers specific certifications for health and fitness exercise specialists and a valuable resource for continuing education on exercise techniques, etc.

Upper Extremity Conditioning Program

www.eatonhand.com/hw/nirschl.htm

Shows strengthening exercises for the upper body.

Rehab Team Site: Passive Stretching

http://calder.med.miami.edu/pointis/upper.html

Passive range of motion exercises.

Body Map

www.athleticadvisor.com/body_map.htm

Describes specific injuries and how to properly rehab with weights.

Physician and Sports Medicine: Weight Training Injuries

www.physsportsmed.com/issues/1998/03mar/laskow2.htm

Article about upper body injuries and how to strengthen the upper body.

NISMAT Exercise Programs

www.nismat.org/orthocor/programs

Step-by-step instructions of strengthening exercises along with diagrams.

Runner Girl.com

www.runnergirl.com

Strengthening and stretching exercises as well as other health and fitness information for women.

Worksheet exercises

As an aid to learning, for in-class and out-of-class assignments, or for testing, tear-out worksheets are found at the end of the text (p. 382).

Upright row exercise worksheet (no. 1)

List the movements that occur in each joint as the subject lifts the weight in performing upright rows. For each joint movement, list the muscles primarily responsible and denote whether they are contracting concentrically or eccentrically.

Dips exercise worksheet (no. 2)

List the movements that occur in each joint as the subject moves the body in performing dips. For each joint movement, list the muscles primarily responsible and denote whether they are contracting concentrically or eccentrically.

LABORATORY AND REVIEW EXERCISES

1. Analyze other conditioning exercises that involve the shoulder areas, such as dips, upright rows, shrugs, dumbbell flys, and inclined presses.
2. Observe and analyze shoulder muscular activities of children on playground equipment.
3. Discuss how you would teach and train boys and girls who cannot perform one chin-up to learn to do chin-ups. Additionally, discuss how you would teach subjects who cannot perform one push-up to do push-ups.
4. Should boys and girls attempt to do chin-ups and push-ups to see whether they have adequate strength in the shoulder area?
5. Test yourself doing chin-ups and push-ups to determine your strength and muscular endurance in the shoulder area.
6. What, if any, benefit would result from doing fingertip push-ups as opposed to push-ups with the hands flat on the floor?

7. Stand slightly farther than arm's length from a wall with your hands facing forward at shoulder-level height. Push your hand straight in front of your shoulder until the elbow is extended fully and the glenohumeral joint is flexed 90 degrees by performing each of the following movements before proceeding to the next:
   - Glenohumeral flexion to 90 degrees
   - Full elbow extension
   - Wrist extension to 70 degrees

   Analyze the movements and the muscles responsible for each movement at the shoulder girdle, glenohumeral joint, elbow, and wrist. Include the type of contraction for each muscle for each movement.
8. Now face a wall and stand about 6 inches from it. Place both hands on the wall at shoulder level and put your nose and chest against the wall. Keeping your palms in place on the wall, slowly push your body from the wall as in a push-up until your chest is as far away from the wall as possible without removing your palms from the wall surface. Analyze the movements and the muscles responsible for each movement at the shoulder girdle, glenohumeral joint, elbow, and wrist. Include the type of contraction for each muscle for each movement.
9. What is the difference between the two exercises in questions 7 and 8? Can you perform exercise 8 one step at a time, as you did in question 7?
10. Analyze each exercise in the exercise analysis chart (p. 214). Use one row for each joint involved that actively moves during the exercise. Do not include joints where there is no active movement or where the joint is maintained in one position isometrically.

Chapter
8

Exercise Analysis Chart

Exercise	Phase	Joint, movement occurring	Force causing movement (muscle or gravity)	Force resisting movement (muscle or gravity)	Functional muscle group, type of contraction
Barbell press (overhead or military press)	Lifting phase				
	Lowering phase				
Chest press (bench press)	Lifting phase				
	Lowering phase				
Chin-up (pull-up)	Pulling up phase				
	Lowering phase				
Latissimus pull (lat pull)	Pull-down phase				
	Return phase				
Push-up	Pushing phase				
	Lowering phase				
Prone row (bent-over row)	Pull-up phase				
	Lowering phase				

References

Adrian M: Isokinetic exercise, *Training and Conditioning* 1:1, June 1991.

Andrews JR, Wilk KE: *The athlete's shoulder,* New York, 1994, Churchill Livingstone.

Andrews JR, Zarins B, Wilk KE: *Injuries in baseball,* Philadelphia, 1998, Lippincott-Raven.

Booher JM, Thibodeau GA: *Athletic injury assessment,* ed 4, Dubuque, IA, 2000, McGraw-Hill.

Ellenbecker TS, Davies GJ: *Closed kinetic chain exercise: a comprehensive guide to multiple-joint exercise,* Champaign, IL, 2001, Human Kinetics.

Fleck SJ: Periodized strength training: a critical review, *Journal of Strength and Conditioning Research,* 13(1) 82–89, 1999.

Geisler P: Kinesiology of the full golf swing—implications for intervention and rehabilitation, *Sports Medicine Update* 11:9, #2, 1996.

Hamilton N, Luttgens K: *Kinesiology: scientific basis of human motion,* ed 10, Boston, 2002, McGraw-Hill.

Matheson O, et al: Stress fractures in athletes, *American Journal of Sports Medicine* 15:46, January–February 1987.

National Strength and Conditioning Association; Baechle TR, Earle RW: *Essentials of strength training and conditioning,* ed 2, Champaign, IL, 2000, Human Kinetics.

Northrip JW, Logan GA, McKinney WC: *Analysis of sport motion: anatomic and biomechanic perspectives,* ed 3, New York, 1983, McGraw-Hill.

Powers SK, Howley ET: *Exercise physiology: theory and application to fitness and performance,* ed 5, New York, 2004, McGraw-Hill.

Smith LK, Weiss EL, Lehmkuhl LD: *Brunnstrom's clinical kinesiology,* ed 5, Philadelphia, 1996, Davis.

Steindler A: *Kinesiology of the human body,* Springfield, IL, 1970, Charles C Thomas.

Chapter 9

The Hip Joint and Pelvic Girdle

Objectives

- To identify on a human skeleton or subject selected bony features of the hip joint and pelvic girdle

- To label on a skeletal chart selected bony features of the hip joint and pelvic girdle

- To draw on a skeletal chart the individual muscles of the hip joint

- To demonstrate, using a human subject, all of the movements of the hip joint and pelvic girdle and list their respective planes of movement and axes of motion

- To palpate on a human subject the muscles of the hip joint and pelvic girdle

- To list and organize the primary muscles that produce movement of the hip joint and pelvic girdle and list their antagonists

---

Online Learning Center Resources

Visit *Manual of Structural Kinesiology's* Online Learning Center at **www.mhhe.com/floyd16e** for additional information and study material for this chapter, including:

- *Self-grading quizzes*
- *Anatomy flashcards*
- *Animations*

The hip, or acetabular femoral joint, is a relatively stable joint due to its bony architecture, strong ligaments, and large, supportive muscles. It functions in weight bearing and locomotion, which is enhanced significantly by its wide range of motion, which provides the ability to run, cross-over cut, side-step cut, jump, and make many other directional changes.

Bones   FIGS. 9.1 TO 9.3

The hip joint is a ball-and-socket joint that consists of the head of the femur connecting with the acetabulum of the pelvic girdle. The femur projects out laterally from its head toward the greater trochanter and then angles back toward the midline as it runs inferiorly to form the proximal bone of the knee. It is the longest bone in the body. The pelvic girdle consists of a right and left pelvic bone joined together posteriorly by the sacrum. The sacrum can be considered an extension of the spinal column with five fused vertebrae. Extending inferior to the sacrum is the coccyx. The pelvic bones are made up of three bones: the ilium, the ischium, and the pubis. At birth and during growth and development, they are three distinct bones. At maturity, they are fused to form one pelvic bone.

The pelvic bone can be divided roughly into three areas, starting from the acetabulum:

Upper two-fifths = ilium
Posterior and lower two-fifths = ischium
Anterior and lower one-fifth = pubis

In studying the muscles of the hip and thigh, it is helpful to focus on the important bony landmarks, keeping in mind their purpose as key attachment points for the muscles. The anterior pelvis serves to provide points of origin for muscles generally involved in flexing the hip. Specifically, the tensor fasciae latae arises from the anterior iliac crest, the sartorius originates on the anterior superior iliac spine, and the rectus femoris originates on the anterior inferior iliac spine. Laterally, the gluteus medius and minimus, which abduct the hip, originate just below the iliac crest. Posteriorly, the gluteus maximus originates on the posterior iliac crest as well as the posterior sacrum and coccyx. Posteroinferiorly, the ischial tuberosity serves as the point of origin for the hamstrings, which extend the hip. Medially, the pubis and its inferior ramus serve as the point of origin for the

hip adductors, which include the adductor magnus, adductor longus, adductor brevis, pectineus, and gracilis.

The proximal thigh generally serves as a point of insertion for some of the short muscles of the hip and as the origin for three of the knee extensors. Most notably, the greater trochanter is the point of insertion for all of the gluteal muscles and most of the six deep external rotators. Although not palpable, the lesser trochanter serves as the bony landmark upon which the iliopsoas inserts. Anteriorly, the three vasti muscles of the quadriceps originate proximally. Posteriorly, the linea aspera serves as the insertion for the hip adductors.

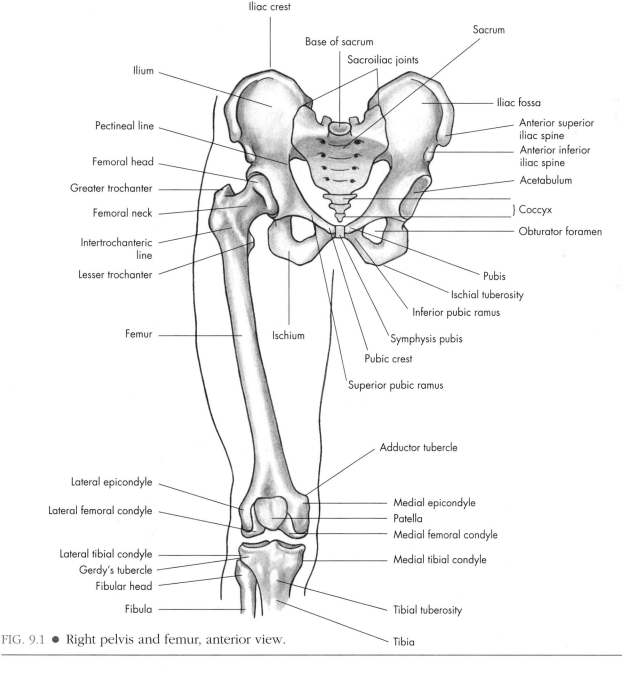

FIG. 9.1 ● Right pelvis and femur, anterior view.

Distally, the patella serves as a major bony landmark to which all four quadriceps muscles insert. The remainder of the hip muscles insert on the proximal tibia or fibula. The sartorius, gracilis, and semitendinosus insert on the upper anteromedial surface of the tibia just below the medial condyle, after crossing the knee posteromedially. The semimembranosus inserts posteromedially on the medial tibial condyle. Laterally, the biceps femoris inserts primarily on the head on the fibula, with some fibers attaching on the lateral tibial condyle. Anterolaterally, Gerdy's tubercle provides the insertion point for the iliotibial tract of the tensor fasciae latae.

Joints FIGS. 9.1 TO 9.5

In the anterior area, the pelvic bones are joined to form the symphysis pubis, an amphiarthrodial joint. In the posterior area, the sacrum is located between the two pelvic bones and forms the sacroiliac joints. Strong ligaments unite these bones to form rigid, slightly movable joints. The bones are large and heavy and for the most part are covered by thick, heavy muscles. Very minimal oscillating-type movements can occur in these joints, as in walking or in hip flexion when lying on one's back. However, movements usually involve the entire pelvic girdle

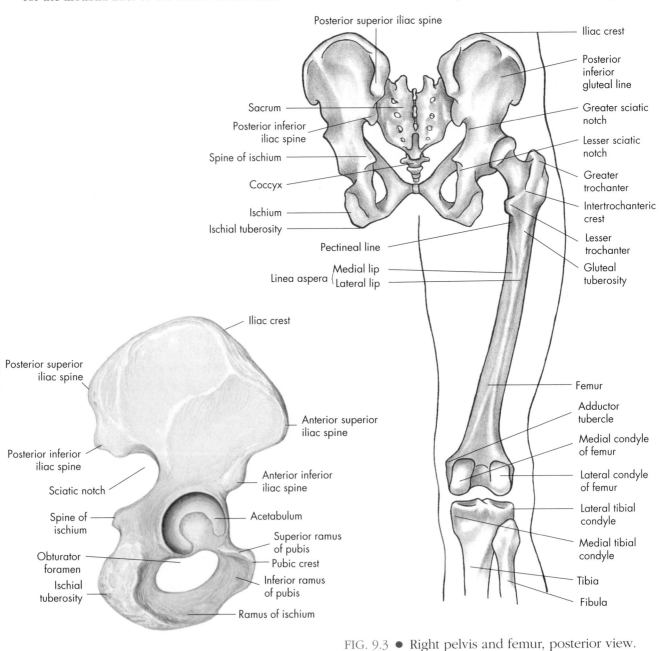

FIG. 9.3 ● Right pelvis and femur, posterior view.

FIG. 9.2 ● Right pelvic bone, lateral view.

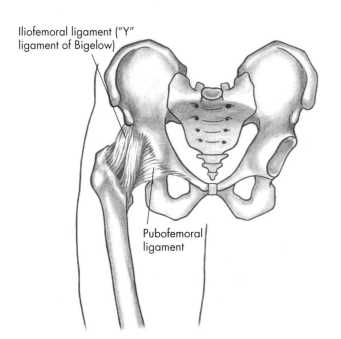

Iliofemoral ligament ("Y" ligament of Bigelow)

Pubofemoral ligament

FIG. 9.4 ● Right hip joint, anterior ligaments.

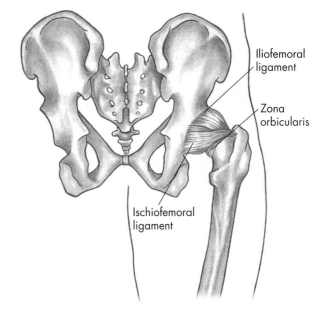

Iliofemoral ligament

Zona orbicularis

Ischiofemoral ligament

FIG. 9.5 ● Right hip joint, posterior ligaments.

and hip joints. In walking, there is hip flexion and extension with rotation of the pelvic girdle, forward in hip flexion and backward in hip extension. Jogging and running result in faster movements and in a greater range of movement.

Sport skills such as kicking a football or soccer ball are other good examples of hip and pelvic movements. Pelvic rotation helps increase the length of the stride in running; in kicking, it can result in a greater range of motion, which translates to a greater distance or more speed to the kick.

Except for the glenohumeral joint, the hip is one of the most mobile joints of the body, largely because of its multiaxial arrangement. Unlike the glenohumeral, the hip joint's bony architecture provides a great deal of stability, resulting in relatively few hip joint subluxations and dislocations.

The hip joint is classified as an enarthrodial-type joint and is formed by the femoral head inserting into the socket provided by the acetabulum of the pelvis. An extremely strong and dense ligamentous capsule, illustrated in Figs. 9.4 and 9.5, reinforces the joint, especially anteriorly. Anteriorly, the iliofemoral, or Y, ligament prevents hip hyperextension. The teres ligament attaches from deep in the acetabulum to a depression in the

head of the femur and slightly limits adduction. The pubofemoral ligament is located anteromedially and inferiorly and limits excessive extension and abduction. Posteriorly, the triangular ischiofemoral ligament extends from the ischium below to the trochanteric fossa of the femur and limits internal rotation.

Because of individual differences, there is some disagreement about the exact possible range of each movement in the hip joint, but the ranges are generally 0 to 130 degrees of flexion, 0 to 30 degrees of extension, 0 to 35 degrees of abduction, 0 to 30 degrees of adduction, 0 to 45 degrees of internal rotation, and 0 to 50 degrees of external rotation (Fig. 9.6).

The pelvic girdle moves back and forth within three planes for a total of six different movements. To avoid confusion, it is important to analyze the pelvic girdle activity to determine the exact location of the movement. All pelvic girdle rotation actually results from motion at one or more of the following locations: the right hip, the left hip, the lumbar spine. Although it is not essential for movement to occur in all three of these areas, it must occur in at least one for the pelvis to rotate in any direction. Table 9.1 lists the motions at the hips and lumbar spine that can often accompany rotation of the pelvic girdle.

Chapter

9

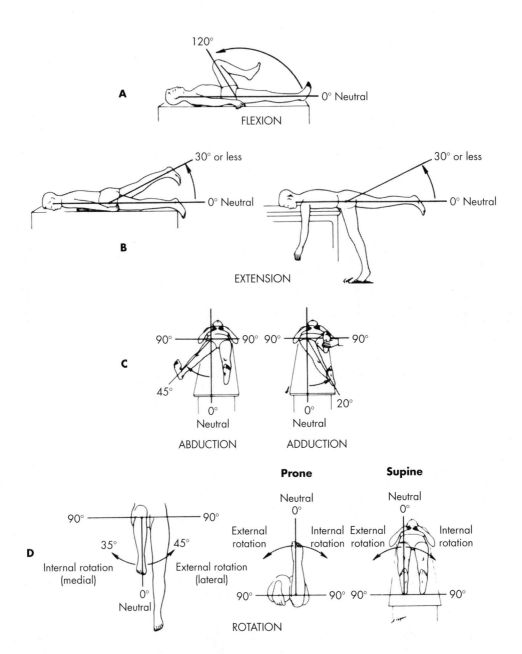

FIG. 9.6 ● Active motion of the hip. **A,** Flexion is measured in degrees from a supine position. The knee can be extended or flexed; **B,** Extension or hyperextension is normally measured with the knee extended; **C,** Abduction can be measured in a supine or side-lying position; adduction is best measured with the athlete lying supine; **D,** Internal and external rotation can be evaluated in a supine or prone position.

TABLE 9.1 • Motions accompanying pelvic rotation

Pelvic rotation	Lumbar spine motion	Right hip motion	Left hip motion
Anterior rotation	Extension	Flexion	Flexion
Posterior rotation	Flexion	Extension	Extension
Right lateral rotation	Left lateral flexion	Abduction	Adduction
Left lateral rotation	Right lateral flexion	Adduction	Abduction
Right transverse rotation	Left lateral rotation	Internal rotation	External rotation
Left transverse rotation	Right lateral rotation	External rotation	Internal rotation

Movements FIGS. 9.7 AND 9.8

Anterior and posterior pelvic rotation occur in the sagittal or anteroposterior plane, whereas right and left lateral rotation occur in the lateral or frontal plane. Right transverse (clockwise) rotation and left transverse (counterclockwise) rotation occur in the horizontal or transverse plane of motion.

Hip flexion: movement of the femur straight anteriorly from any point in the sagittal plane toward the pelvis

Hip extension: movement of the femur straight posteriorly from any point in the sagittal plane away from the pelvis; sometimes referred to as hyperextension

Hip abduction: movement of the femur in the frontal plane laterally to the side away from the midline

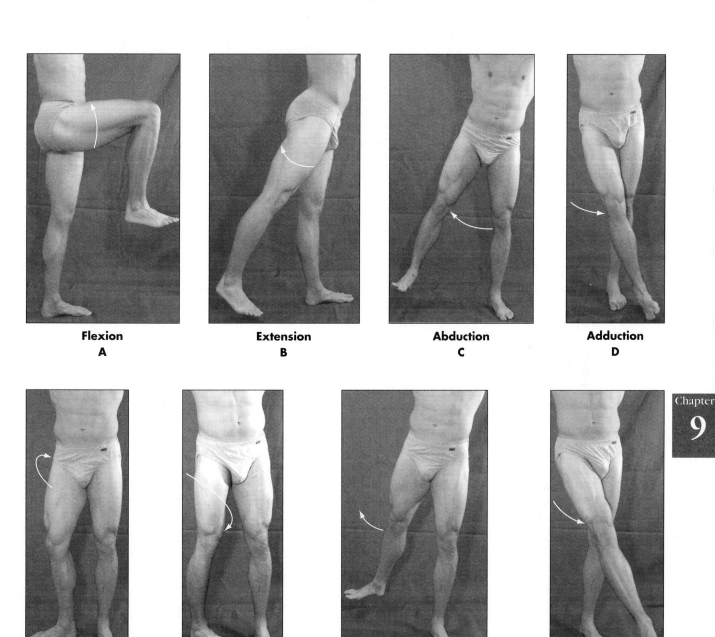

| Flexion A | Extension B | Abduction C | Adduction D |

| External rotation E | Internal rotation F | Diagonal abduction G | Diagonal adduction H |

FIG. 9.7 ● Movements of the hip.

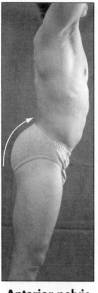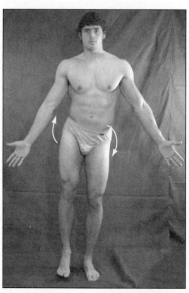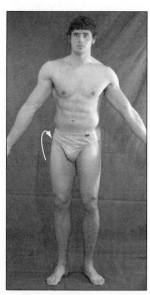

**Anterior pelvic
rotation
A**

**Posterior pelvic
rotation
B**

**Left lateral pelvic rotation
C**

**Right transverse pelvic
rotation
D**

FIG. 9.8 ● Pelvic girdle motions.

Hip adduction: movement of the femur in the frontal plane medially toward the midline

Hip external rotation: lateral rotary movement of the femur in the transverse plane around its longitudinal axis away from the midline; lateral rotation

Hip internal rotation: medial rotary movement of the femur in the transverse plane around its longitudinal axis toward the midline; medial rotation

Hip diagonal abduction: movement of the femur in a diagonal plane away from the midline of the body

Hip diagonal adduction: movement of the femur in a diagonal plane toward the midline of the body

Anterior pelvic rotation: anterior movement of the upper pelvis; the iliac crest tilts forward in a sagittal plane; anterior tilt; accomplished by hip flexion and/or lumbar extension

Posterior pelvic rotation: posterior movement of the upper pelvis; the iliac crest tilts backward in a sagittal plane; posterior tilt; accomplished by hip extension and/or lumbar flexion

Left lateral pelvic rotation: in the frontal plane, the left pelvis moves inferiorly in relation to the right pelvis; either the left pelvis rotates downward or the right pelvis rotates upward; left lateral tilt; accomplished by left hip abduction, right hip adduction, and/or right lumbar lateral flexion

Right lateral pelvic rotation: in the frontal plane, the right pelvis moves inferiorly in relation to the left pelvis; either the right pelvis rotates downward or the left pelvis rotates upward; right lateral tilt; accomplished by the right hip abduction, left hip adduction, and/or left lumbar lateral flexion

Left transverse pelvic rotation: in a horizontal plane of motion, rotation of the pelvis to the body's left; the right iliac crest moves anteriorly in relation to the left iliac crest, which moves posteriorly; accomplished by right hip external rotation, left hip internal rotation, and/or right lumbar rotation

Right transverse pelvic rotation: in a horizontal plane of motion, rotation of the pelvis to the body's right; the left iliac crest moves anteriorly in relation to the right iliac crest, which moves posteriorly; accomplished by left hip external rotation, right hip internal rotation, and/or left lumbar rotation

Muscles FIGS. 9.9 TO 9.11

At the hip joint, there are seven two-joint muscles that have one action at the hip and another at the knee. The muscles actually involved in hip and pelvic girdle motions depend largely on the direction of the movement and the position of the body in relation to the earth and its gravita-

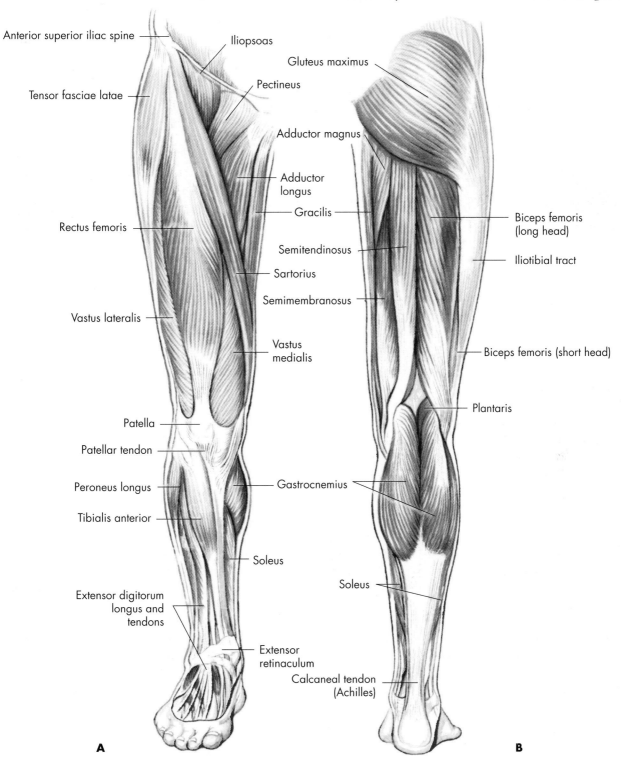

Anterior superior iliac spine
Tensor fasciae latae
Rectus femoris
Vastus lateralis
Patella
Patellar tendon
Peroneus longus
Tibialis anterior
Extensor digitorum longus and tendons
Extensor retinaculum

Iliopsoas
Pectineus
Gluteus maximus
Adductor magnus
Adductor longus
Gracilis
Semitendinosus
Sartorius
Semimembranosus
Vastus medialis
Gastrocnemius
Soleus

Biceps femoris (long head)
Iliotibial tract
Biceps femoris (short head)
Plantaris
Soleus
Calcaneal tendon (Achilles)

A

B

FIG. 9.9 ● **A,** Superficial muscles of the right upper leg, anterior surface; **B,** Superficial muscles of the right upper leg, posterior surface.

Modified from Anthony CP, Kolthoff NJ: *Textbook of anatomy and physiology,* ed 9, St. Louis, 1975, Mosby.

Chapter
9

tional forces. In addition, it should be noted that the body part that moves the most will be the part least stabilized. For example, when standing on both feet and contracting the hip flexors, the trunk and pelvis will rotate anteriorly, but, when lying supine and contracting the hip flexors, the thighs will move forward into flexion on the stable pelvis.

In another example, the hip flexor muscles are used in moving the thighs toward the trunk, but the extensor muscles are used eccentrically when the pelvis and the trunk move downward slowly on the femur and concentrically when the trunk is raised on the femur—this, of course, in rising to the standing position.

In the downward phase of the knee-bend exercise, the movement at the hips and knees is flexion. The muscles primarily involved are the hip and knee extensors in eccentric contraction.

Hip joint and pelvic girdle muscles—location

Muscle location largely determines the muscle action. Seventeen or more muscles are found in the area (the six external rotators are counted as one muscle). Most hip joint and pelvic girdle muscles are large and strong.

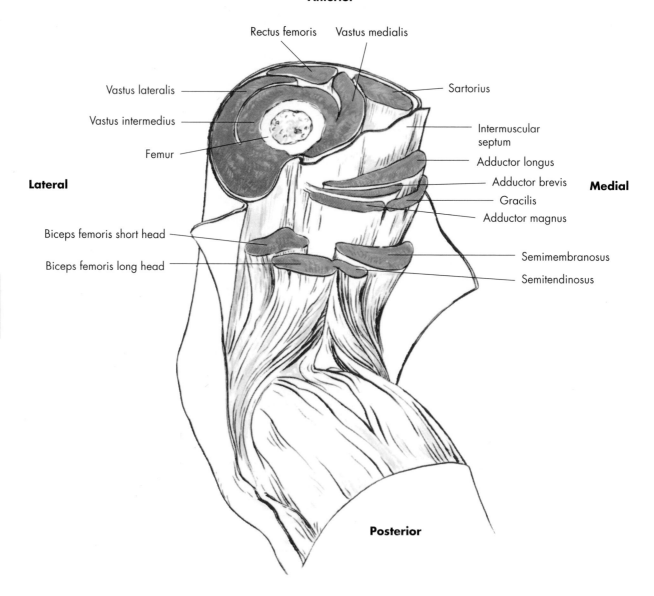

FIG. 9.10 ● Cross-section of the left thigh at the midsection.

Anterior
  Primarily hip flexion
    Iliopsoas (iliacus and psoas)
    Pectineus
    Rectus femoris*†
    Sartorius†
Lateral
  Primarily hip abduction
    Gluteus medius
    Gluteus minimus
    External rotators
    Tensor fasciae latae†

Posterior
  Primarily hip extension
    Gluteus maximus
    Biceps femoris*†
    Semitendinosus*†
    Semimembranosus*†
    External rotators (six deep)
Medial
  Primarily hip adduction
    Adductor brevis
    Adductor longus
    Adductor magnus
    Gracilis†

*Two-joint muscles, knee actions are discussed in Chapter 8.
†Two-joint muscles.

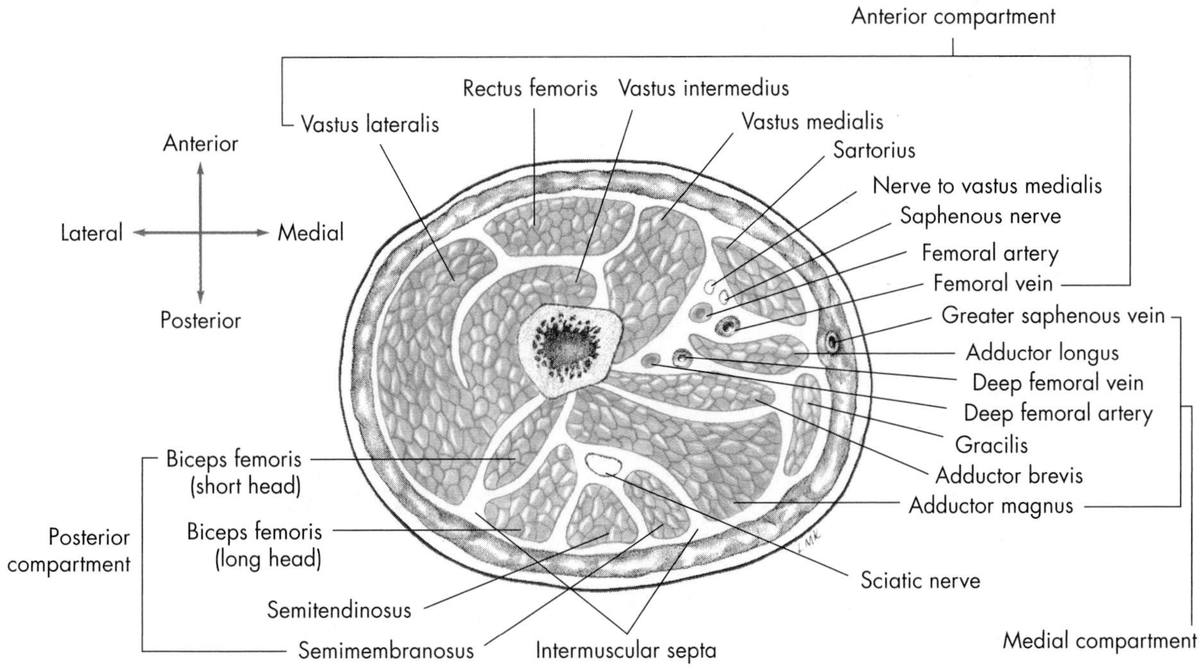

FIG. 9.11 ● Transverse section of the left midthigh, detailing the anterior, posterior, and medial compartments.

Muscle identification

In developing a thorough and practical knowledge of the muscular system, it is essential that individual muscles be understood. Figs. 9.9, 9.10, 9.12, 9.13, 9.14, and 9.15 illustrate groups of muscles that work together to produce joint movement. While viewing the muscles in these figures, correlate them with Table 9.2.

The muscles of the pelvis that act on the hip joint may be divided into two regions—the iliac and gluteal regions. The iliac region contains the iliopsoas muscle, which flexes the hip. The iliopsoas actually is three different muscles—the iliacus, the psoas major, and the psoas minor. The 10 muscles of the gluteal region function primarily to extend and rotate the hip. Located in the gluteal region are the gluteus maximus, gluteus medius, gluteus minimi, and tensor fasciae latae, and the six deep external rotators—piriformis, obturator externus, obturator internus, gemellus superior, gemellus inferior, and quadratus femoris.

The thigh is divided into three compartments by the intermuscular septa (Fig. 9.11). The anterior compartment contains the rectus femoris, vastus medialis, vastus intermedius, vastus lateralis, and sartorius. The hamstring muscle group, consisting of the biceps femoris, semitendinosus,

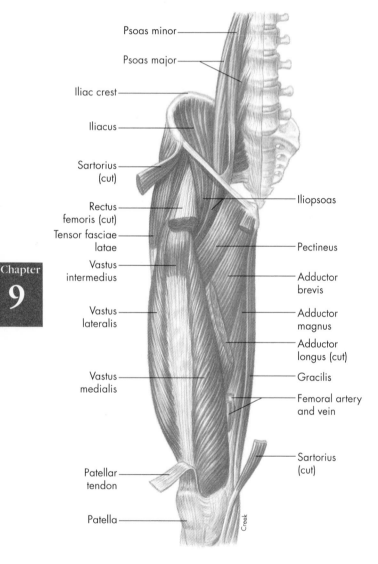

FIG. 9.12 ● Muscles of the right anterior pelvic and thigh regions.

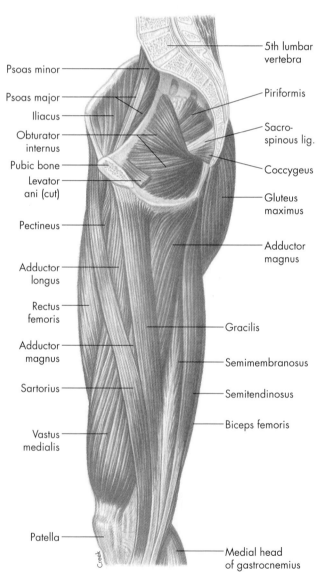

FIG. 9.13 ● Muscles of the right medial thigh.

and semimembranosus, is located in the posterior compartment. The medial compartment contains the thigh muscles primarily responsible for adduction of the hip, which are the adductor brevis, adductor longus, adductor magnus, pectineus, and gracilis.

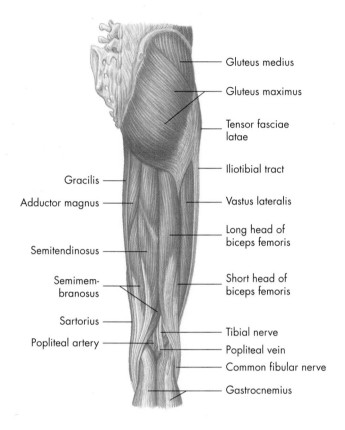

FIG. 9.14 ● Muscles of the right posterior thigh.

Gluteus medius
Gluteus maximus
Tensor fasciae latae
Iliotibial tract
Gracilis
Adductor magnus
Vastus lateralis
Long head of biceps femoris
Semitendinosus
Semimem-branosus
Short head of biceps femoris
Sartorius
Popliteal artery
Tibial nerve
Popliteal vein
Common fibular nerve
Gastrocnemius

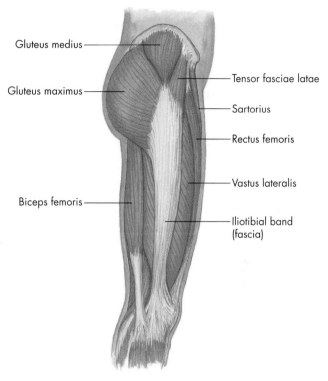

FIG. 9.15 ● Muscles of the lateral right thigh.

Gluteus medius
Gluteus maximus
Biceps femoris
Tensor fasciae latae
Sartorius
Rectus femoris
Vastus lateralis
Iliotibial band (fascia)

TABLE 9.2 • Agonist muscles of the hip joint

	Muscle	Origin	Insertion	Action	Plane of motion	Palpation	Innervation
Anterior	Iliacus	Inner surface of the ilium	Lesser trochanter of the femur and shaft just below	Flexion of the hip	Sagittal	Difficult to palpate; deep against posterior abdominal wall; with subject seated and leaning slightly forward to relax abdominal muscles, palpate psoas major deeply between iliac crest and 12th rib about halfway between ASIS and umbilicus with active hip flexion; palpate iliopsoas distal tendon on anterior aspect of hip approximately 1 1/2 inches below center of inguinal ligament with active hip flexion/extension of supine subject, immediately lateral to pectineus and medial to sartorius	Lumbar nerve and femoral nerve (L2–L4)
				External rotation of the femur	Transverse		
				Transverse pelvic rotation contralaterally when ipsilateral femur is stabilized			
	Psoas major and minor	Lower borders of the transverse processes (L1–L5), sides of the bodies of the last thoracic vertebrae (T12), the lumbar vertebrae (L1–L5), intervertebral fibrocartilages, and base of the sacrum	Lesser trochanter of the femur and shaft just below Psoas minor: Pectineal line and ilopectineal eminence	Flexion of the hip	Sagittal		
				External rotation of the femur	Transverse		
				Transverse pelvic rotation contralaterally when ipsilateral femur is stabilized			
	Rectus femoris	Anterior inferior iliac spine of the ilium and groove (posterior) above the acetabulum	Superior aspect of the patella tendon to the tibial tuberosity	Flexion of the hip	Sagittal	Straight down anterior thigh from anterior inferior iliac spine to patella with resisted hip flexion/knee extension	Femoral nerve (L2–L4)
				Extension of the knee			
				Anterior pelvic rotation			
	Sartorius	Anterior superior iliac spine and notch just below the spine	Anterior medial surface of the tibia just below the condyle	Flexion of the hip	Sagittal	Located proximally just medial to tensor fascia latae and lateral to iliopsoas; palpate superficially from anterior superior iliac spine to medial tibial condyle during combined resistance of hip flexion/external rotation/abduction and knee flexion in a supine position	Femoral nerve (L2, L3)
				Flexion of the knee			
				Anterior pelvic rotation			
				External rotation of the thigh as it flexes the hip and knee	Transverse		
				Abduction of the hip	Frontal		
	Pectineus	Space 1 inch wide on the front of the pubis just above the crest	Rough line leading from the lesser trochanter down to the linea aspera	Flexion of the hip	Sagittal	Difficult to distinguish from other adductors; anterior aspect of hip approximately 1 1/2 inches below center of inguinal ligament; just lateral and slightly proximal to adductor longus and medial to iliopsoas during flexion and adduction of supine subject	Femoral nerve (L2–L4)
				Adduction of the hip	Frontal		
				External rotation of the hip	Transverse		

Chapter 9

TABLE 9.2 (continued) • **Agonist muscles of the hip joint**

	Muscle	Origin	Insertion	Action	Plane of motion	Palpation	Innervation
Medial	Adductor brevis	Front of the inferior pubic ramus just below the origin of the adductor longus	Lower 2/3 of the pectineal line of the femur and the upper half of the medial lip of the linea aspera	Adduction of the hip	Frontal	Deep to adductor longus and superficial to adductor magnus; very difficult to palpate and differentiate from adductor longus, which is immediately inferior; proximal portion is just lateral to adductor longus	Obturator nerve (L3, L4)
				External rotation as it adducts the hip	Transverse		
				Assists in flexion of the hip	Sagittal		
	Adductor longus	Anterior pubis just below its crest	Middle 1/3 of the linea aspera	Adduction of the hip	Frontal	Most prominent muscle proximally on anteromedial thigh just inferior to the pubic bone with resisted adduction	Obturator nerve (L3, L4)
				Assists in flexion of the hip	Sagittal		
	Adductor magnus	Edge of the entire pubic ramus and the ischium and ischial tuberosity	Whole length of the linea aspera, inner condyloid ridge, and adductor tubercle	Adduction of the hip	Frontal	Medial aspect of thigh between gracilis and medial hamstrings from ischial tuberosity to adductor tubercle with resisted adduction from abducted position	Anterior: obturator nerve (L2–L4) Posterior: sciatic nerve (L4, L5, S1–S3)
				External rotation as the hip adducts	Transverse		
				Extension of the hip	Sagittal		
	Gracilis	Anteromedial edge of the descending ramus of the pubis	Anterior medial surface of the tibia below the condyle	Adduction of the hip	Frontal	A thin superficial tendon on anteromedial thigh with knee flexion and resisted adduction; just posterior to adductor longus and medial to the semitendinosus	Obturator nerve (L2–L4)
				Weak flexion of the knee	Sagittal		
				Assists with flexion of the hip			
				Internal rotation of the hip	Transverse		

TABLE 9.2 (continued) • Agonist muscles of the hip joint

	Muscle	Origin	Insertion	Action	Plane of motion	Palpation	Innervation
Posterior	Biceps femoris	Long head: ischial tuberosity. Short head: lower 1/2 of the linea aspera, and lateral condyloid ridge	Head of the fibula and lateral condyle of the tibia	Flexion of the knee	Sagittal	Posterolateral aspect of distal thigh with combined knee flexion and external rotation against resistance; just distal to the ischial tuberosity in a prone position with hip internally rotated during active knee flexion	Long head: sciatic nerve—tibial division (S1–S3) Short head: sciatic nerve—peroneal division (L5, S1, S2)
				Extension of the hip			
				Posterior pelvic rotation			
				External rotation of the hip	Transverse		
				External rotation of the knee			
	Semi-tendinosus	Ischial tuberosity	Upper anterior medial surface of the tibia just below the condyle	Flexion of the knee	Sagittal	Posteromedial aspect of distal thigh with combined knee flexion and internal rotation against resistance; just distal to ischial tuberosity in a prone position with hip internally rotated during active knee flexion	Sciatic nerve—tibial division (L5, S1, S2)
				Extension of the hip			
				Posterior pelvic rotation			
				Internal rotation of the hip	Transverse		
				Internal rotation of the flexed knee			
	Semi-membranosus	Ischial tuberosity	Postermedial surface of the medial tibial condyle	Flexion of the knee	Sagittal	Largely covered by other muscles, tendon can be felt at posteromedial aspect of knee just deep to semitendinosus tendon with combined knee flexion and internal rotation against resistance	Sciatic nerve—tibial division (L5, S1, S2)
				Extension of the hip			
				Posterior pelvic rotation			
				Internal rotation of the hip	Transverse		
				Internal rotation of the knee			
	Gluteus maximus	Posterior 1/4 of the crest of the ilium, posterior surface of the sacrum and coccyx near the ilium, and fascia of the lumbar area	Oblique ridge on the lateral surface of the greater trochanter and the iliotibial band of the fasciae latae	Extension of the hip	Sagittal	Running downward and laterally between posterior iliac crest superiorly, anal cleft medially, and gluteal fold inferiorly, emphasized with hip extension, external rotation, and abduction	Inferior gluteal nerve (L5, S1, S2)
				Posterior pelvic rotation			
				External rotation of the hip	Transverse		
				Upper fibers: assist in hip abduction	Frontal		
				Lower fibers: assist in hip adduction			

Chapter 9

TABLE 9.2 (continued) • **Agonist muscles of the hip joint**

	Muscle	Origin	Insertion	Action	Plane of motion	Palpation	Innervation
Lateral	Gluteus medius	Lateral surface of the ilium just below the crest	Posterior and middle surfaces of the greater trochanter of the femur	Abduction of the hip	Frontal	Slightly in front of and a few inches above the greater trochanter with active elevation of opposite pelvis from a standing position or active abduction when side-lying on contralateral pelvis	Superior gluteal nerve (L4, L5, S1)
				Anterior fibers: internal rotation of the hip	Transverse		
				Posterior fibers: external rotation of the hip			
				Anterior fibers: flexion of the hip	Sagittal		
				Posterior fibers: extension of the hip			
	Gluteus minimus	Lateral surface of the ilium just below the origin of the gluteus medius	Anterior surface of the greater trochanter of the femur	Abduction of the hip	Frontal	Deep to the gluteus medius; covered by tensor fasciae latae between anterior iliac crest and greater trochanter during internal rotation and abduction	Superior gluteal nerve (L4, L5, S1)
				Hip internal rotation as femur is abducted	Transverse		
				Flexion of the hip	Sagittal		
	Tensor fasciae latae	Anterior iliac crest and surface of the ilium just below the crest	One fourth of the way down the thigh into the iliotibial tract, which in turn inserts onto Gerdy's tubercle of the anterolateral tibial condyle	Abduction of the hip	Frontal	Anterolaterally, between anterior iliac crest and greater trochanter during flexion, internal rotation and abduction	Superior gluteal nerve (L4, L5, S1)
				Flexion of the hip	Sagittal		
				Anterior pelvic rotation			
				Internal rotation of the hip as it flexes	Transverse		

Chapter **9**

TABLE 9.2 (continued) • Agonist muscles of the hip joint

	Muscle	Origin	Insertion	Action	Plane of motion	Palpation	Innervation
Deep posterior	Piriformis	Anterior sacrum, posterior portions of the ischium, and obturator foramen	Superior and posterior aspect of the greater trochanter	Hip external rotation	Transverse	With subject prone and gluteus maximus relaxed, palpate deeply between posterior superior greater trochanter and sacrum while passively internally/externally rotating femur	First and second sacral nerve (S1, S2)
	Gemellus superior	Ischial spine	Posterior aspect of the greater trochanter immediately below piriformis	Hip external rotation	Transverse	With subject prone and gluteus maximus relaxed, palpate deeply between posterior superior greater trochanter and ischial spine while passively internally/externally rotating femur	Sacral nerve (L5, S1, S2)
	Gemellus inferior	Ischial tuberosity	Posterior aspect of the greater trochanter with obturator internus	Hip external rotation	Transverse	With subject prone and gluteus maximus relaxed, palpate deeply between posterior greater trochanter and ischial tuberosity while passively internally/externally rotating femur	Branches from sacral plexus (L4, L5, S1, S2)
	Obturator internus	Margin of obturator foramen	Posterior aspect of the greater trochanter with gemellus superior	Hip external rotation	Transverse	With subject prone and gluteus maximus relaxed, palpate deeply between posterior superior greater trochanter and obturator foramen while passively internally/externally rotating femur	Branches from sacral plexus (L4, L5, S1, S2)
	Obturator externus	Inferior margin of obturator foramen	Posterior aspect of the greater trochanter immediately below obturator internus	Hip external rotation	Transverse	With subject prone and gluteus maximus relaxed, palpate deeply between inferior posterior greater trochanter and obturator foramen while passively internally/externally rotating femur	Obturator nerve (L3, L4)
	Quadratus femoris	Ischial tuberosity	Intertrochanteric ridge of femur	Hip external rotation	Transverse	With subject prone and gluteus maximus relaxed, palpate deeply between inferior posterior greater trochanter and ischial tuberosity while passively internally/externally rotating femur	Branches from sacral plexus (L4, L5, S1)

Nerves

The muscles of the hip and pelvic girdle are all innervated from the lumbar and sacral plexus, known collectively as the lumbosacral plexus. The lumbar plexus (Fig. 9.16) is formed by the anterior rami of spinal nerves L1 through L4 and some fibers from T12. The lower abdomen and the anterior and medial portions of the lower extremity are innervated by nerves arising from the lumbar plexus. The sacral plexus (Fig. 9.17) is formed by the anterior rami of L4, L5, and S1 through S4. The lower back, pelvis, perineum, posterior surface of the thigh and leg, and dorsal and plantar surfaces of the foot are innervated by nerves arising from the sacral plexus.

The major nerves of significance arising from the lumbar plexus to innervate the muscles of the hip are the femoral and obturator nerves. The femoral nerve (Fig. 9.18) arises from the posterior division of the lumbar plexus and innervates the anterior muscles of the thigh, including the iliopsoas, rectus femoris, vastus medialis, vastus intermedius, vastus lateralis, pectineus, and sartorius. It also pro-vides sensation to the anterior and lateral thigh and the medial leg and foot. The obturator nerve (Fig. 9.19) arises from the anterior division of the lumbar plexus and provides innervation to the hip adductors, such as the adductor brevis, adductor longus, adductor magnus, and gracilis, as well as the obturator externus. The obturator nerve provides sensation to the medial thigh.

The nerves arising from the sacral plexus that innervate the muscles of the hip are the superior gluteal, inferior gluteal, sciatic, and branches from the sacral plexus. The superior gluteal nerve arises from L4, L5, and S1 to innervate the gluteus medius, gluteus minimus, and tensor fasciae latae. The inferior gluteal nerve arises from L5, S1, and S2 to supply the gluteus maximus. Branches from the sacral plexus innervate the piriformis (S1, S2), gemellus superior (L5, S1, S2), gemellus inferior and obturator internus (L4, L5, S1, S2), and quadratus femoris (L4, L5, S1).

The sciatic nerve is composed of the tibial and common peroneal (fibular) nerves, which are wrapped together in a connective tissue sheath until reaching approximately midway down the

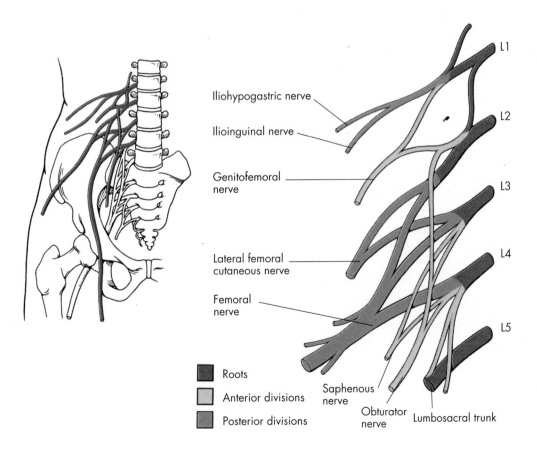

Iliohypogastric nerve

Ilioinguinal nerve

Genitofemoral nerve

Lateral femoral cutaneous nerve

Femoral nerve

Roots
Anterior divisions
Posterior divisions

Saphenous nerve
Obturator nerve
Lumbosacral trunk

L1
L2
L3
L4
L5

FIG. 9.16 ● The lumbar plexus.

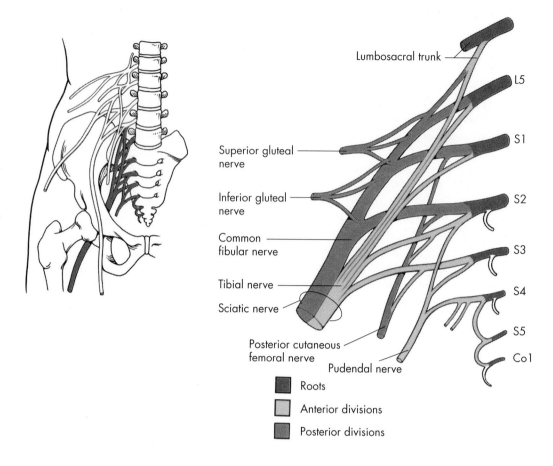

FIG. 9.17 ● The sacral plexus.

posterior thigh. The sciatic nerve tibial division (Fig. 9.20) innervates the semitendinosus, semimembranosus, biceps femoris (long head), and adductor magnus. The sciatic nerve supplies sensation to the anterolateral and posterolateral lower leg as well as most of the dorsal and plantar aspect of the foot. The tibial division provides sensation for the posterolateral lower leg and plantar aspect of the foot, while the peroneal division provides sensation to the anterolateral lower leg and dorsum of the foot. Both of these nerves continue down the lower extremity to provide motor and sensory function to the muscles of the lower leg; this will be addressed in Chapters 10 and 11.

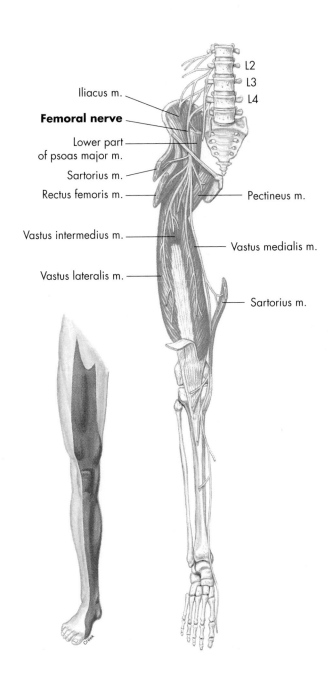

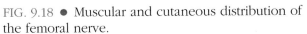

FIG. 9.18 ● Muscular and cutaneous distribution of the femoral nerve.

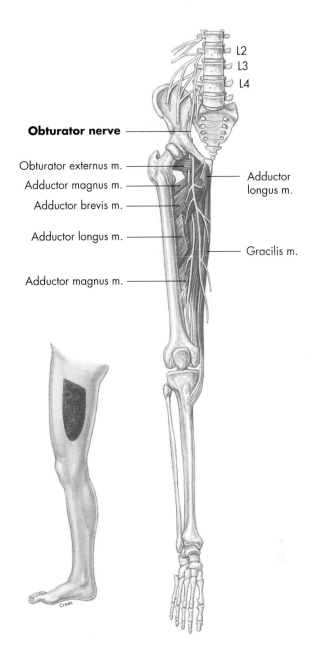

FIG. 9.19 ● Muscular and cutaneous distribution of the obturator nerve.

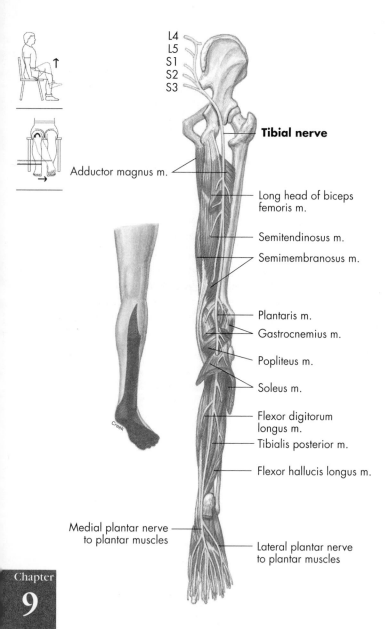

L4
L5
S1
S2
S3

Tibial nerve

Adductor magnus m.

Long head of biceps femoris m.

Semitendinosus m.

Semimembranosus m.

Plantaris m.

Gastrocnemius m.

Popliteus m.

Soleus m.

Flexor digitorum longus m.

Tibialis posterior m.

Flexor hallucis longus m.

Medial plantar nerve to plantar muscles

Lateral plantar nerve to plantar muscles

Creek

FIG. 9.20 ● Muscular and cutaneous distribution of the tibial nerve.

Iliopsoas muscle FIG. 9.21

(il´e-o-so´as)

Origin

Iliacus: inner surface of the ilium

Psoas major and minor: lower borders of the transverse processes (L1–L5), sides of the bodies of the last thoracic vertebrae (T12), the lumbar vertebrae (L1–L5), intervertebral fibrocartilages, and base of the sacrum

Insertion

Iliacus and psoas major: lesser trochanter of the femur and the shaft just below

Psoas minor: pectineal line and iliopectineal eminence

Action

Flexion of the hip

External rotation of the femur

Transverse pelvis rotation contralaterally when ipsilateral femur is stabilized

Palpation

Difficult to palpate; deep against posterior abdominal wall; with subject seated and leaning slightly forward to relax abdominal muscles, palpate psoas major deeply between iliac crest and 12th rib about halfway between ASIS and umbilicus with active hip flexion; palpate iliopsoas distal tendon on anterior aspect of hip approximately 1 1/2 inches below center of inguinal ligament with active hip flexion/extension of supine subject, immediately lateral to pectineus and medial to sartorius

Innervation

Lumbar nerve and femoral nerve (L2–L4)

Application, strengthening, and flexibility

The iliopsoas is commonly referred to as if it were one muscle, but it is actually composed of the iliacus, the psoas major, and the psoas minor. Some anatomy texts make this distinction and list each muscle individually.

The iliopsoas muscle is powerful in actions such as raising the lower extremity from the floor while in a supine position. The psoas major's origin in the lower back tends to move the lower back anteriorly or, in the supine position, pulls the lower back up as it raises the legs. For this reason, lower back problems are often aggravated by this activity, and bilateral 6-inch leg raises are usually not recommended. The abdominals are the muscles that can be used to prevent this lower back strain by pulling up on the front of the pelvis, thus flattening the back. Leg raising is primarily hip flexion and not abdominal action. Backs may be injured by strenuous and prolonged leg-raising exercises.

The iliopsoas contracts strongly, both concentrically and eccentrically, in sit-ups, particularly if the hip is not flexed. The more flexed and/or abducted the hips are, the less the iliopsoas will be activated with abdominal strengthening exercises.

The iliopsoas may be exercised by supporting the arms on a dip bar or parallel bars and then flexing the hips to lift the legs. This may be done initially with the knees flexed in a tucked position to lessen the resistance. As the muscle becomes more developed, the knees can be straightened, which increases the resistance arm length to add more resistance. This concept of increasing or decreasing the resistance by modifying the resistance arm is explained further in Chapter 3.

To stretch the iliopsoas, which often becomes tight with excessive straight leg sit-ups and contributes to anterior pelvic tilting, the hip must be extended so that the femur is behind the plane of the body. In order to somewhat isolate the iliopsoas, full knee flexion should be avoided. Slight additional stretch may be applied by internally rotating the hip while it is extended.

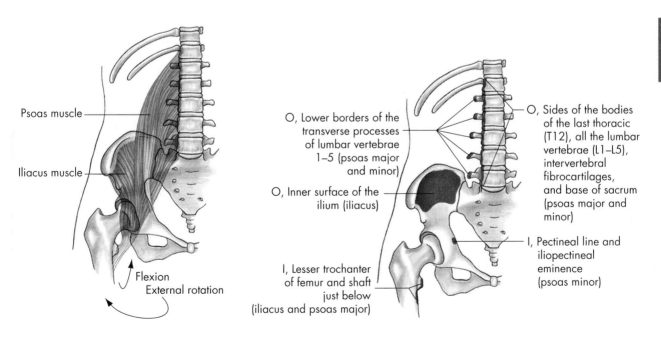

Psoas muscle

Iliacus muscle

Flexion
External rotation

O, Lower borders of the transverse processes of lumbar vertebrae 1–5 (psoas major and minor)

O, Inner surface of the ilium (iliacus)

I, Lesser trochanter of femur and shaft just below (iliacus and psoas major)

O, Sides of the bodies of the last thoracic (T12), all the lumbar vertebrae (L1–L5), intervertebral fibrocartilages, and base of sacrum (psoas major and minor)

I, Pectineal line and iliopectineal eminence (psoas minor)

FIG. 9.21 ● Iliopsoas muscle, anterior view. O, Origin; I, Insertion.

Sartorius muscle   FIG. 9.22

(sar-to´ri-us)

Origin

Anterior superior iliac spine and notch just below the spine

Insertion

Anterior medial surface of the tibia just below the condyle

Action

Flexion of the hip

Flexion of the knee

External rotation of the thigh as it flexes the hip and knee

Abduction of the hip

Anterior pelvic rotation

Palpation

Located proximally just medial to tensor fascia latae and lateral to iliopsoas; palpate superficially from the anterior superior iliac spine to the medial tibial condyle during combined resistance of hip flexion/ external rotation/abduction and knee flexion in a supine position

Innervation

Femoral nerve (L2–L3)

Application, strengthening, and flexibility

Pulling from the anterior superior iliac spine and the notch just below it, the tendency again is to tilt the pelvis anteriorly (down in front) as this muscle contracts. The abdominal muscles must prevent this tendency by posteriorly rotating the pelvis (pulling up in front), thus flattening the lower back.

The sartorius, a two-joint muscle, is effective as a hip flexor or as a knee flexor. It is weak when both actions take place at the same time. Observe that, in attempting to cross the knees when in a sitting position, one customarily leans well back, thus raising the origin to lengthen this muscle, making it more effective in flexing and crossing the knees. With the knees held extended, the sartorius becomes a more effective hip flexor. It is the longest muscle in the body and is strengthened when hip flexion activities are performed as described for developing the iliopsoas. Stretching may be accomplished by a partner passively taking the hip into extreme extension, adduction, and internal rotation with the knee extended.

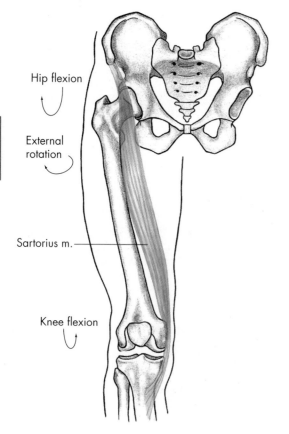

Hip flexion

External rotation

Sartorius m.

Knee flexion

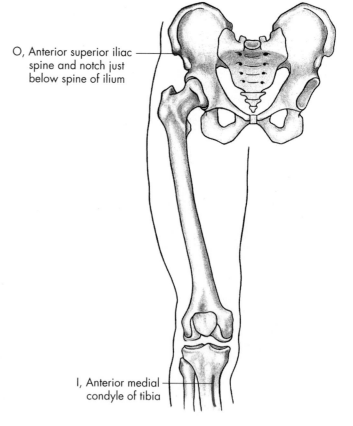

O, Anterior superior iliac spine and notch just below spine of ilium

I, Anterior medial condyle of tibia

FIG. 9.22 ● Sartorius muscle, anterior view. *O,* Origin; *I,* Insertion.

Rectus femoris muscle FIG. 9.23

(rek´tus fem´or-is)

Origin

Anterior inferior iliac spine of the ilium and groove
(posterior) above the acetabulum

Insertion

Superior aspect of the patella and patellar tendon to
the tibial tuberosity

Action

Flexion of the hip
Extension of the knee
Anterior pelvic rotation

Palpation

Straight down anterior thigh from anterior inferior
iliac spine to patella with resisted knee extension
and hip flexion

Innervation

Femoral nerve (L2–L4)

Application, strengthening, and flexibility

Pulling from the anterior inferior iliac spine of
the ilium, the rectus femoris muscle has the same

tendency to anteriorly rotate the pelvis (down in
front and up in back). Only the abdominal muscles can prevent this from occurring. In speaking
of the hip flexor group in general, it may be said
that many people permit the pelvis to be perma-
nently tilted forward as they get older. The relaxed abdominal wall does not hold the pelvis up;
therefore, an increased lumbar curve results.

Generally, a muscle's ability to exert force decreases as it shortens. This explains why the rectus
femoris muscle is a powerful extensor of the knee
when the hip is extended but weaker when the
hip is flexed. This muscle is exercised, along with
the vastus group, in running, jumping, hopping,
and skipping. In these movements, the hips are extended powerfully by the gluteus maximus and the
hamstring muscles, which counteract the tendency
of the rectus femoris muscle to flex the hip while it
extends the knee. It can be remembered as one of
the quadriceps muscle group. The rectus femoris is
developed by performing hip flexion exercises or
knee extension exercises against manual resistance.

The rectus femoris is best stretched in a side-lying
position by having a partner take the knee into full flexion and simultaneously take the hip into extension.

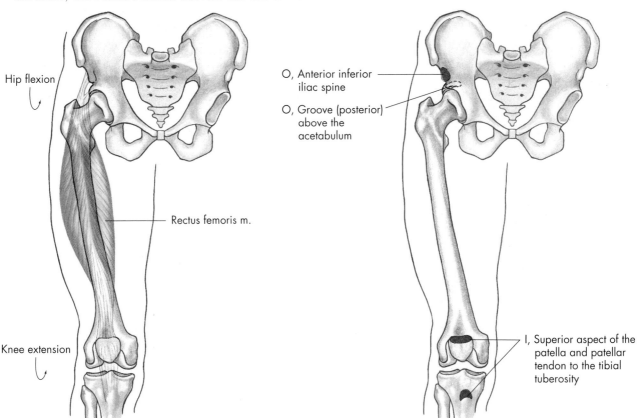

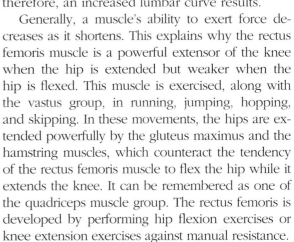

FIG. 9.23 ● Rectus femoris muscle, anterior view. *O,* Origin; *I,* Insertion.

Tensor fasciae latae muscle FIG. 9.24

(ten´sor fas´i-e la´te)

Origin

Anterior iliac crest and surface of the ilium just below the crest

Insertion

One-fourth of the way down the thigh into the ilio-tibial tract, which in turn inserts onto Gerdy's tubercle of the anterolateral tibial condyle

Action

Abduction of the hip
Flexion of the hip
Tendency to rotate the hip internally as it flexes
Anterior pelvic rotation

Palpation

Anterolaterally, between anterior iliac crest and greater trochanter during internal rotation, flexion, and abduction

Innervation

Superior gluteal nerve (L4–L5, S1)

Application, strengthening, and flexibility

The tensor fasciae latae muscle aids in preventing external rotation of the hip as it is flexed by other flexor muscles.

The tensor fasciae latae muscle is used when flexion and internal rotation take place. This is a weak movement but is important in helping direct the leg forward so that the foot is placed straight forward in walking and running. Thus, from the supine position, raising the leg with definite internal rotation of the femur will call it into action.

The tensor fasciae latae may be developed by performing hip abduction exercises against gravity and resistance while in a side-lying position. This is done simply by abducting the hip that is up and then slowly lowering it back to rest against the other leg. Stretch may be applied by remaining on the side and having a partner passively move the downside hip into full extension, adduction, and external rotation.

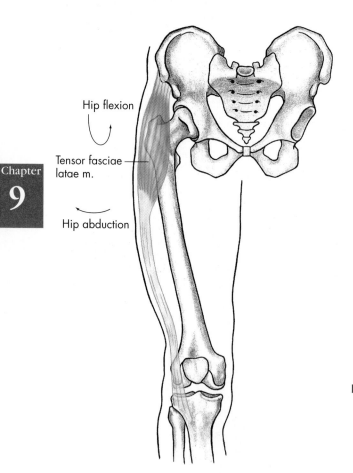

Hip flexion

Tensor fasciae latae m.

Hip abduction

Chapter 9

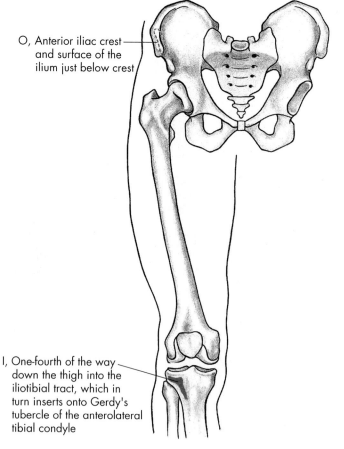

O, Anterior iliac crest and surface of the ilium just below crest

I, One-fourth of the way down the thigh into the iliotibial tract, which in turn inserts onto Gerdy's tubercle of the anterolateral tibial condyle

FIG. 9.24 ● Tensor fasciae latae muscle, anterior view. *O,* Origin; *I,* Insertion.

Gluteus maximus muscle   FIG. 9.25

(glu´te-us maks´i-mus)

Origin

Posterior one-fourth of the crest of the ilium, posterior surface of the sacrum and coccyx near the ilium, and fascia of the lumbar area

Insertion

Oblique ridge on the lateral surface of the greater trochanter and the iliotibial band of the fasciae latae

Action

Extension of the hip
External rotation of the hip
Upper fibers: assist in hip abduction
Lower fibers: assist in hip adduction
Posterior pelvic rotation

Palpation

Running downward and laterally between posterior iliac crest superiorly, anal cleft medially, and gluteal fold inferiorly, emphasized with hip extension, external rotation, and abduction

Innervation

Inferior gluteal nerve (L5, S1–S2)

Application, strengthening, and flexibility

The gluteus maximus muscle comes into action when movement between the pelvis and the femur approaches and goes beyond 15 degrees of extension. As a result, it is not used extensively in ordinary walking. It is important in extension of the thigh with external rotation.

Strong action of the gluteus maximus muscle is seen in running, hopping, skipping, and jumping. Powerful extension of the thigh is secured in the return to standing from a squatting position, especially if a barbell with weights is placed on the shoulders.

Hip extension exercises from a forward-leaning or prone position may be used to develop this muscle. This muscle is most emphasized when the hip starts from a flexed position and moves to full extension and abduction, with the knee flexed 30 degrees or more to reduce the hamstrings' involvement in the action.

The gluteus maximus is stretched in the supine position with full hip flexion to the ipsilateral axilla and then to the contralateral axilla with the knee in flexion. Simultaneous internal hip rotation accentuates this stretch.

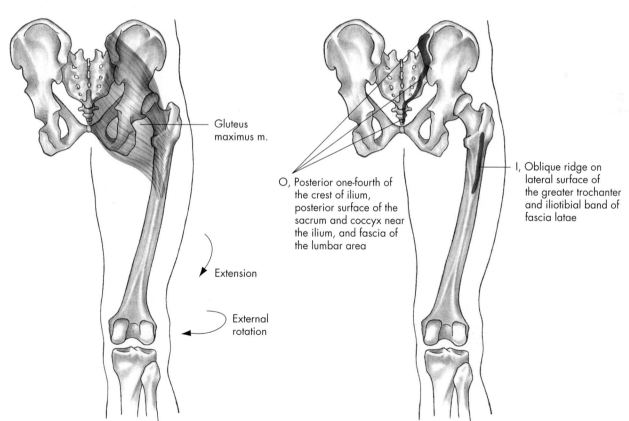

Gluteus maximus m.

O, Posterior one-fourth of the crest of ilium, posterior surface of the sacrum and coccyx near the ilium, and fascia of the lumbar area

Extension

External rotation

I, Oblique ridge on lateral surface of the greater trochanter and iliotibial band of fascia latae

Chapter 9

FIG. 9.25 ● Gluteus maximus muscle, posterior view. *O,* Origin; *I,* Insertion.

Gluteus medius muscle FIG. 9.26

(glu´te-us me´di-us)

Origin
Lateral surface of the ilium just below the crest

Insertion
Posterior and middle surfaces of the greater trochanter of the femur

Action
Abduction of the hip

Anterior fibers: internal rotation and flexion of the hip

Posterior fibers: external rotation and extension of the hip

Palpation
Slightly in front of and a few inches above the greater trochanter with active elevation of opposite pelvis from a standing position or active abduction when side-lying on contralateral pelvis

Innervation
Superior gluteal nerve (L4–L5, S1)

Application, strengthening, and flexibility

Typical action of the gluteus medius and gluteus minimus muscles is seen in walking. As the weight of the body is suspended on one leg, these muscles prevent the opposite pelvis from sagging. Weakness in the gluteus medius and gluteus minimus can result in a Trendelenburg gait. With this weakness, the individual's opposite pelvis will sag on weight bearing because the hip abductors cannot maintain proper alignment.

Hip external rotation exercises performed against resistance can provide some strengthening for the gluteus medius, but it is best strengthened by performing the side-lying leg raises or hip abduction exercises as described for the tensor fasciae latae. The gluteus medius is best stretched by moving the hip into extreme adduction in front of the opposite extremity and then behind it.

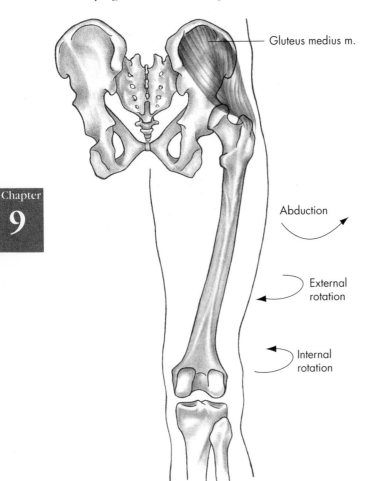
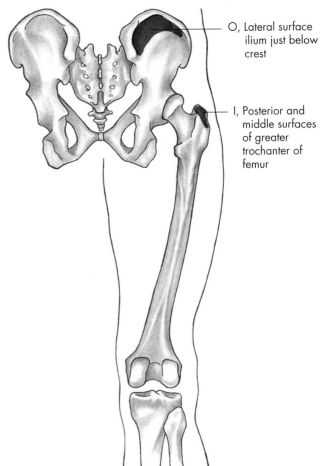

Gluteus medius m.

Abduction

External rotation

Internal rotation

O, Lateral surface ilium just below crest

I, Posterior and middle surfaces of greater trochanter of femur

FIG. 9.26 ● Gluteus medius muscle, posterior view. *O*, Origin; *I*, Insertion.

Gluteus minimus muscle FIG. 9.27

(glu´te-us min´i-mus)

Origin

Lateral surface of the ilium just below the origin of the gluteus medius

Insertion

Anterior surface of the greater trochanter of the femur

Action

Abduction of the hip

Internal rotation as the femur abducts

Flexion of the hip

Palpation

Deep to the gluteus medius; covered by tensor fasciae latae between anterior iliac crest and greater trochanter during internal rotation and abduction

Innervation

Superior gluteal nerve (L4–L5, S1)

Application, strengthening, and flexibility

Both the gluteus minimus and the gluteus medius are used in powerfully maintaining proper hip abduction while running. As a result, both of these muscles are exercised effectively in running, hopping, and skipping, in which weight is transferred forcefully from one foot to the other. As the body ages, the gluteus medius and gluteus minimus muscles tend to lose their effectiveness. The spring of youth, as far as the hips are concerned, resides in these muscles. To have great drive in the legs, these muscles must be fully developed.

The gluteus minimus is best strengthened by performing hip abduction exercises similar to the ones described for the tensor fasciae latae and gluteus medius muscles. It may also be developed by performing hip internal rotation exercises against manual resistance. Stretching of this muscle is accomplished by extreme hip adduction with slight external rotation.

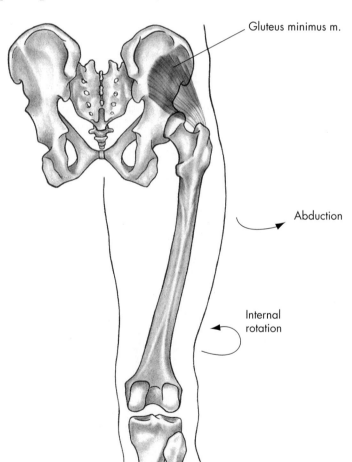

Gluteus minimus m.

Abduction

Internal rotation

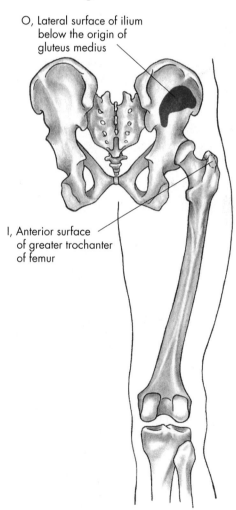

O, Lateral surface of ilium below the origin of gluteus medius

I, Anterior surface of greater trochanter of femur

FIG. 9.27 ● Gluteus minimus muscle, posterior view. *O,* Origin; *I,* Insertion.

The six deep lateral rotator muscles—
FIG. 9.28

piriformis   (pi-ri-for´mis)
gemellus superior   (je-mel´us su-pe´ri-or)
gemellus inferior   (je-mel´us in-fe´ri-or)
obturator externus   (ob-tu-ra´tor eks-ter´nus)
obturator internus   (ob-tu-ra´tor in-ter´nus)
quadratus femoris   (kwad-ra´tus fem´or-is)

Origin
Anterior sacrum, posterior portions of the ischium, and obturator foramen

Insertion
Superior and posterior aspect of the greater trochanter

Action
External rotation of the hip

Palpation
Although not directly palpable, deep palpation is possible between posterior superior greater trochanter and obturator foramen with the subject prone during relaxation of the gluteus maximus while passively using the lower leg flexed at the knee to passively internally and externally rotate the femur or alternating contracting/relaxing the external rotators slightly

Innervation
Piriformis: first or second sacral nerve (S1–S2)
Gemellus superior: sacral nerve (L5, S1–S2)

Gemellus inferior: branches from sacral plexus (L4–L5, S1–S2)
Obturator externus: obturator nerve (L3–L4)
Obturator internus: branches from sacral plexus (L4–L5, S1–S2)
Quadratus femoris: branches from sacral plexus (L4–L5, S1)

Application, strengthening, and flexibility
The six lateral rotators are used powerfully in movements of external rotation of the femur, as in sports in which the individual takes off on one leg from preliminary internal rotation. Throwing a baseball and swinging a baseball bat, in which there is rotation of the hip, are typical examples.

Standing on one leg and forcefully turning the body away from that leg is accomplished by contraction of these muscles, and it may be repeated for strengthening purposes. A partner may provide resistance as development progresses. The six deep lateral rotators may be stretched in the supine position with a partner passively internally rotating and slightly flexing the hip.

Of special note, the sciatic nerve usually passes just inferior to the piriformis muscle but may pass through it. As a result, tightness in the piriformis muscle may contribute to compression on the sciatic nerve. The piriformis may be stretched by having the subject lie on the uninvolved side with a partner passively taking the hip into full internal rotation combined with hip adduction and slight to moderate hip flexion.

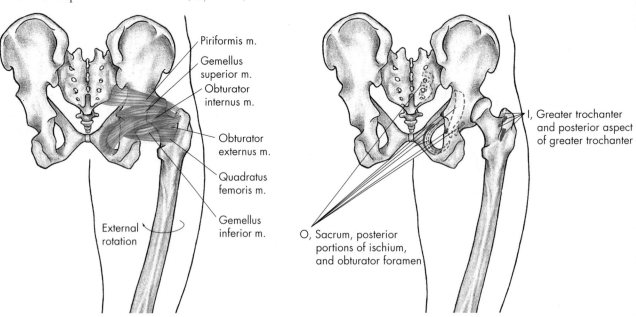

FIG. 9.28 ● The six deep lateral rotator muscles, posterior view: piriformis, gemellus superior, gemellus inferior, obturator externus, obturator internus, and quadratus femoris.

Semitendinosus muscle   FIG. 9.29

(sem´i-ten-di-no´sus)

Origin
Ischial tuberosity

Insertion
Anterior medial surface of the tibia just below the condyle

Action
Flexion of the knee
Extension of the hip
Internal rotation of the hip
Internal rotation of the flexed knee
Posterior pelvic rotation

Palpation
Posteromedial aspect of distal thigh with combined knee flexion and internal rotation against resistance just distal to the ischial tuberosity in a prone position with hip internally rotated during active knee flexion

Innervation
Sciatic nerve—tibial division (L5, S1–S2)

Application, strengthening, and flexibility

This two-joint muscle is most effective when contracting to either extend the hip or flex the knee. When there is extension of the hip and flexion of the knee at the same time, both movements are weak. When the trunk is flexed forward with the knees straight, the hamstring muscles have a powerful pull on the rear pelvis and tilt it down in back by full contraction. If the knees are flexed when this movement takes place, one can observe that the work is done chiefly by the gluteus maximus muscle.

On the other hand, when the muscles are used in powerful flexion of the knees, as in hanging by the knees from a bar, the flexors of the hip come into play to raise the origin of these muscles and make them more effective as knee flexors. By full extension of the hips in this movement, the knee flexion movement is weakened. These muscles are used in ordinary walking as extensors of the hip and allow the gluteus maximus to relax in the movement.

The semitendinosus is best developed through knee flexion exercises against resistance. Commonly known as hamstring curls or leg curls, they may be performed in a prone position on a knee table or standing with ankle weights attached. This muscle is emphasized when performing hamstring curls while attempting to maintain the knee joint in internal rotation. This internally rotated position brings its insertion in alignment with its origin.

The semitendinosus is stretched by maximally extending the knee while flexing the internally rotated and slightly abducted hip.

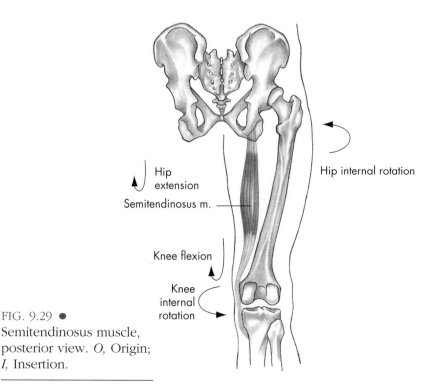

FIG. 9.29 ●
Semitendinosus muscle, posterior view. *O,* Origin; *I,* Insertion.

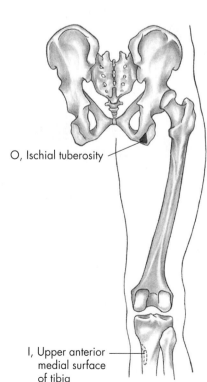

Semimembranosus muscle FIG. 9.30

(sem´i-mem´bra-no´sus)

Origin

Ischial tuberosity

Insertion

Posteromedial surface of the medial tibial condyle

Action

Flexion of the knee
Extension of the hip
Internal rotation of the hip
Internal rotation of the flexed knee
Posterior pelvic rotation

Palpation

Largely covered by other muscles, the tendon can be felt at the posteromedial aspect of the knee just deep to the semitendinosus tendon with combined knee flexion and internal rotation against resistance

Innervation

Sciatic nerve—tibial division (L5, S1–S2)

Application, strengthening, and flexibility

Both the semitendinosus and semimembranosus are responsible for internal rotation of the knee, along with the popliteus muscle, which is discussed in the next chapter. Because of the manner in which they cross the joint, the muscles are very important in providing dynamic medial stability to the knee joint.

The semimembranosus is best developed by performing leg curls. Internal rotation of the knee throughout the range accentuates the activity of this muscle. The semimembranosus is stretched in the same manner as the semitendinosus.

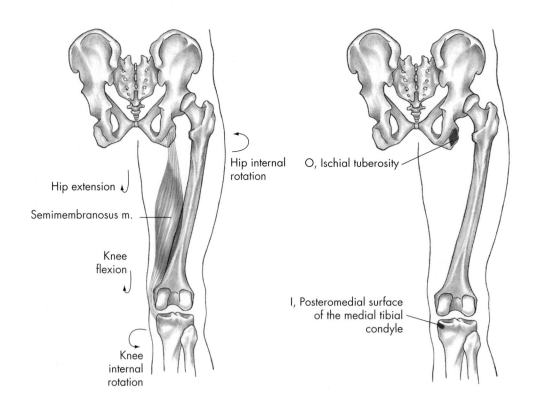

FIG. 9.30 ● Semimembranosus muscle, posterior view. *O,* Origin; *I,* Insertion.

Biceps femoris muscle   FIG. 9.31

(bi´seps fem´or-is)

Origin

Long head: ischial tuberosity
Short head: lower half of the linea aspera, and lateral
  condyloid ridge

Insertion

Lateral condyle of the tibia and head of the fibula

Action

Flexion of the knee
Extension of the hip
External rotation of the hip
External rotation of the flexed knee
Posterior pelvic rotation

Palpation

Posterolateral aspect of distal thigh with combined
  knee flexion and external rotation against resistance and just distal to the ischial tuberosity in a
  prone position with hip internally rotated during
  active knee flexion

Innervation

Long head: sciatic nerve—tibial division (S1–S3)
Short head: sciatic nerve—peroneal division (L5, S1–S2)

Application, strengthening, and flexibility

The semitendinosus, semimembranosus, and biceps femoris muscles are known as the hamstrings.
These muscles, together with the gluteus maximus
muscle, are used in extension of the hip when the
knees are straight or nearly so. Thus, in running,
jumping, skipping, and hopping, these muscles are
used together. The hamstrings are used without the
aid of the gluteus maximus, however, when one is
hanging from a bar by the knees. Similarly, the gluteus maximus is used without the aid of the hamstrings when the knees are flexed while the hips
are being extended. This occurs when rising from
a knee-bend position to a standing position.

The biceps femoris is best developed through
hamstring curls as described for the semitendinosus,
but it is emphasized more if the knee is maintained
in external rotation throughout the range of motion,
which brings the origin and insertion more in line
with each other. The biceps femoris is stretched by
maximally extending the knee while flexing the externally rotated and slightly adducted hip.

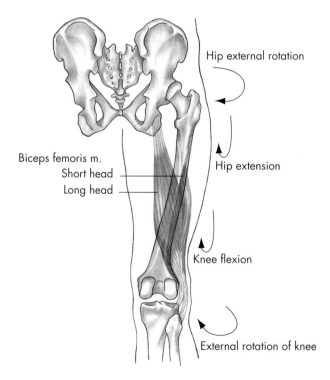

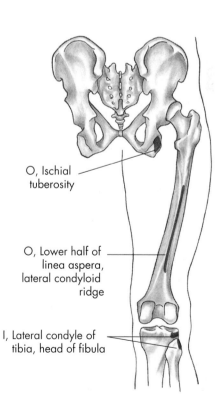

FIG. 9.31 ● Biceps femoris muscle, posterior view. *O,* Origin; *I,* Insertion.

Adductor brevis muscle FIG. 9.32

(ad-duk´tor bre´vis)

Origin
Front of the inferior pubic ramus just below the origin of the longus

Insertion
Lower two-thirds of the pectineal line of the femur and the upper half of the medial lip of the linea aspera

Action
Adduction of the hip
External rotation as it adducts the hip
Assists in flexion of the hip

Palpation
Deep to the adductor longus and superficial to the adductor magnus; very difficult to palpate and differentiate from adductor longus, which is immediately inferior; proximal portion is just lateral to adductor longus

Innervation
Obturator nerve (L3–L4)

Application, strengthening, and flexibility
The adductor brevis muscle, along with the other adductor muscles, provides powerful movement of the thighs toward each other. Squeezing the legs together toward each other against resistance is effective in strengthening the adductor brevis. Abducting the extended and internally rotated hip provides stretching of the adductor brevis.

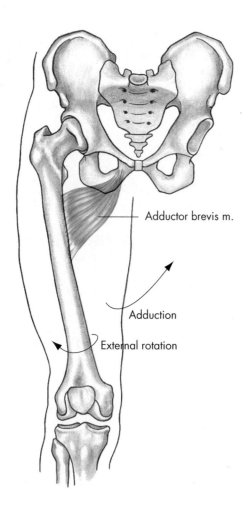

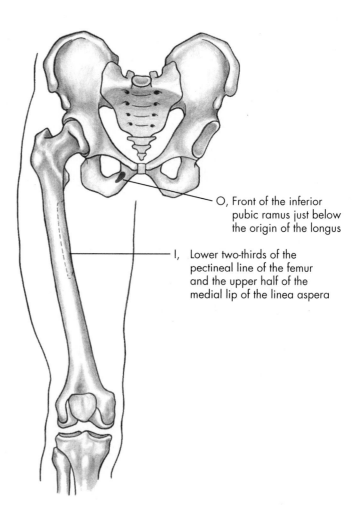

Adductor brevis m.

Adduction

External rotation

O, Front of the inferior pubic ramus just below the origin of the longus

I, Lower two-thirds of the pectineal line of the femur and the upper half of the medial lip of the linea aspera

FIG. 9.32 ● Adductor brevis muscle, anterior view. *O*, Origin; *I*, Insertion.

Adductor longus muscle FIG. 9.33

(ad-duk´tor long´gus)

Origin
Anterior pubis just below its crest

Insertion
Middle third of the linea aspera

Action
Adduction of the hip
Assists in flexion of the hip

Palpation
Most prominent muscle proximally on anteromedial thigh just inferior to the pubic bone with resisted adduction

Innervation
Obturator nerve (L3–L4)

Application, strengthening, and flexibility
The muscle may be strengthened by using the scissors exercise, which requires the subject to sit on the floor with the legs spread wide while a partner puts his or her legs or arms inside each lower leg to provide resistance. As the subject attempts to adduct his or her legs together, the partner provides manual resistance throughout the range of motion. This exercise may be used for either one or both legs. The adductor longus is stretched in the same manner as the adductor brevis.

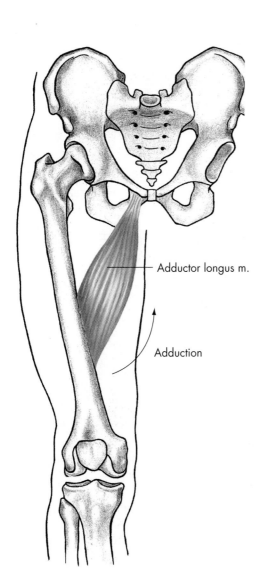
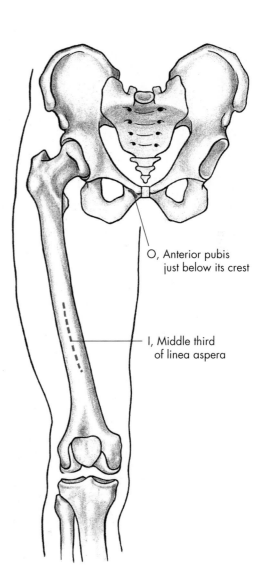

Adductor longus m.

Adduction

O, Anterior pubis just below its crest

I, Middle third of linea aspera

FIG. 9.33 ● Adductor longus muscle, anterior view. *O,* Origin; *I,* Insertion.

Adductor magnus muscle FIG. 9.34

(ad-duk´tor mag´nus)

Origin

Edge of the entire ramus of the pubis and the ischium and ischial tuberosity

Insertion

Whole length of the linea aspera, inner condyloid ridge, and adductor tubercle

Action

Adduction of the hip
External rotation as the hip adducts
Extension of the hip

Palpation

Medial aspect of thigh between gracilis and medial hamstrings from ischial tuberosity to adductor tubercle with resisted adduction from abducted position

Innervation

Anterior: obturator nerve (L2–L4)
Posterior: sciatic nerve (L4–L5, S1–S3)

Application, strengthening, and flexibility

The adductor magnus muscle is used in the breaststroke kick in swimming or in horseback riding. Since the adductor muscles (adductor magnus, adductor longus, adductor brevis, and gracilis muscles) are not heavily used in ordinary movement, some prescribed activity for them should be provided. Some modern exercise equipment is engineered to provide resistance for hip adduction movement. Hip adduction exercises such as those described for the adductor brevis and the adductor longus may be used for strengthening the adductor magnus as well. The adductor magnus is stretched in the same manner as the adductor brevis and adductor longus.

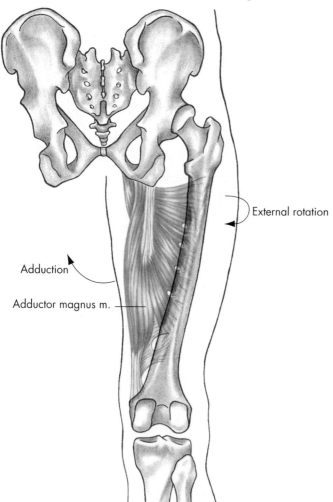
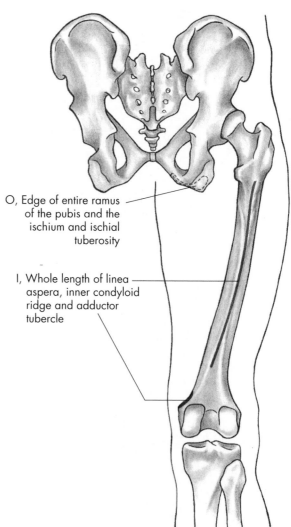

External rotation

Adduction

Adductor magnus m.

O, Edge of entire ramus of the pubis and the ischium and ischial tuberosity

I, Whole length of linea aspera, inner condyloid ridge and adductor tubercle

FIG. 9.34 ● Adductor magnus muscle, posterior view. *O*, Origin; *I*, Insertion.

Chapter
9

Pectineus muscle FIG. 9.35

(pek-tin´e-us)

Origin
Space 1-inch wide on the front of the pubis just above the crest

Insertion
Rough line leading from the lesser trochanter down to the linea aspera

Action
Flexion of the hip
Adduction of the hip
External rotation of the hip

Palpation
Difficult to distinguish from other adductors; anterior aspect of hip approximately 1 1/2 inches below center of inguinal ligament; just lateral and slightly proximal to adductor longus and medial to iliopsoas during flexion and adduction of supine subject

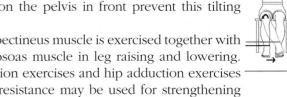

Innervation
Femoral nerve (L2–L4)

Application, strengthening, and flexibility
As the pectineus contracts, it also tends to rotate the pelvis anteriorly. The abdominal muscles pulling up on the pelvis in front prevent this tilting action.

The pectineus muscle is exercised together with the iliopsoas muscle in leg raising and lowering. Hip flexion exercises and hip adduction exercises against resistance may be used for strengthening this muscle.

The pectineus is stretched by fully abducting the extended and internally rotated hip.

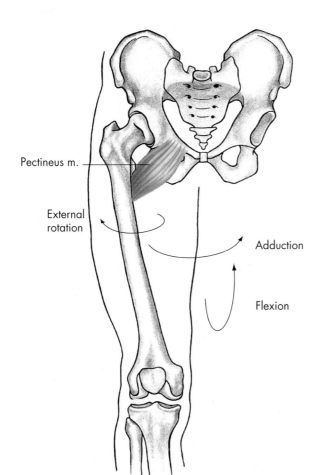

Pectineus m.

External rotation

Adduction

Flexion

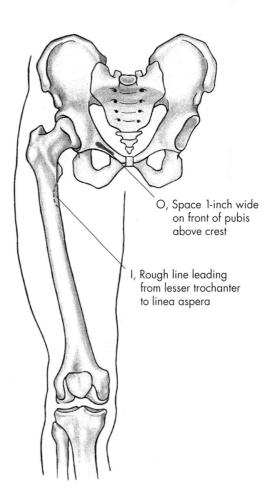

O, Space 1-inch wide on front of pubis above crest

I, Rough line leading from lesser trochanter to linea aspera

FIG. 9.35 ● Pectineus muscle, anterior view. *O,* Origin; *I,* Insertion.

Gracilis muscle   FIG. 9.36

(gras´-il-is)

Origin

Anteromedial edge of the descending ramus of the pubis

Insertion

Anterior medial surface of the tibia just below the condyle

Action

Adduction of the hip
Weak flexion of the knee
Internal rotation of the hip
Assists with flexion of the hip

Palpation

A thin tendon on anteromedial thigh with knee flexion and resisted adduction just posterior to adductor longus and medial to the semitendinosus

Innervation

Obturator nerve (L2–L4)

Application, strengthening, and flexibility

Also known as the adductor gracilis, this muscle performs the same function as the other adductors but adds some weak assistance to knee flexion.

The adductor muscles as a group (adductor magnus, adductor longus, adductor brevis, and gracilis) are called into action in horseback riding and in doing the breaststroke kick in swimming. Proper development of the adductor group prevents soreness after participation in these sports. The gracilis is strengthened with the same exercises as described for the other hip adductors. The gracilis may be stretched in a manner similar to the adductors, except that the knee must be extended.

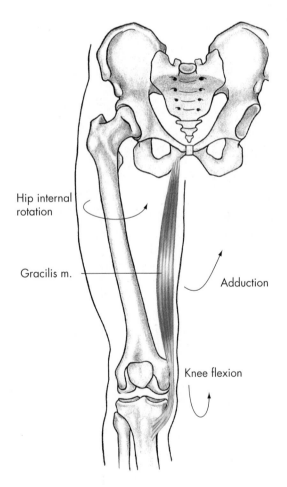

Hip internal rotation

Gracilis m.

Adduction

Knee flexion

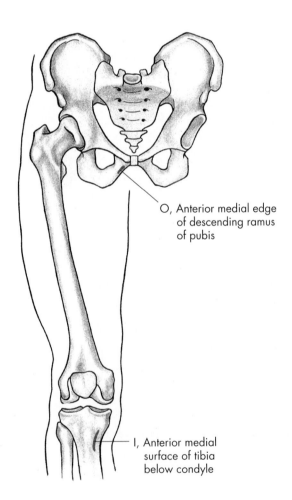

O, Anterior medial edge of descending ramus of pubis

I, Anterior medial surface of tibia below condyle

FIG. 9.36 ● Gracilis muscle, anterior view. *O,* Origin; *I,* Insertion.

Web sites

Radiologic Anatomy Browser

http://radlinux1.usuf1.usuhs.mil/rad/iong

Numerous radiological views of the musculoskeletal system.

University of Arkansas Medical School Gross Anatomy for Medical Students

http://anatomy.uams.edu/anatomyhtml/gross.html

Dissections, anatomy tables, atlas images, links, etc.

Loyola University Medical Center: Structure of the Human Body

www.meddean.luc.edu/lumen/meded/grossanatomy/index.htm

Excellent site with many slides, dissections, tutorials, etc., for the study of human anatomy.

Wheeless' Textbook of Orthopaedics

www.wheelessonline.com

An extensive index of links to the fractures, joints, muscles, nerves, trauma, medications, medical topics, and lab tests, as well as links to orthopaedic journals and other orthopaedic and medical news.

Premiere Medical Search Engine

www.medsite.com

This site allows the reader to enter any medical condition and it will search the net to find relevant articles.

Virtual Hospital

www.vh.org

Numerous slides, patient information, etc.

Arthroscopy.com

www.arthroscopy.com/sports.htm

Patient information on various musculoskeletal problems of the lower extremity.

Human Anatomy Online

www.innerbody.com/image/musc08.html

Interactive musculoskeletal anatomy.

The Hip and Knee Institute

www.hipsandknees.com/hip/contents.htm

Arthritis of the hip joint.

Adam Healthcare Center

http://adam.about.com/surgery/100006.htm

Hip joint replacement.

American Academy of Orthopaedic Surgeons

http://orthoinfo.aaos.org/category.cfm?topcategory=Hip

Patient education library on the hip.

Sports Injury Bulletin

www.sportsinjurybulletin.com/archive/1054-groin-strain.htm

Groin strain causes.

HealthGate Data Corp

http://healthgate.partners.org/browsing/browseContent.asp?fileName=11822.xml&title=Groin%20Strain

Groin strain.

The Physician and Sportsmedicine

www.physsportsmed.com/issues/2004/0104/meislin.htm

Symptomatic snapping hip.

Neurography Institute

www.neurography.com/Images/Piriformis/Piriformis1.htm

Piriformis syndrome.

Worksheet exercises

As an aid to learning, for in-class or out-of-class assignments, or for testing, tear-out worksheets are found at the end of the text (pp. 383 and 384).

Anterior skeletal worksheet (no. 1)

Draw and label on the worksheet the anterior hip joint and pelvic girdle muscles.

Posterior skeletal worksheet (no. 2)

Draw and label on the worksheet the posterior hip joint and pelvic girdle muscles.

LABORATORY AND REVIEW EXERCISES

1. Locate the following parts of the pelvic girdle and hip joint on a human skeleton and on a subject:

Chapter 9

a. Skeleton
   1. Ilium
   2. Ischium
   3. Pubis
   4. Symphysis pubis
   5. Acetabulum
   6. Rami (ascending and descending)
   7. Obturator foramen
   8. Ischial tuberosity
   9. Anterior superior iliac spine
   10. Greater trochanter
   11. Lesser trochanter
b. Subject
   1. Crest of ilium
   2. Anterior superior iliac spine
   3. Ischial tuberosity
   4. Greater trochanter

2. How and where can the following muscles be palpated on a human subject?
   a. Gracilis
   b. Sartorius
   c. Gluteus maximus
   d. Gluteus medius
   e. Gluteus minimus
   f. Biceps femoris
   g. Rectus femoris
   h. Semimembranosus
   i. Semitendinosus
   j. Adductor magnus
   k. Adductor longus
   l. Adductor brevis

3. Be prepared to indicate on a human skeleton, using a long rubber band, where each muscle has its origin and insertion.

4. Distinguish between hip flexion and trunk flexion.

5. Demonstrate the movement and list the muscles primarily responsible for the following hip movements:
   a. Flexion
   b. Extension
   c. Adduction
   d. Abduction
   e. External rotation
   f. Internal rotation

6. List the planes in which each of the following hip joint movements occur. List the respective axis of rotation for each movement in each plane.
   a. Flexion
   b. Extension
   c. Adduction
   d. Abduction

e. External rotation
f. Internal rotation

7. How is walking different from running in relation to the use of the hip joint muscle actions and the range of motion?

8. How may the walking gait be affected by a weakness in the gluteus medius muscle? Have a laboratory partner demonstrate the gait pattern associated with gluteus medius weakness. What is the name of this dysfunctional gait?

9. How might bilateral iliopsoas tightness affect the posture and movement of the lumbar spine in the standing position? Demonstrate and discuss this effect with a laboratory partner.

10. How might bilateral hamstring tightness affect the posture and movement of the lumbar spine in the standing position? Demonstrate and discuss this effect with a laboratory partner.

11. Research common hip disorders such as osteoarthritis, groin strains, hamstring strains, greater trochanteric bursitis, and slipped capital femoral epiphysis. Report your findings in class.

12. Fill in the muscle analysis chart (p. 255) by listing the muscles primarily involved in each movement.

13. Fill in the antagonistic muscle action chart (p. 255) by listing the muscle(s) or parts of muscles that are antagonistic in their actions to the muscles in the left column.

14. After analyzing each of the exercises in the hip joint movement analysis chart (p. 256), break each into two primary movement phases such as a lifting phase and lowering phase. For each phase, determine the hip joint movements occurring, and then list the hip joint muscles primarily responsible for causing/controlling those movements. Beside each muscle in each movement indicate the type of contraction as follows: I-isometric; C-concentric; E-eccentric.

15. Analyze each skill in the hip joint sport skill analysis chart (p. 256), and list the movements of the right and left hip joint in each phase of the skill. You may prefer to list the initial position the hip joint is in for the stance phase. After each movement, list the hip joint muscle(s) primarily responsible for causing/ controlling those movements. Beside each muscle in each movement indicate the type of contraction as follows: I-isometric, C-concentric; E-eccentric. It may be desirable to review the concepts for analysis in Chapter 8 for the various phases.

Muscle analysis chart • Hip joint

Flexion	Extension
Abduction	**Adduction**
External rotation	**Internal rotation**

Antagonistic muscle action chart • Hip joint and pelvic girdle

Agonist	Antagonist
Gluteus maximus	
Gluteus medius	
Gluteus minimus	
Biceps femoris	
Semimembranosus/ semitendinosus	
Adductor magnus/ adductor brevis	
Adductor longus	
Gracilis	
Lateral rotators	
Rectus femoris	
Sartorius	
Pectineus	
Iliopsoas	
Tensor fasciae latae	

Chapter

9

Hip joint movement analysis chart

Exercise	Initial movement phase		Secondary movement phase	
	Movement(s)	Agonist(s)–(contraction type)	Movement(s)	Agonist(s)–(contraction type)
Push-up				
Squat				
Dead lift				
Hip sled				
Forward lunge				
Rowing exercise				
Stair machine				

Hip joint sport skill analysis chart

Exercise		Stance phase	Preparatory phase	Movement phase	Follow-through phase
Baseball pitch	(R)				
	(L)				
Football punting	(R)				
	(L)				
Walking	(R)				
	(L)				
Softball pitch	(R)				
	(L)				
Soccer pass	(R)				
	(L)				
Batting	(R)				
	(L)				
Bowling	(R)				
	(L)				
Basketball jump shot	(R)				
	(L)				

References

Field D: *Anatomy: palpation and surface markings,* ed 3, Oxford, 2001, Butterworth-Heinemann.

Hamilton N, Luttgens K: *Kinesiology: scientific basis of human motion,* ed 10, Boston, 2002, McGraw-Hill.

Hislop HJ, Montgomery J: *Daniels and Worthingham's muscle testing: techniques of manual examination,* ed 7, Philadelphia, 2002, Saunders.

Kendall FP, McCreary EK, Provance, PG: *Muscles: testing and function,* ed 4, Baltimore, 1993, Lippincott Williams & Wilkins.

Lindsay DT: *Functional human anatomy,* St. Louis, 1996, Mosby.

Lysholm J, Wikland J: Injuries in runners, *American Journal of Sports Medicine* 15:168, September–October 1986.

Magee DJ: *Orthopedic physical assessment,* ed 4, Philadelphia 2002, Saunders.

Muscolino JE: *The muscular system manual: the skeletal muscles of the human body,* ed 2, St. Louis, 2003, Elsevier Mosby.

Noahes TD, et al: Pelvic stress fractures in long distance runners, *American Journal of Sports Medicine* 13:120, March–April 1985.

Oatis CA: *Kinesiology: the mechanics and pathomechanics of human movement,* Philadelphia, 2004, Lippincott Williams & Wilkins.

Chapter 9

Prentice WE: *Arnheim's principles of athletic training,* ed 12, New York, 2006, McGraw-Hill.

Seeley RR, Stephens TD, Tate P: *Anatomy & physiology,* ed 7, New York, 2006, McGraw-Hill.

Sieg KW, Adams SP: *Illustrated essentials of musculoskeletal anatomy,* ed 2, Gainesville, FL, 1985, Megabooks.

Stone RJ, Stone JA: *Atlas of the skeletal muscles,* New York, 1990, McGraw-Hill.

Thibodeau GA, Patton KT: *Anatomy & physiology,* ed 9, St. Louis, 1993, Mosby.

Van De Graaff KM: *Human anatomy,* ed 6, Dubuque, IA, 2002, McGraw-Hill.

Chapter

9

The Knee Joint

Objectives

- To identify on a human skeleton selected bony features of the knee

- To explain the cartilaginous and ligamentous structures of the knee joint

- To draw and label on a skeletal chart muscles and ligaments of the knee joint

- To palpate the superficial knee joint structures and muscles on a human subject

- To demonstrate and palpate with a fellow student all the movements of the knee joint and list their respective planes of motion and axes of rotation

- To name and explain the actions and importance of the quadriceps and hamstring muscles

- To list and organize the muscles that produce the movements of the knee joint and list their antagonists

Online Learning Center Resources

Visit *Manual of Structural Kinesiology's* Online Learning Center at **www.mhhe.com/floyd16e** for additional information and study material for this chapter, including:

- *Self-grading quizzes*
- *Anatomy flashcards*
- *Animations*

The knee joint is the largest joint in the body and is very complex. It is primarily a hinge joint. The combined functions of weight bearing and locomotion place considerable stress and strain on the knee joint. Powerful knee joint extensor and flexor muscles, combined with a strong ligamentous structure, provide a strong functioning joint in most instances.

Bones   FIG. 10.1

The enlarged femoral condyles articulate on the enlarged condyles of the tibia, somewhat in a horizontal line. Since the femur projects downward at an oblique angle toward the midline, its medial condyle is slightly larger than the lateral condyle.

The top of the medial and lateral tibial condyles, known as the medial and lateral tibial plateaus, serve as receptacles for the femoral condyles. The tibia is the medial bone in the leg and bears much more of the body's weight than the fibula. The fibula serves as the attachment for some very important knee joint structures, although it does not articulate with the femur or patella and is not part of the knee joint.

The patella is a sesamoid (floating) bone contained within the quadriceps muscle group and patellar tendon. Its location allows it to serve the quadriceps in a fashion similar to the work of a pulley by creating an improved angle of pull. This results in a greater mechanical advantage when performing knee extension.

Key bony landmarks of the knee include the superior and inferior poles of the patella, the tibial tuberosity, Gerdy's tubercle, the medial and lateral

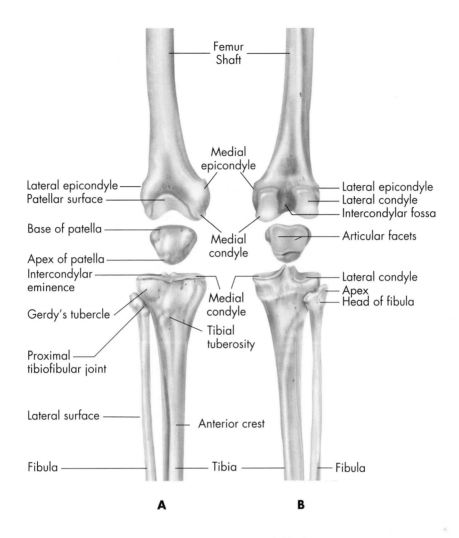

FIG. 10.1 ● Bones of the right knee—femur, patella, tibia, and fibula. **A,** Anterior view; **B,** Posterior view.

Modified for Prentice WE: *Arnheim's principles of athletic training,* ed 12, New York, 2006, McGraw-Hill; from Saladin, KS: *Anatomy & physiology: the unity of form and function,* ed 2, New York, 2001, McGraw-Hill.

femoral condyles, the upper anterior medial surface of the tibia, and the head of the fibula. The three vasti muscles of the quadriceps originate on the proximal femur and insert along with rectus femoris on the superior pole of the patella. Their specific insertion into the patella varies in that the vastus medialis and the vastus lateralis insert into the patella from a superomedial and superolateral angle, respectively. The superficial rectus femoris and the vastus intermedius, which lies directly beneath it, both attach to the patella from superior direction. From here their insertion is ultimately on the tibial tuberosity by way of the large patella tendon, which runs from the inferior patellar pole to the tibial tuberosity. Gerdy's tubercle, located on the anterolateral aspect of the lateral tibial condyle is the insertion point for the iliotibial tract of the tensor fasciae latae.

The upper anteromedial surface of the tibia just below the medial condyle serves as the insertion for the sartorius, gracilis, and semitendinosus. The semimembranosus inserts posteromedially on the medial tibial condyle. The head of the fibula is the primary location of the biceps femoris insertion, although some of its fibers insert on the lateral tibial condyle. The popliteus origin is located on the lateral aspect of the lateral femoral condyle.

Additionally, the tibial collateral ligament originates on the medial aspect of the upper medial femoral condyle and inserts on the medial surface of the tibia. Laterally, the shorter fibula collateral originates on the lateral femoral condyle very close to the popliteus origin and inserts on the head of the fibula.

Chapter

10

Joints FIGS. 10.2 and 10.3

The knee joint proper, or tibiofemoral joint, is classified as a ginglymus joint because it functions like a hinge. It moves between flexion and extension without side-to-side movement into abduction or adduction. However, it is sometimes referred to as a trochoginglymus joint because of the internal and external rotation movements that can occur during flexion. Some authorities argue that it should be classified as a condyloid or "double condyloid" joint due to its bicondylar structure. The patellofemoral joint is classified as an arthrodial joint due to the gliding nature of the patella on the femoral condyles.

The ligaments provide static stability to the knee joint, and contractions of the quadriceps and hamstrings produce dynamic stability. The surfaces between the femur and tibia are protected by articular cartilage, as is true of all diarthrodial joints. In addition to the articular cartilage covering the ends of the bones, there are specialized cartilages (see Fig. 10.2), known as the menisci, that form cushions between the bones. These menisci are attached to the tibia and deepen the tibial plateaus, thereby enhancing stability.

The medial semilunar cartilage or, more technically, the medial meniscus, is located on the medial tibial plateau to form a receptacle for the medial femoral condyle. The lateral semilunar cartilage (lateral meniscus) sits on the lateral tibial plateau to receive the lateral femoral condyle. Both of these menisci are thicker on the outside border and taper down to be very thin on the inside border. They can slip about slightly and are held in place by various small ligaments. The medial meniscus is the larger of the two and has a much more open C appearance than the rather closed C lateral meniscus configuration. Either or both of the menisci may be torn in several different areas from a variety of mechanisms, resulting in varying degrees of severity and problems. These injuries often occur due to the significant compression and shear forces that develop as the knee rotates while flexing or extending during quick directional changes in running.

Two very important ligaments of the knee are the anterior and posterior cruciate, so named because they cross within the knee between the tibia and the femur. These ligaments are vital in respectively maintaining the anterior and posterior stability of the knee joint, as well as the rotatory stability (see Fig. 10.2).

The anterior cruciate ligament (ACL) tear is one of the most common serious injuries to the knee and has been shown to be significantly more common in females than males during similar sports such as basketball or soccer. The mechanism of this injury often involves noncontact rotary forces associated with planting and cutting. Studies have also shown that the ACL maybe disrupted in a hyperextension mechanism or solely by a violent contraction of the quadriceps that pulls the tibia forward on the femur. Recent studies suggest that ACL injury prevention programs incorporating detailed conditioning exercises and techniques designed to improve neuromuscular coordination and control among the hamstrings and quadriceps, maintain proper knee alignment, and utilize proper landing techniques may be effective in reducing the likelihood of injury.

Fortunately, the posterior cruciate ligament (PCL) is not often injured. Injuries of the posterior cruciate usually come about through direct contact with an opponent or with the playing surface. Many of the PCL injuries that do occur are partial tears with minimal involvement of other knee structures. In many cases, even with complete tears, athletes may remain fairly competitive at a high level after a brief nonsurgical treatment and rehabilitation program.

On the medial side of the knee is the tibial (medial) collateral ligament (MCL) (see Fig. 10.2), which maintains medial stability by resisting valgus forces or preventing the knee joint from being abducted. Injuries to the tibial collateral occur quite commonly, particularly in contact or collision sports in which a teammate or an opponent falls against the lateral aspect of the knee or leg, causing medial opening of the knee joint and stress to the medial ligamentous structures. Its deeper fibers are attached to the medial meniscus, which may be affected with injuries to the ligament.

On the lateral side of the knee, the fibular (lateral) collateral ligament (LCL) joins the fibula and the femur. Injuries to this ligament are infrequent.

In addition to the other intraarticular ligaments detailed in Fig. 10.2, there are numerous other ligaments not shown that are contiguous with the joint capsule. Generally, these ligaments are of lesser importance and will not be discussed further.*

The knee joint, as shown in Fig. 10.3, is well supplied with synovial fluid from the synovial cavity, which lies under the patella and between the

---

*Most detailed discussion of the knee is found in anatomy texts and athletic training manuals.

Chapter **10**

Joints FIGS. 10.2 and 10.3

The knee joint proper, or tibiofemoral joint, is classified as a ginglymus joint because it functions like a hinge.

Chapter **10**

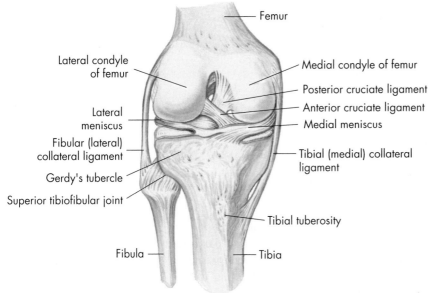

Anterior view with patella removed

Femur

Medial condyle of femur

Posterior cruciate ligament

Anterior cruciate ligament

Medial meniscus

Tibial (medial) collateral ligament

Tibial tuberosity

Tibia

Fibula

Superior tibiofibular joint

Gerdy's tubercle

Fibular (lateral) collateral ligament

Lateral meniscus

Lateral condyle of femur

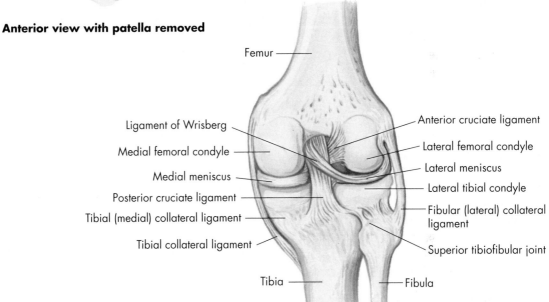

Posterior view

Femur

Ligament of Wrisberg

Medial femoral condyle

Medial meniscus

Posterior cruciate ligament

Tibial (medial) collateral ligament

Tibial collateral ligament

Tibia

Anterior cruciate ligament

Lateral femoral condyle

Lateral meniscus

Lateral tibial condyle

Fibular (lateral) collateral ligament

Superior tibiofibular joint

Fibula

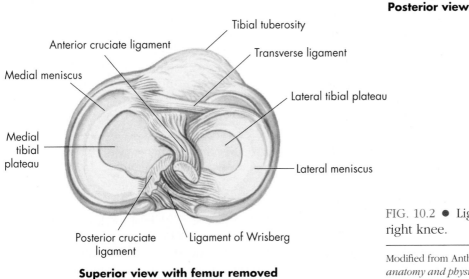

Anterior cruciate ligament

Medial meniscus

Medial tibial plateau

Tibial tuberosity

Transverse ligament

Lateral tibial plateau

Lateral meniscus

Posterior cruciate ligament

Ligament of Wrisberg

Superior view with femur removed

FIG. 10.2 ● Ligaments and menisci of the right knee.

Modified from Anthony CP, Kolthoff NJ: *Textbook of anatomy and physiology,* ed 9, St. Louis, 1975, Mosby.

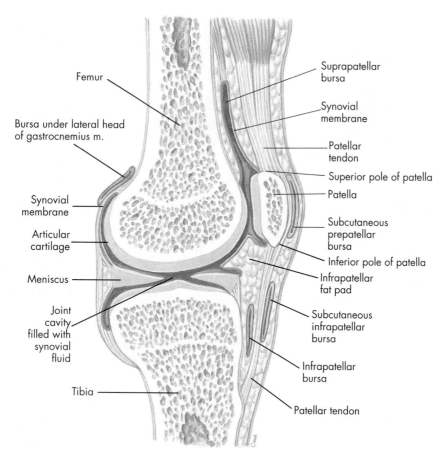

FIG. 10.3 ● Knee joint, sagittal view.

surfaces of the tibia and the femur. Commonly, this synovial cavity is called the capsule of the knee. Just posterior to the patellar tendon is the infrapatellar fat pad, which is often an insertion point for synovial folds of tissue known as **plica**. A plica is an anatomical variant among some individuals that may be irritated or inflamed with injuries or overuse of the knee. More than 10 bursae are located in the knee, some of which are connected to the synovial cavity. Bursae are located where they can absorb shock or prevent friction.

The knee can usually extend to 180 degrees or a straight line, although it is not uncommon for some knees to hyperextend up to 10 degrees or more. When the knee is in full extension, it can move from there to about 140 degrees of flexion. With the knee flexed 30 degrees or more, approximately 30 degrees of internal rotation and 45 degrees of external rotation can occur (Fig. 10.4).

Due to the shape of the medial femoral condyle, the knee must "screw home" to fully extend. As the knee approaches full extension, the tibia must externally rotate approximately 10 degrees to achieve proper alignment of the tibial and femoral condyles. In full extension, due to the

close congruency of the articular surfaces, there is no appreciable rotation of the knee. During initial flexion from a fully extended position, the knee "unlocks" by the tibia rotating internally, to a degree, from its externally rotated position to achieve flexion.

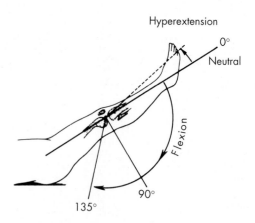

FIG. 10.4 ● Active motion of the knee. Flexion is measured in degrees from the zero starting position, which is an extended straight leg with the athlete either prone or supine. Hyperextension is measured with degrees opposite the zero starting point.

Movements FIG. 10.5

Flexion and extension of the knee occur in the sagittal plane, whereas internal and external rotation occur in the horizontal plane. The knee will not allow rotation unless it is flexed 20 to 30 degrees or more.

Flexion: bending or decreasing the angle between the femur and lower leg; characterized by the heel moving toward the buttocks

Extension: straightening or increasing the angle between the femur and the lower leg

External rotation: rotary movement of the lower leg laterally away from the midline

Internal rotation: rotary movement of the lower leg medially toward the midline

Muscles FIG 9.9

Some of the muscles involved in knee joint movements were discussed in Chapter 9 because of their biarticular arrangement with both the hip and knee joints. As a result, they will not be covered again fully in this chapter. The knee joint muscles that have already been addressed are

Knee extensor: rectus femoris

Knee flexors: sartorius, biceps femoris, semitendinosus, semimembranosus, and gracilis

The gastrocnemius muscle, discussed in Chapter 11, also assists minimally with knee flexion.

The muscle group that extends the knee is located in the anterior compartment of the thigh and is known as the quadriceps. It consists of four muscles: the rectus femoris, the vastus lateralis, the vastus intermedius, and the vastus medialis. All four muscles work together to pull the patella superiorly, which in turn pulls the leg into extension at the knee by its attachment to the tibial tuberosity via the patellar tendon.

The central line of pull for the entire quadriceps runs from the anterior superior iliac spine (ASIS) to the center of the patella. The line of pull of the patella tendon runs from the center of the patella to the center of the tibial tuberosity. The angle formed by the intersection of these two lines at the patella is known as the **Q angle** or quadriceps angle (Fig. 10.6). Normally, this angle will be 15 degrees or less for males and 20 degrees or less in females. Generally, females have higher angles due to a wider pelvis. Higher Q angles generally predispose people, in varying degrees, to a variety of potential knee problems including patellar subluxation or dislocation, patellar compression syndrome, chondromalacia, and ligamentous injuries.

The hamstring muscle group is located in the posterior compartment of the thigh and is responsible for knee flexion. The hamstrings consist of three muscles: the semitendinosus, the semimembranosus, and the

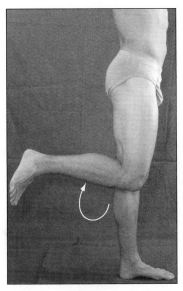
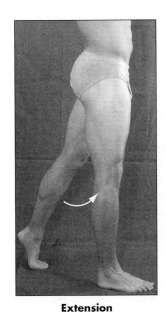
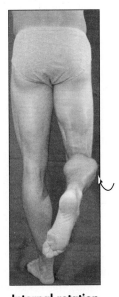
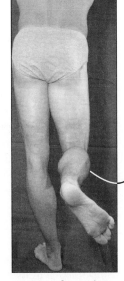

| **Flexion** | **Extension** | **Internal rotation** | **External rotation** |
| A | B | C | D |

FIG. 10.5 ● Movements of the knee.

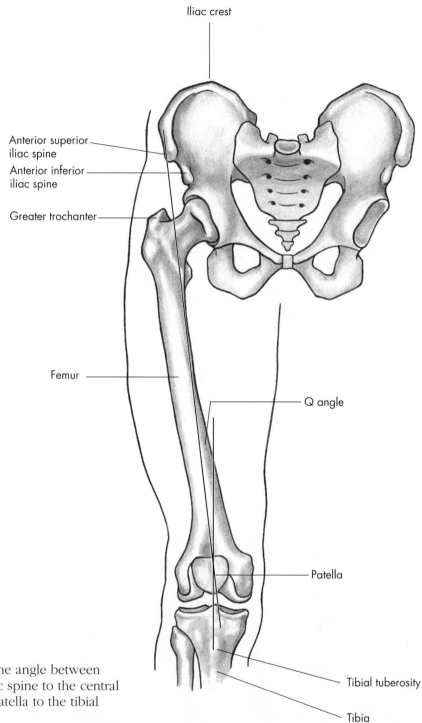

Iliac crest

Anterior superior
iliac spine

Anterior inferior
iliac spine

Greater trochanter

Femur

Q angle

Patella

Tibial tuberosity

Tibia

FIG. 10.6 ● Q angle; represented by the angle between the line from the anterior superior iliac spine to the central patella and the line from the central patella to the tibial tuberosity.

biceps femoris. The semimembranosus and semitendinosus muscles (medial hamstrings) are assisted by the popliteus in internally rotating the knee, whereas the biceps femoris (lateral hamstring) is responsible for knee external rotation.

Two-joint muscles are most effective when either the origin or the insertion is stabilized to prevent movement in the direction of the mus-

cle when it contracts. Additionally, muscles are able to exert greater force when lengthened than when shortened. All of the hamstring muscles, as well as the rectus femoris, sartorius, and gracilis, are biarticular (two-joint) muscles.

As an example, the sartorius muscle becomes a better flexor at the knee when the pelvis is rotated posteriorly and stabilized by the abdominal muscles, thus increasing its total length by moving its origin farther from its insertion. This is exemplified by trying to flex the knee and cross the legs in the sitting position. One usually leans backward to flex the legs at the knees. Again, this is illustrated by kicking a football. The kicker invariably leans well backward to raise and fix the origin of the rectus femoris muscle to make it more effective as an extensor of the leg at the knee. Also, when youngsters hang by the knees, they flex the hips to fix or raise the origin of the hamstrings to make the latter more effective flexors of the knees.

The sartorius, gracilis, and semitendinosus all join together distally to form a tendinous expansion, known as the **pes anserinus**, which attaches to the anteromedial aspect of the proximal tibia below the level of the tibial tuberosity. This attachment and the line of pull these muscles have posteromedially to the knee enable them to assist with knee flexion, particularly once the knee is flexed and the hip is externally rotated. The medial and lateral heads of the gastrocnemius attach posteriorly on the medial and lateral femoral condyles, respectively. This relationship to the knee provides the gastrocnemius a line of pull to assist with knee flexion.

Knee joint muscles—location

Muscle location closely relates to muscle function with the knee. While viewing the muscles, in Fig. 9.9, correlate them with Table 10.1.

Anterior
  Primarily knee extension
    Rectus femoris*†
    Vastus medialis
    Vastus intermedius
    Vastus lateralis

---

*Two-joint muscles; hip actions are discussed in Chapter 9 and ankle actions are discussed in Chapter 11.
†Two-joint muscles.

TABLE 10.1 • Agonist muscles of the knee joint

	Muscle	Origin	Insertion	Action	Plane of motion	Palpation	Innervation
Anterior aspect	Rectus femoris	Anterior inferior iliac spine of the ilium and groove (posterior) above the acetabulum	Superior aspect of the patella tendon to the tibial tuberosity	Extension of the knee / Flexion of the hip / Anterior pelvic rotation	Sagittal	Straight down anterior thigh from anterior inferior iliac spine to patella with resisted hip flexion/knee extension	Femoral nerve (L2–L4)
	Vastus intermedius	Upper 2/3 of anterior surface of femur	Upper border of patella and patellar tendon to tibial tuberosity	Extension of the knee	Sagittal	Anteromedial distal 1/3 of thigh just above the superomedial patella and deep to the rectus femoris, with extension of the knee, particularly full extension against resistance	Femoral nerve (L2–L4)
	Vastus lateralis (externus)	Intertrochanteric line, anterior and inferior borders of the greater trochanter, gluteal tuberosity, upper 1/2 of the linea aspera and entire lateral intermuscular septum	Lateral border of the patella and patellar tendon to the tibial tuberosity	Extension of the knee	Sagittal	Slightly distal to greater trochanter down the anterolateral aspect of the thigh to the superolateral patella, with extension of the knee, particularly full extension against resistance	Femoral nerve (L2–L4)
	Vastus medalis (internus)	Whole length of linea aspera and medial condyloid ridge	Medial 1/2 of upper border of patella and patellar tendon to tibial tuberosity	Extension of the knee	Sagittal	Anterior medial side of the thigh just above the superomedial patella, with extension of the knee, particularly full extension against resistance	Femoral nerve (L2–L4)

Chapter
10

TABLE 10.1 (continued) • Agonist muscles of the knee joint

	Muscle	Origin	Insertion	Action	Plane of motion	Palpation	Innervation
Posterior muscles	Biceps femoris	Long head: ischial tuberosity. Short head: lower half of the linea aspera, and lateral condyloid ridge	Head of the fibula and lateral condyle of the tibia	Flexion of the knee	Sagittal	Posterolateral aspect of distal thigh with combined knee flexion and external rotation against resistance; just distal to the ischial tuberosity in a prone position with hip internally rotated during active knee flexion	Long head: sciatic nerve—tibial division (S1–S3). Short head: sciatic nerve—peroneal division (L5, S1, S2)
				Extension of the hip			
				Posterior pelvic rotation			
				External rotation of the knee	Transverse		
				External rotation of the hip			
	Popliteus	Posterior surface of lateral condyle of femur	Upper posterior medial surface of tibia	Internal rotation of the knee as it flexes	Transverse	With subject sitting, knee flexed 90 degrees, palpate deep to the gastrocnemius medially on the posterior proximal tibia and proceed superolaterally toward lateral epicondyle of tibia just deep to fibular collateral ligament, while subject internally rotates knee	Tibial nerve (L5, S1)
				Flexion of the knee	Sagittal		
	Semi-membranosus	Ischial tuberosity	Postero-medial surface of the medial tibial condyle	Extension of the hip	Sagittal	Largely covered by other muscles, tendon can be felt at posteromedial aspect of knee just deep to semitendinosus tendon with combined knee flexion and internal rotation against resistance	Sciatic nerve—tibial division (L5, S1, S2)
				Flexion of the knee			
				Posterior pelvic rotation			
				Internal rotation of the hip	Transverse		
				Internal rotation of the knee			
	Semi-tendinosus	Ischial tuberosity	Upper anterior medial surface of the tibia just below the condyle	Extension of the hip	Sagittal	Posteromedial aspect of the distal thigh with combined knee flexion and internal rotation against resistance; just distal to ischial tuberosity in a prone position with hip internally rotated during active knee flexion	Sciatic nerve—tibial division (L5, S1, S2)
				Flexion of the knee			
				Posterior pelvic rotation			
				Internal rotation of the hip	Transverse		
				Internal rotation of the knee			

Posterior
    Primarily knee flexion
        Biceps femoris*†
        Semimembranosus*†
        Semitendinosus*†
        Sartorius*†
        Gracilis*†
        Popliteus
        Gastrocnemius*†

---

*Two-joint muscles; hip actions are discussed in Chapter 9 and ankle actions are discussed in Chapter 11.
†Two-joint muscles.

Nerves

The femoral nerve (Fig. 9.18) innervates the knee extensors—rectus femoris, vastus medialis, vastus intermedius, and vastus lateralis. The knee flexors, consisting of the semitendinosus, semimembranosus, biceps femoris (long head), and popliteus are innervated by the tibial division of the sciatic nerve (Fig. 9.20). The biceps femoris short head is supplied by the peroneal nerve (Fig. 11.9).

Quadriceps muscles   FIG. 10.7

(kwod´ri-seps)

The ability to jump is essential in nearly all sports. Individuals who have good jumping ability always have strong quadriceps muscles that extend the leg at the knee. The quadriceps function as a decelerator when it is necessary to decrease speed for changing direction or to prevent falling when landing. This deceleration function is also evident in stopping the body when coming down from a jump. The contraction that occurs in the quadriceps during braking or decelerating actions is eccentric. This eccentric action of the quadriceps controls the slowing of movements initiated in previous phases of the sports skill.

The muscles are the rectus femoris (the only two-joint muscle of the group), vastus lateralis (the largest muscle of the group), vastus intermedius, and vastus medialis. All attach to the patella and by the patellar tendon to the tuberosity of the tibia. All are superficial and palpable, except the vastus intermedius, which is under the rectus femoris. The vertical jump is a simple test that may be used to indicate the strength or power of the quadriceps. This muscle group is generally desired to be 25% to 33% stronger than the hamstring muscle group (knee flexors).

Development of strength and endurance of the quadriceps or "quads" is essential for maintenance of patellofemoral stability, which is often a problem in many physically active individuals. This problem is further perpetuated due to the quads being particularly prone to atrophy when injuries occur. The muscle of the quadriceps may be developed by resisted knee extension activities from a seated position. Performing functional weight-bearing activities such as step-ups or squats is particularly useful for strengthening and endurance.

Rectus femoris muscle   FIG. 9.23

(rek´tus fem´o-ris)

Origin
Anterior inferior iliac spine of the ilium and superior margin of the acetabulum

Insertion
Superior aspect of the patella and patellar tendon to the tibial tuberosity

Action
Flexion of the hip
Extension of the knee
Anterior pelvic rotation

Palpation
Straight down anterior thigh from anterior inferior iliac spine to patella, with resisted knee extension and hip flexion

Innervation
Femoral nerve (L2–L4)

Application, strengthening, and flexibility
When the hip is flexed, the rectus femoris becomes shorter, which reduces its effectiveness as an extensor of the knee. The work is then done primarily by the three vasti muscles.

Also see the rectus femoris discussion in Chapter 9, p. 239 (Fig. 9.23).

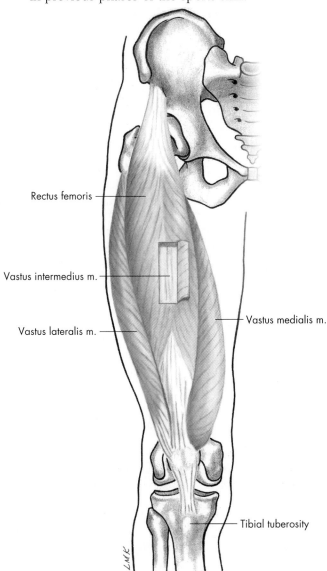

Rectus femoris

Vastus intermedius m.

Vastus lateralis m.

Vastus medialis m.

Tibial tuberosity

FIG. 10.7 ● Quadriceps muscle group.

Chapter
10

Vastus lateralis (externus) muscle

FIG. 10.8

(vas´tus lat-er-a´lis)

Origin

Intertrochanteric line, anterior and inferior borders of the greater trochanter, gluteal tuberosity, upper half of the linea aspera and entire lateral intermuscular septum

Insertion

Lateral border of the patella and patellar tendon to the tibial tuberosity

Action

Extension of the knee

Palpation

Slightly distal to greater trochanter down the anterolateral aspect of the thigh to the superolateral patella, with extension of the knee, particularly full extension against resistance

Innervation

Femoral nerve (L2–L4)

Application, strengthening, and flexibility

All three of the vasti muscles function with the rectus femoris in knee extension. They are typically used in walking and running and must be used to keep the knees straight, as in standing. The vastus lateralis has a slightly superior lateral pull on the patella and, as a result, is occasionally blamed in part for common lateral patellar subluxation and dislocation problems.

The vastus lateralis is strengthened through knee extension activities against resistance. Stretching occurs by pulling the knee into maximum flexion, such as by standing on one leg and pulling the heel of the other leg to the buttocks.

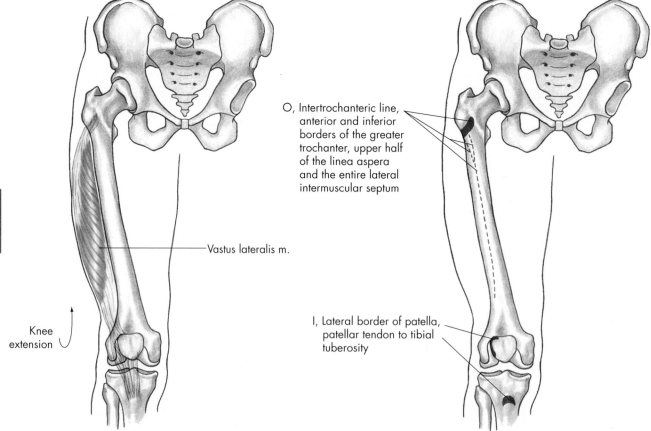

O, Intertrochanteric line, anterior and inferior borders of the greater trochanter, upper half of the linea aspera and the entire lateral intermuscular septum

Vastus lateralis m.

Knee extension

I, Lateral border of patella, patellar tendon to tibial tuberosity

FIG. 10.8 ● Vastus lateralis muscle, anterior view. *O,* Origin; *I,* Insertion.

Chapter
10

Vastus intermedius muscle FIG. 10.9

(vas´tus in´ter-me´di-us)

Origin
Upper two-thirds of the anterior surface of the femur

Insertion
Upper border of the patella and patellar tendon to the tibial tuberosity

Action
Extension of the knee

Palpation
Anteromedial distal one-third of thigh just above the superomedial patella and deep to the rectus femoris, with extension of the knee, particularly full extension against resistance

Innervation
Femoral nerve (L2–L4)

Application, strengthening, and flexibility

The three vasti muscles all contract in knee extension. They are used together with the rectus femoris in running, jumping, hopping, skipping, and walking. The vasti muscles are primarily responsible for extending the knee while the hip is flexed or being flexed. Thus, in doing a knee bend with the trunk bent forward at the hip, the vasti are exercised with little involvement of the rectus femoris. These natural activities mentioned develop the quadriceps.

Squats with a barbell of varying weights on the shoulders, depending on strength, are an excellent exercise for developing the quadriceps if done properly. Caution should be used, along with strict attention to proper technique, to avoid injuries to the knees and lower back. Leg press exercises and knee extensions with weight machines are other good exercises. Full knee flexion stretches all of the quadriceps musculature.

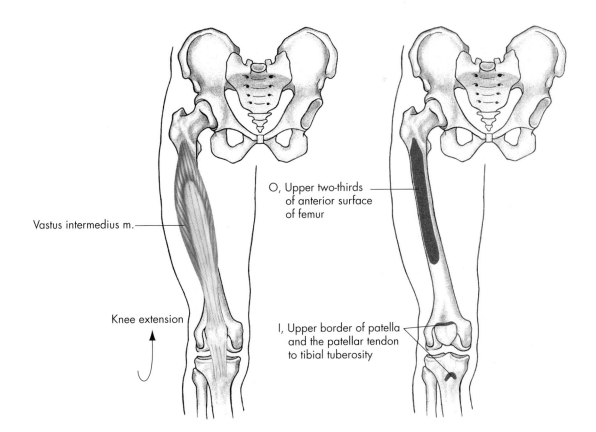

Vastus intermedius m.

Knee extension

O, Upper two-thirds of anterior surface of femur

I, Upper border of patella and the patellar tendon to tibial tuberosity

FIG. 10.9 ● Vastus intermedius muscle, anterior view. *O,* Origin; *I,* Insertion.

Vastus medialis (internus) muscle

FIG. 10.10

(vas′tus me-di-a′lis)

Origin

Whole length of the linea aspera and the medial condyloid ridge

Insertion

Medial half of the upper border of the patella and patellar tendon to the tibial tuberosity

Action

Extension of the knee

Palpation

Anterior medial side of the thigh just above the superomedial patella, with extension of the knee, particularly full extension against resistance

Innervation

Femoral nerve (L2–L4)

Application, strengthening, and flexibility

The vastus medialis is thought to be very important in maintaining patellofemoral stability because of the oblique attachment of its distal fibers to the superior medial patella. This portion of the vastus medialis is referred to as the vastus medialis obliquus (VMO). The vastus medialis is strengthened similarly to the other quadriceps muscles by squats, knee extensions, and leg presses, but the VMO is not really emphasized until the last 10 to 20 degrees of knee extension. Full knee flexion stretches all of the quadriceps muscles.

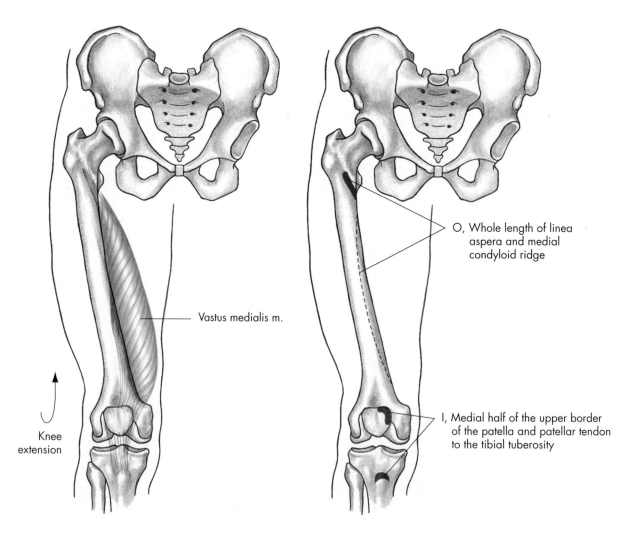

Vastus medialis m.

Knee extension

O, Whole length of linea aspera and medial condyloid ridge

I, Medial half of the upper border of the patella and patellar tendon to the tibial tuberosity

FIG. 10.10 ● Vastus medialis muscle, anterior view. *O,* Origin; *I,* Insertion.

Hamstring muscles   FIG. 10.11

The hamstring muscle group, consisting of the biceps femoris, semimembranosus, and semitendinosus, is covered in complete detail in Chapter 9, but further discussion is included here because of its importance in knee function.

Muscle strains involving the hamstrings are very common in football and other sports that require explosive running. This muscle group is often referred to as the running muscle because of its function in acceleration. The hamstring muscles are antagonists to the quadriceps muscles at the knee and are named for their cordlike attachments at the knee. All of the hamstring muscles originate on the ischial tuberosity of the pelvic bone. The semitendinosus and semimembranosus insert on the anteromedial and posteromedial side of the tibia, respectively. The biceps femoris inserts on the lateral tibial condyle and head of the fibula—hence the saying "Two to the inside and one to the outside." The short head of the biceps femoris originates on the linea aspera of the femur.

Special exercises to improve the strength and flexibility of this muscle group are important in decreasing knee injuries. Inability to touch the floor with the fingers when the knees are straight is largely a result of a lack of flexibility of the hamstrings. The hamstrings may be strengthened by performing knee or hamstring curls on a knee table against resistance. Additionally, tight or inflexible hamstrings are also contributing factors in painful conditions involving the low back and knee. The flexibility of these muscles may be improved by performing slow, static stretching exercises, such as flexing the hip slowly while maintaining knee extension in a long sitting position.

The hamstrings are primarily knee flexors in addition to serving as hip extensors. Rotation of the knee can occur when it is in a flexed position. Knee rotation is brought about by the hamstring muscles. The biceps femoris externally rotates the lower leg at the knee. The semitendinosus and semimembranosus perform internal rotation. Rotation of the knee permits pivoting movements and change in direction of the body. This rotation of the knee is vital in accommodating to forces developing at the hip or ankle during directional changes in order to make the total movement more functional as well as more fluid in appearance.

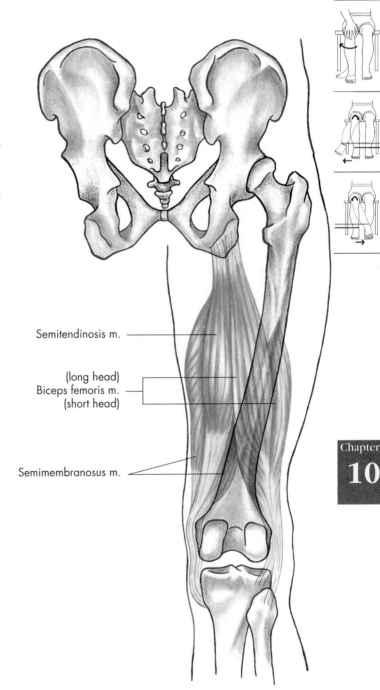

Semitendinosis m.

(long head)
Biceps femoris m.
(short head)

Semimembranosus m.

FIG. 10.11 ● The hamstring muscle group.

Popliteus muscle FIG. 10.12

(pop´li-te´us)

Origin

Posterior surface of the lateral condyle of the femur

Insertion

Upper posterior medial surface of the tibia

Action

Flexion of the knee

Internal rotation of the knee as it flexes

Palpation

With subject sitting, knee flexed 90 degrees, palpate deep to the gastrocnemius medially on the posterior proximal tibia and proceed superolaterally toward lateral epicondyle of tibia just deep to fibular collateral ligament, while subject internally rotates knee.

Innervation

Tibial nerve (L5, S1)

Application, strengthening, and flexibility

The popliteus muscle is the only true flexor of the leg at the knee. All other flexors are two-joint muscles. The popliteus is vital in providing posterolateral stability to the knee. It assists the medial hamstrings in internal rotation of the lower leg at the knee and is crucial in internally rotating the knee to unlock it from the "screwed home" full extension position.

Hanging from a bar with the legs flexed at the knee strenuously exercises the popliteus muscle. Also, the less strenuous activities of walking and running exercise this muscle. Specific efforts to strengthen this muscle combine knee internal rotation and flexion exercises against resistance. Stretching of the popliteus is difficult but may be done through passive full-knee extension without flexing the hip. Passive maximum external rotation with the knee flexed approximately 20 to 30 degrees also stretches the popliteus.

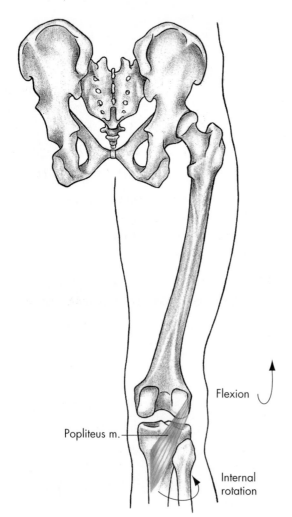

Popliteus m.

Flexion

Internal rotation

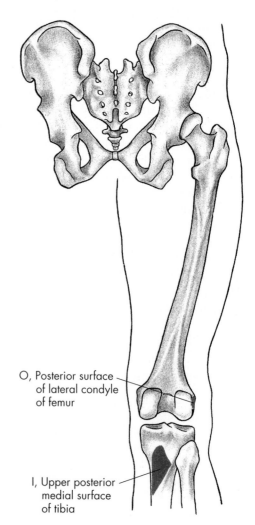

O, Posterior surface of lateral condyle of femur

I, Upper posterior medial surface of tibia

FIG. 10.12 ● Popliteus muscle, posterior view. *O,* Origin; *I,* Insertion.

Web sites

Radiologic Anatomy Browser

http://radlinux1.usuf1.usuhs.mil/rad/iong

    Numerous radiological views of the musculoskeletal system.

University of Arkansas Medical School Gross Anatomy for Medical Students

http://anatomy.uams.edu/anatomyhtml/gross.html

    Dissections, anatomy tables, atlas images, links, etc.

Loyola University Medical Center: Structure of the Human Body

www.meddean.luc.edu/lumen/meded/grossanatomy/index.htm

    An excellent site with many slides, dissections, tutorials, etc., for the study of human anatomy.

Wheeless' Textbook of Orthopaedics

www.wheelessonline.com

    Extensive index of links to the fractures, joints, muscles, nerves, trauma, medications, medical topics and lab tests, as well as links to orthopaedic journals and other orthopaedic, and medical news.

Premiere Medical Search Engine

www.medsite.com

    Allows the reader to enter any medical condition, and it will search the net to find relevant articles.

Arthroscopy.com

www.arthroscopy.com/sports.htm

    Patient information on various musculoskeletal problems of the upper and lower extremity.

Virtual Hospital

www.vh.org

    Numerous slides, patient information, etc.

Human Anatomy Online

www.innerbody.com/image/musc08.html

    Interactive musculoskeletal anatomy.

The Hip and Knee Institute

www.hipsandknees.com/knee/index.html

    Arthritis of the knee joint.

Adam Healthcare Center

http://adam.about.com/surgery/100088.htm#

    Knee joint replacement.

American Academy of Orthopaedic Surgeons

http://orthoinfo.aaos.org/category.cfm?topcategory=Knee

    Patient education library on the knee.

Edheads Activities

www.edheads.org/activities/knee

    Allows you to perform virtual knee surgery.

Gross Anatomy: The Functional Anatomy of the Knee Joint

www.upstate.edu/cdb/grossanat/limbs8.shtml

    Functional anatomy of the knee.

Knee Ligament Anatomy and Injury

www.orthoassociates.com/knee_lig.htm

    Anatomy and injuries of the knee and its ligaments.

Duke Orthopaedics

www.wheelessonline.com/ortho/anatomy_and_kinematics_of_the_knee_joint

    Anatomy and kinematics of the knee joint.

Knee Injury: Meniscus

www.patient.co.uk/showdoc/27000672

    Understanding the knee joint and purpose of meniscus.

Smart Play: The Knee

www.smartplay.net/ouch/bodybits/b_bitsknee.html

    Anatomy, functions, injuries, etc. of the knee.

Patellofemoral Instability

www.massgeneral.org/ortho/PatellofemoralInstability.htm

    Patella femoral alignment.

Chiroweb.com

www.chiroweb.com/archives/21/24/03.html

    Abnormal Q angle and orthotic support.

The Physician and Sportsmedicine

www.physsportsmed.com/issues/1997/05may/bach.htm

    Acute knee injuries: when to refer.

The Physician and Sportsmedicine

www.physsportsmed.com/issues/1999/10_01_99/laprade.htm

    Acute knee injuries: On-the-field and sideline evaluation.

Worksheet exercise

As an aid to learning, for in-class or out-of-class assignments, or for testing, a tear-out worksheet is found at the end of the text (p. 385).

Skeletal worksheet (no. 1)

Draw and label on the worksheet the knee joint muscles.

LABORATORY AND REVIEW EXERCISES

1. Locate the following parts of bones on a human skeleton and on a subject:
   a. Skeleton
      1. Head and neck of femur
      2. Greater trochanter
      3. Shaft of femur
      4. Lesser trochanter
      5. Linea aspera
      6. Adductor tubercle
      7. Medial femoral condyle
      8. Lateral femoral condyle
      9. Patella
      10. Fibula head
      11. Medial tibial condyle
      12. Lateral tibial condyle
      13. Tibial tuberosity
      14. Gerdy's tubercle
   b. Subject
      1. Greater trochanter
      2. Adductor tubercle
      3. Medial femoral condyle
      4. Lateral femoral condyle
      5. Patella
      6. Fibula head
      7. Medial tibial condyle
      8. Lateral tibial condyle
      9. Tibial tuberosity
      10. Gerdy's tubercle

2. How and where can the following muscles be palpated on a human subject?
   NOTE: Palpate the previously studied hip joint muscles while they are performing actions at the knee.
   a. Gracilis   f. Rectus femoris
   b. Sartorius   g. Vastus lateralis
   c. Biceps femoris   h. Vastus intermedius
   d. Semitendinosus   i. Vastus medialis
   e. Semimembranosus   j. Popliteus

3. Be prepared to indicate on a human skeleton, by using a long rubber band, the origin and insertion of the muscles just listed.

4. Demonstrate the following movements and list the muscles primarily responsible for each.
   a. Extension of the leg at the knee
   b. Flexion of the leg at the knee
   c. Internal rotation of the leg at the knee
   d. External rotation of the leg at the knee

5. List the planes in which each of the following movements occurs. List the respective axis of rotation for each movement in each plane.
   a. Extension of the leg at the knee
   b. Flexion of the leg at the knee
   c. Internal rotation of the leg at the knee
   d. External rotation of the leg at the knee

6. With a laboratory partner, determine how and why maintaining the position of full-knee extension limits the ability to maximally flex the hip both actively and passively. Does maintaining excessive hip flexion limit your ability to accomplish full knee extension?

7. With a laboratory partner, determine how and why maintaining the position of full-knee flexion limits the ability to maximally extend the hip both actively and passively. Does maintaining excessive hip extension limit your ability to accomplish full-knee flexion?

8. Compare and contrast the bony, ligamentous, articular, and cartilaginous aspects of the medial knee joint versus the lateral knee joint.

9. Research the acceptability of deep knee-bends and duck-walk activities in a physical education program and report your findings in class.

10. Prepare a report on the knee on one of the following topics: anterior cruciate ligament injuries, meniscal injuries, medial collateral ligament injuries, patellar tendon tendonitis, plica syndrome, anterior knee pain, osteochondritis dissecans, patella subluxation/dislocation, knee bracing, quadriceps rehabilitation.

11. Research preventive and rehabilitative exercises to strengthen the knee joint and report your findings in class.

12. Which muscle group about the knee would be most important to develop for an athlete with a torn anterior cruciate ligament and why? A torn posterior cruciate ligament? Why?

13. Fill in the muscle analysis chart by listing the muscles primarily involved in each joint movement.

14. Fill in the antagonistic muscle action chart (p. 275) by listing the muscle(s) or parts of muscles that are antagonistic in their actions to the muscles in the left column.

Muscle analysis chart • Knee joint

Flexion	Extension
Internal rotation	External rotation

Antagonistic muscle action chart • Knee joint

Agonist	Antagonist
Biceps femoris	
Semitendinosus	
Semimembranosus	
Rectus femoris	
Vastus lateralis	
Vastus intermedius	
Vastus medialis	
Rectus femoris	
Popliteus	

15. After analyzing each of the exercises in the knee joint movement analysis chart below, break each into two primary movement phases such as a lifting phase and lowering phase. For each phase, determine the knee joint movements occurring, and then list the knee joint muscles primarily responsible for causing/controlling those movements. Beside each muscle in each movement indicate the type of contraction as follows: I-isometric; C-concentric; E-eccentric.

Knee joint movement analysis chart

Exercise	Initial movement phase		Secondary movement phase	
	Movement(s)	Agonist(s)–(contraction type)	Movement(s)	Agonist(s)–(contraction type)
Push-up				
Squat				
Dead lift				
Hip sled				
Forward lunge				
Rowing exercise				
Stair machine				

16. Analyze each skill in the knee joint sport skill analysis chart and list the movements of the right and left knee joint in each phase of the skill. You may prefer to list the initial position the knee joint is in for the stance phase. After each movement, list the knee joint muscle(s) primarily responsible for causing/controlling those movements. Beside each muscle in each movement indicate the type of contraction as follows: I-isometric; C-concentric; E-eccentric. It may be desirable to review the concepts for analysis in Chapter 8 for the various phases.

Knee joint sport skill analysis chart

Exercise		Stance phase	Preparatory phase	Movement phase	Follow-through phase
Baseball pitch	(R)				
	(L)				
Football punting	(R)				
	(L)				
Walking	(R)				
	(L)				
Softball pitch	(R)				
	(L)				
Soccer pass	(R)				
	(L)				
Batting	(R)				
	(L)				
Bowling	(R)				
	(L)				
Basketball jump shot	(R)				
	(L)				

References

Baker BE, et al: Review of meniscal injury and associated sports, *American Journal of Sports Medicine* 13:1, January–February 1985.

Field D: *Anatomy: palpation and surface markings,* ed 3, Oxford, 2001, Butterworth-Heinemann.

Garrick JG, Regna RK: Prophylactic knee bracing, *American Journal of Sports Medicine* 15:471, September–October 1987.

Hamilton N, Luttgens K: *Kinesiology: scientific basis of human motion,* ed 10, Boston, 2002, McGraw-Hill.

Hislop HJ, Montgomery J: *Daniels and Worthingham's muscle testing: techniques of manual examination,* ed 7, Philadelphia, 2002, Saunders.

Kelly DW, et al: Patellar and quadriceps tendon ruptures—jumping knee, *American Journal of Sports Medicine* 12:375, September–October 1984.

Lysholm J, Wikland J: Injuries in runners, *American Journal of Sports Medicine* 15:168, September–October 1986.

Magee DJ: *Orthopedic physical assessment,* ed 4, Philadelphia, 2002, Saunders.

Muscolino JE: *The muscular system manual: the skeletal muscles of the human body,* ed 2, St. Louis, 2003, Elsevier Mosby.

Oatis CA: *Kinesiology: the mechanics and pathomechanics of human movement,* Philadelphia, 2004, Lippincott Williams & Wilkins.

Prentice WE: *Arnheim's principles of athletic training,* ed 12, New York, 2006, McGraw-Hill.

Seeley RR, Stephens TD, Tate P: *Anatomy & physiology,* ed 7, New York, 2006, McGraw-Hill.

Sieg KW, Adams SP: *Illustrated essentials of musculoskeletal anatomy,* ed 2, Gainesville, FL, 1985, Megabooks.

Stone RJ, Stone JA: *Atlas of the skeletal muscles,* New York, 1990, McGraw-Hill.

Van De Graaff KM: *Human anatomy,* ed 6, Dubuque, IA, 2002, McGraw-Hill.

Wroble RR, et al: Pattern of knee injuries in wrestling, a six-year study, *American Journal of Sports Medicine* 14:55, January–February 1986.

Chapter

10

Chapter 11

The Ankle and Foot Joints

Objectives

- To identify on a human skeleton the most important bone features, ligaments, and arches of the ankle and foot

- To draw and label on a skeletal chart the muscles of the ankle and foot

- To demonstrate and palpate with a fellow student the movements of the ankle and foot and list their respective planes of motion and axes of rotation

- To palpate the superficial joint structures and muscles of the ankle and foot on a human subject

- To list and organize the muscles that produce movement of the ankle and foot and list their antagonists

The complexity of the foot is evidenced by the 26 bones, 19 large muscles, many small (intrinsic) muscles, and more than 100 ligaments that make up its structure.

Support and propulsion are the two functions of the foot. Proper functioning and adequate development of the muscles of the foot and practice of proper foot mechanics are essential for everyone. In our modern society, foot trouble is one of the most common ailments. Quite often, people develop poor foot mechanics or gait abnormalities secondary to improper shoe wear or other relatively minor problems. Poor foot mechanics early in life inevitably leads to foot discomfort in later years.

Walking and running may be divided into stance and swing phases (Fig. 11.1). The **stance** phase is further divided into three components—heel-strike, midstance, and toe-off. Midstance may be further separated into loading response, midstance, and terminal stance. Normally, **heel-strike** is characterized by landing on the heel with the foot in supination and the leg in external rotation, followed immediately by pronation and internal rotation of the foot and leg, respectively, during **midstance**. The foot returns to supination, and the leg returns to external rotation immediately prior to and during **toe-off**. The **swing** phase occurs when the foot leaves the ground, and the leg moves forward to another point of contact. The swing phase may be divided into initial swing, midswing, and terminal swing. Problems often arise when the foot is too rigid and does not pronate adequately or when the foot remains in pronation past midstance. If the foot remains too rigid and does not pronate adequately, then impact forces will not be absorbed through the gait, resulting in shock being transmitted up the kinetic chain. If the foot overpronates or remains in pronation too much past midstance, then propulsive forces are diminished and additional stresses are placed on the kinetic chain. Walking differs from running in that one foot is always in contact with the ground, and there is a point at which both feet contact the ground, whereas in running there is a point when neither foot is in contact with the ground, and both feet are never in contact with the ground at the same time.

The fitness revolution that has occurred during the past three decades has resulted in great improvements in shoes available for sports and recreational activities. In the past, a pair of sneakers would suffice for most activities. Now there are basketball, baseball, football, jogging, soccer, tennis, walking, and cross-training shoes. Good shoes are important, but there is no substitute for adequate muscular development, strength, and proper foot mechanics.

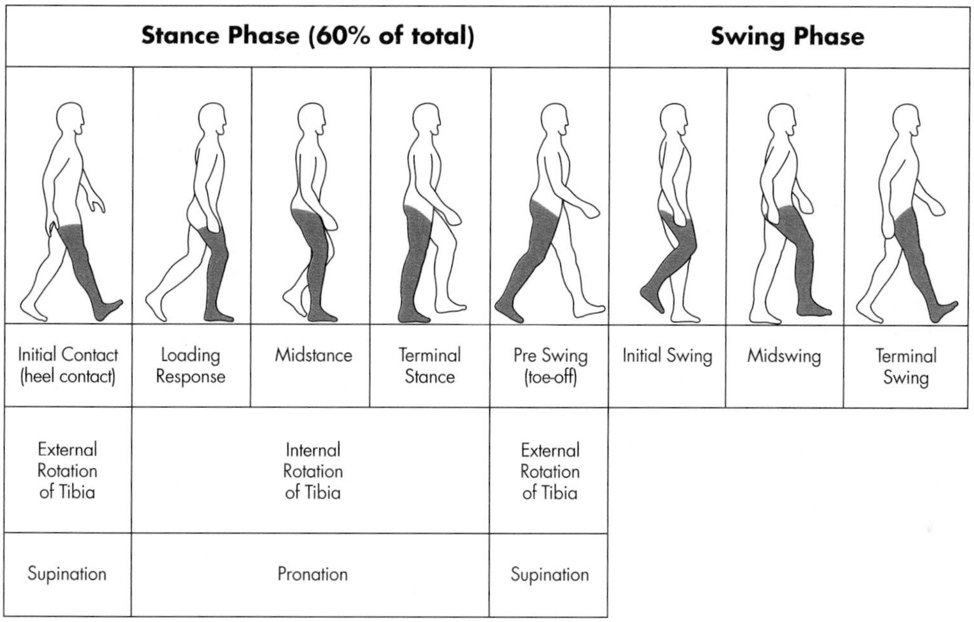

Stance Phase (60% of total)					Swing Phase		
Initial Contact (heel contact)	Loading Response	Midstance	Terminal Stance	Pre Swing (toe-off)	Initial Swing	Midswing	Terminal Swing
External Rotation of Tibia		Internal Rotation of Tibia		External Rotation of Tibia			
Supination		Pronation		Supination			

FIG. 11.1 ● Walking gait cycle.

Prentice WE: *Arnheim's principles of athletic training,* ed 12, New York, 2006, McGraw-Hill.

Chapter

11

Bones

Each foot has 26 bones, which collectively form the shape of an arch. They connect with the thigh and the remainder of the body through the fibula and tibia (Figs. 11.2 and 11.3). Body weight is transferred from the tibia to the talus and the calcaneus.

In addition to the talus and calcaneus, there are five other bones in the rear foot and midfoot known as the tarsals. Between the talus and the three cuneiform bones lies the navicular. The cuboid is located between the calcaneus and the fourth and fifth metatarsals. Distal to the tarsals are the five metatarsals, which in turn correspond to each of the five toes. The toes are known as the phalanges. There are three individual bones in each phalange, except for the great toe, which has only two. Each of these bones is known as a phalanx. Finally, there are two sesamoid bones located beneath the first metatarsophalangeal joint and contained within the flexor hallucis longus tendons.

The distal end of the tibia and fibula are enlarged and protrude horizontally and inferiorly. These bony protrusions, known as malleoli, serve as a sort of pulley for the tendons of the muscles that run directly posterior to them. Specifically, the peroneus brevis and peroneus longus are immediately behind the lateral malleolus. The muscles immediately posterior to the medial malleolus may be remembered by the phrase "Tom, Dick and Harry" with the "T" for the tibialis posterior, the "D" for the flexor digitorum longus, and the "H" for the flexor hallucis longus. This bony arrangement increases the mechanical advantage of these muscles in performing their actions of inversion and eversion. The base of the fifth metatarsal is enlarged and prominent to serve as an attachment point for the peroneus brevis and tertius.

The inner surface of the medial cuneiform and base of the first metatarsal provide insertion points for the tibialis anterior, while the undersurface of the same bones serves as the insertion for the peroneus longus. The tibialis posterior has multiple insertions on the lower inner surfaces of the navicular, cuneiform, and second through fifth metatarsal bases. The tops and undersurfaces of the bases of the second through fifth distal phalanxes are the insertion points for the extensor digitorum longus and the flexor digitorum longus, respectively. Similarly, the top and undersurface

of the base of the first distal phalanx provide insertions for the extensor hallucis longus and flexor hallucis longus, respectively.

The posterior surface of the calcaneus is very prominent and serves as the attachment point for the Achilles tendon of the gastrocnemius-soleus complex.

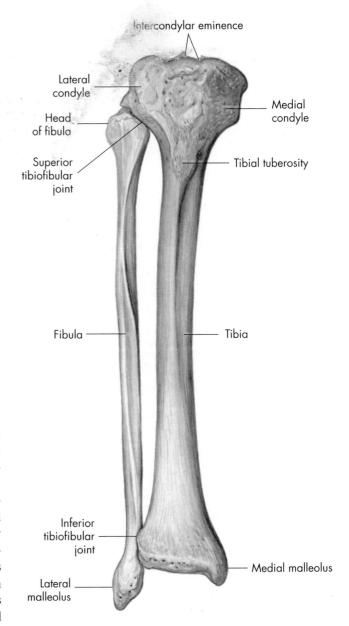

FIG. 11.2 ● Right fibula and tibia.

From Anthony CP, Kolthoff NJ: *Textbook of anatomy and physiology,* ed 9, St. Louis, 1975, Mosby.

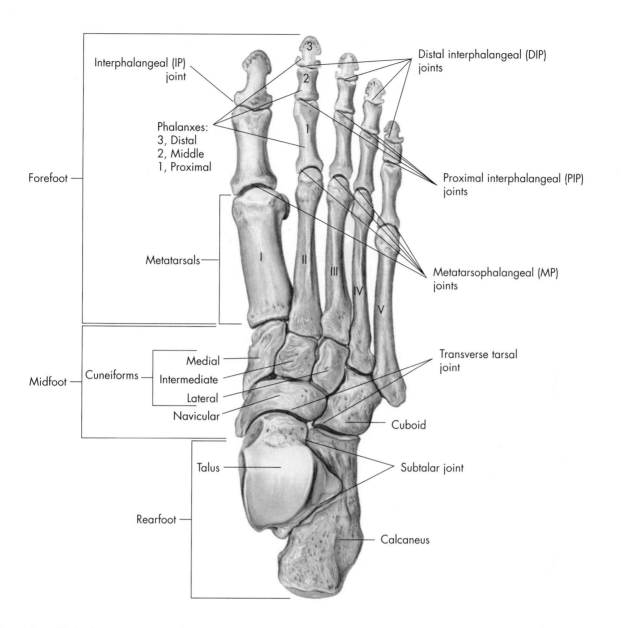

FIG. 11.3 ● Right foot, superior view.

Modified from Anthony CP, Kolthoff NJ: *Textbook of anatomy and physiology*, ed 9, St. Louis, 1975, Mosby.

Joints

The tibia and fibula form the tibiofibular joint, a syndesmotic amphiarthrodial joint (see Fig. 11.2). The bones are joined at both the proximal and distal tibiofibular joints. In addition to the ligaments supporting both of these joints, there is a strong, dense interosseus membrane between the shafts of these two bones. Although only minimal movement is possible between these bones, the distal joint does become sprained occasionally in heavy contact sports such as football. This injury, a sprain of the syndesmosis joint, is commonly referred to as a high ankle sprain and primarily involves the anterior inferior tibiofibular ligament. Secondarily, and with more severe injuries, the posterior tibiofibular ligament, interosseus ligament, and interosseus membrane may be involved.

The ankle joint, technically known as the talocrural joint, is a hinge or ginglymus-type joint (Fig. 11.4). Specifically, it is the joint made up of the talus, the distal tibia, and the distal fibula. The ankle joint allows approximately 50 degrees of plantar flexion and 15 to 20 degrees of dorsiflexion (Fig. 11.5). Greater range of dorsiflexion, particularly in weight bearing, is possible when the knee is flexed, which reduces the tension of the biarticular gastrocnemius muscle. The fibula rotates on its axis 3 to 5 degrees externally with dorsiflexion of the ankle and 3 to 5 degrees internally during plantar flexion. The syndesmosis joint widens by approximately 1 to 2 millimeters during full dorsiflexion.

Inversion and eversion, although commonly thought to be ankle joint movements, technically occur in the subtalar and transverse tarsal joints. These joints, classified as gliding or arthrodial, combine to allow approximately 20 to 30 degrees of inversion and 5 to 15 degrees of eversion. There is minimal movement within the remainder of the intertarsal and tarsometatarsal arthrodial joints.

The phalanges join the metatarsals to form the metatarsophalangeal joints, which are classified as condyloid-type joints. The metatarsophalangeal (MP) joint of the great toe flexes 45 degrees and extends 70 degrees, whereas the interphalangeal (IP) joint can flex from 0 degrees of full extension to 90 degrees of flexion. The MP joints of the four lesser toes allow approximately 40 degrees of flexion and 40 degrees of extension. The MP joints also abduct and adduct minimally. The proximal interphalangeal (PIP) joints in the lesser toes flex from 0 degrees of extension to 35 degrees of flexion. The distal interphalangeal (DIP) joints flex 60 degrees and extend 30 degrees. There is much variation from joint to joint and from person to person in all of these joints.

Ankle sprains are one of the most common injuries among physically active people. Sprains involve the stretching or tearing of one or more ligaments. There are far too many ligaments in the foot and ankle to discuss in this text, but a few of the ankle ligaments are shown in Fig. 11.4. Far and away the most common ankle sprain results from excessive inversion, which causes damage to the lateral ligamentous structures, primarily the anterior talofibular ligament and the calcaneofibular ligament. Excessive eversion forces causing injury to the deltoid ligament on the medial aspect of the ankle occur less commonly.

Ligaments in the foot and the ankle maintain the position of an arch. All 26 bones in the foot are connected with ligaments. This brief discussion is focused on the longitudinal and transverse arches.

There are two longitudinal arches (Fig. 11.6). The medial longitudinal arch is located on the medial side of the foot and extends from the calcaneus bone to the talus, the navicular, the three cuneiforms, and the distal ends of the three medial metatarsals. The lateral longitudinal arch is located on the lateral side of the foot and extends from the calcaneus to the cuboid and distal ends of the fourth and fifth metatarsals. Individual long arches can be high, medium, or low, but a low arch is not necessarily a weak arch.

The transverse arch (see Fig. 11.6) extends across the foot from one metatarsal bone to the other.

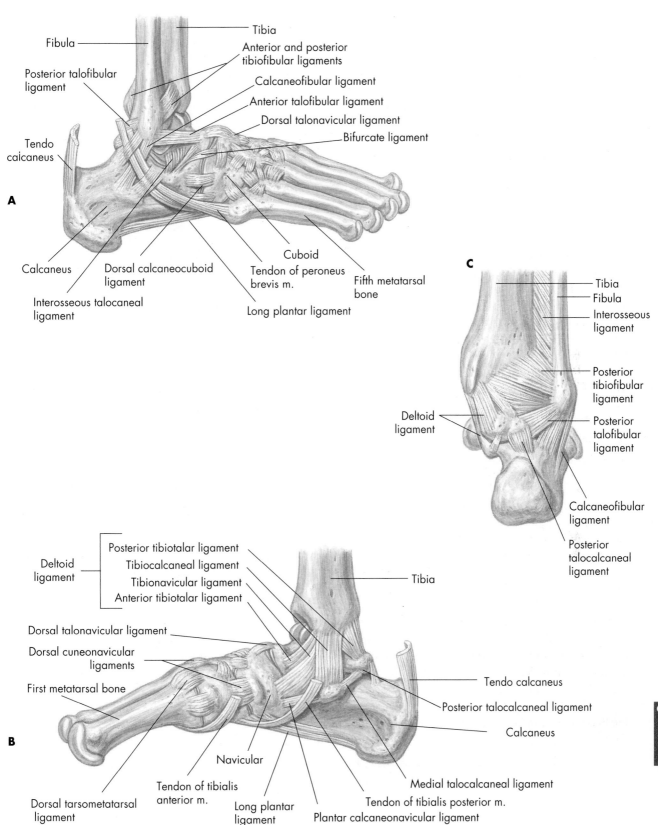

A

Fibula
Posterior talofibular ligament
Tendo calcaneus
Calcaneus
Dorsal calcaneocuboid ligament
Interosseous talocaneal ligament

Tibia
Anterior and posterior tibiofibular ligaments
Calcaneofibular ligament
Anterior talofibular ligament
Dorsal talonavicular ligament
Bifurcate ligament
Cuboid
Tendon of peroneus brevis m.
Fifth metatarsal bone
Long plantar ligament

C

Tibia
Fibula
Interosseous ligament
Posterior tibiofibular ligament
Posterior talofibular ligament
Calcaneofibular ligament
Posterior talocalcaneal ligament
Deltoid ligament

B

Deltoid ligament
Posterior tibiotalar ligament
Tibiocalcaneal ligament
Tibionavicular ligament
Anterior tibiotalar ligament
Dorsal talonavicular ligament
Dorsal cuneonavicular ligaments
First metatarsal bone
Dorsal tarsometatarsal ligament
Tendon of tibialis anterior m.
Navicular
Long plantar ligament
Plantar calcaneonavicular ligament
Tendon of tibialis posterior m.
Medial talocalcaneal ligament
Calcaneus
Posterior talocalcaneal ligament
Tendo calcaneus
Tibia

FIG. 11.4 ● Right ankle joint. A, Lateral view; **B,** Medial view; **C,** Posterior view.

Modified from Van De Graff KM: *Human anatomy,* ed 6, New York, 2002, McGraw-Hill.

Chapter
11

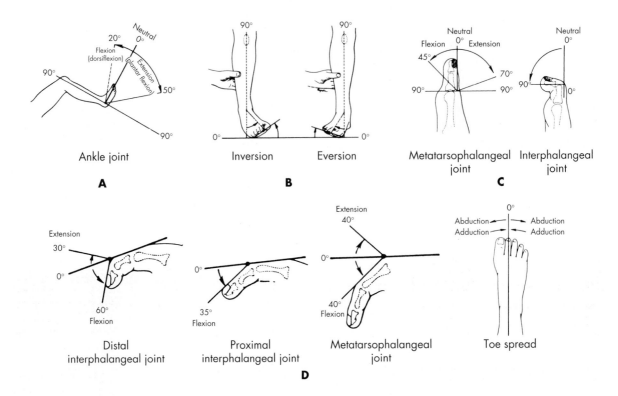

FIG. 11.5 ● Active motion of the ankle, foot, and toes. **A,** Dorsiflexion and plantar flexion are measured in degrees from the right-angle neutral position or in percentages of motion as compared to the opposite ankle; **B,** Inversion and eversion are normally estimated in degrees or expressed in percentages as compared to the opposite foot; **C,** Flexion and extension of the great toe; **D,** ROM for the lateral four toes.

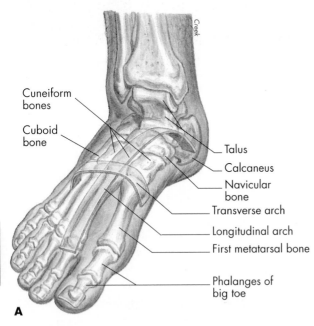

FIG. 11.6 ● Longitudinal and transverse arches. **A,** Medial view of the right foot showing both arches. **B,** Transverse view through the bases of the metatarsal bones showing a portion of the transverse arch.

Chapter

11

Movements   FIG. 11.7

Dorsiflexion (flexion): dorsal flexion; movement of the top of the ankle and foot toward the anterior tibia bone

Plantar flexion (extension): movement of the ankle and foot away from the tibia

Eversion: turning the ankle and foot outward; abduction, away from the midline; weight is on the medial edge of the foot

Inversion: turning the ankle and foot inward; adduction, toward the midline; weight is on the lateral edge of the foot

Toe flexion: movement of the toes toward the plantar surface of the foot

Toe extension: movement of the toes away from the plantar surface of the foot

Pronation: a combination of ankle dorsiflexion, subtalar eversion, and forefoot abduction (toe-out)

Supination: a combination of ankle plantar flexion, subtalar inversion, and forefoot adduction (toe-in)

A
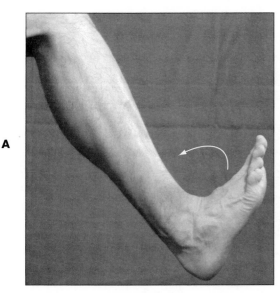

Dorsiflexion

B
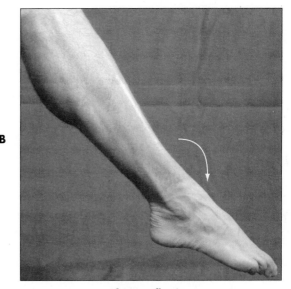

Plantar flexion

C
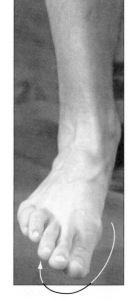

Transverse tarsal and subtalar eversion

D
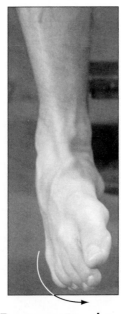

Transverse tarsal and subtalar inversion

FIG. 11.7 ● Movements of the ankle and foot (continued next page).

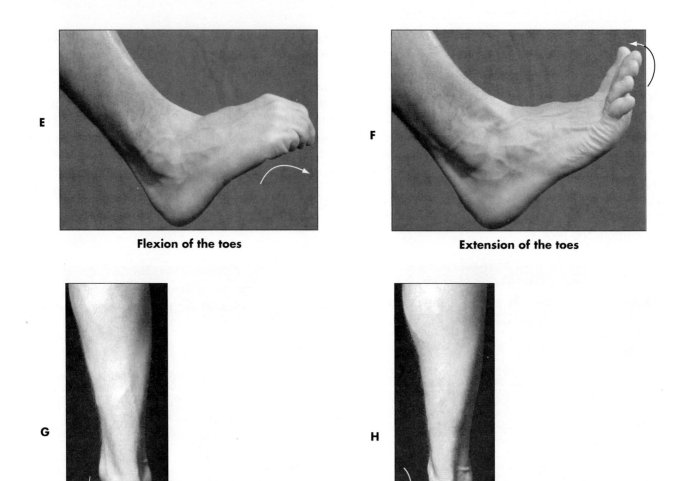

E Flexion of the toes

F Extension of the toes

G Pronation

H Supination

FIG. 11.7 (continued) ● Movements of the ankle and foot.

Ankle and foot muscles   FIGS 11.8, 11.9

The large number of muscles in the ankle and foot may be easier to learn if grouped according to location and function. In general, the muscles located on the anterior aspect of the ankle and foot are the dorsal flexors. Those on the posterior aspect are plantar flexors. Specifically, the gastrocnemius and the soleus collectively are known as the triceps surae, due to their three heads, which join together to the Achilles tendon. Muscles that are evertors are located more to the lateral side, whereas the invertors are located medially.

The lower leg is divided into four compartments, each containing specific muscles (Fig. 11.9). Tightly surrounding and binding each compartment is a dense fascia, which facilitates venous return and prevents excessive swelling of the muscles during exercise. The anterior compartment contains the dorsiflexor group, consisting of the tibialis anterior, peroneus tertius, extensor digitorum longus, and extensor hallucis

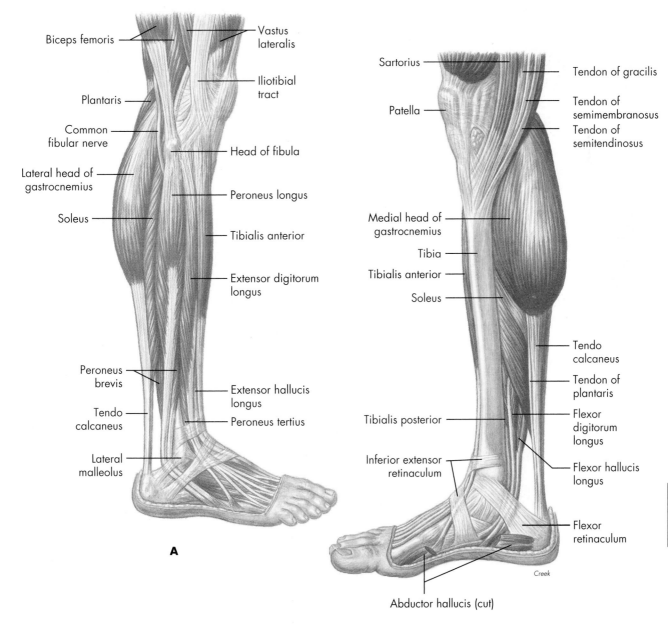

FIG. 11.8 ● Lower leg, ankle, and foot muscles. **A,** Lateral view; **B,** Medial view (continued next page).

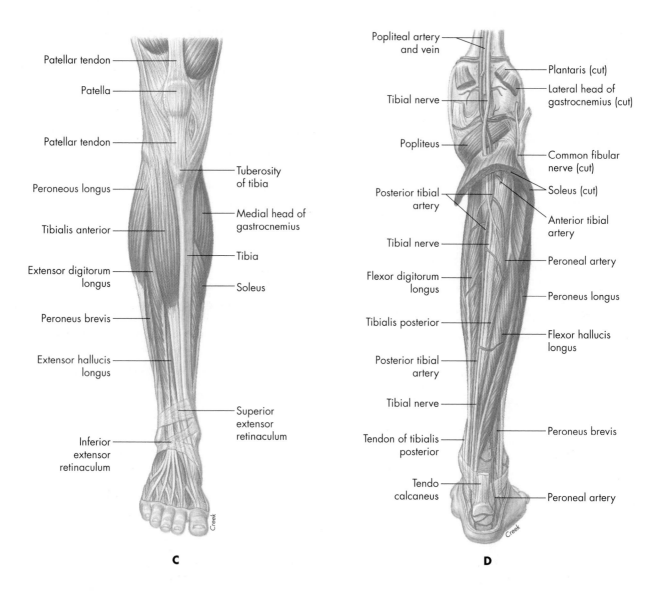

Patellar tendon
Patella
Patellar tendon
Peroneous longus
Tibialis anterior
Extensor digitorum longus
Peroneus brevis
Extensor hallucis longus
Inferior extensor retinaculum
Tuberosity of tibia
Medial head of gastrocnemius
Tibia
Soleus
Superior extensor retinaculum

C

Popliteal artery and vein
Tibial nerve
Popliteus
Posterior tibial artery
Tibial nerve
Flexor digitorum longus
Tibialis posterior
Posterior tibial artery
Tibial nerve
Tendon of tibialis posterior
Tendo calcaneus
Plantaris (cut)
Lateral head of gastrocnemius (cut)
Common fibular nerve (cut)
Soleus (cut)
Anterior tibial artery
Peroneal artery
Peroneus longus
Flexor hallucis longus
Peroneus brevis
Peroneal artery

D

FIG. 11.8 (continued) ● Lower leg, ankle, and foot muscles. **C,** Anterior view; **D,** Deep posterior view.

longus. The lateral compartment contains the peroneus longus and peroneus brevis—the two most powerful evertors. The posterior compartment is divided into deep and superficial compartments. The gastrocnemius, soleus, and plantaris are located in the superficial posterior compartment, while the deep posterior compartment is composed of the flexor digitorum longus, flexor hallucis longus, popliteus, and tibialis posterior. All of the muscles of the superficial posterior compartment are primarily plantar flexors. The plantaris, absent in some humans, is a vestigial biarticular muscle that contributes minimally to ankle plantar flexion. The deep posterior compartment muscles, except for the popliteus, are plantar flexors but also function as invertors.

Due to heavy demands placed on the musculature of the legs in the running activities of most sports, both acute and chronic injuries are common. "Shin splints" is a common term used to describe a painful condition of the leg that is often associated with running activities. This condition is not a specific diagnosis but, rather, is attributed to a number of specific musculotendinous injuries. Most often the tibialis posterior, medial soleus, or tibialis anterior is involved, but the extensor digitorum longus may also be involved. Shin splints may be prevented in part by stretching the plantar flexors and strengthening the dorsiflexors.

Additionally, painful cramps caused by acute muscle spasm in the gastrocnemius and soleus occur somewhat commonly and may be relieved

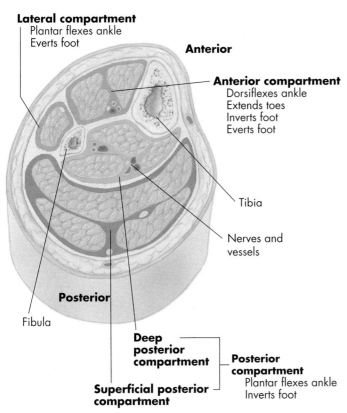

Lateral compartment
Plantar flexes ankle
Everts foot

Anterior

Anterior compartment
Dorsiflexes ankle
Extends toes
Inverts foot
Everts foot

Tibia

Nerves and
vessels

Posterior

Fibula

**Deep
posterior
compartment**

**Posterior
compartment**
Plantar flexes ankle
Inverts foot

**Superficial posterior
compartment**

FIG. 11.9 ● Cross-section of the leg, demonstrating
the muscular compartments.

From Seeley RR, Stephens TD, Tate P: *Anatomy and physiology,* ed
3, St. Louis, 1995, Mosby.

through active and passive dorsiflexion. Also, a
very disabling injury involves the complete rupture of the strong Achilles tendon, which connects
these two plantar flexors to the calcaneus.

NOTE: A number of the ankle and foot muscles are capable of helping produce more than
one movement.

Ankle and foot muscles by function

Plantar flexors
Gastrocnemius
Flexor digitorum longus
Flexor hallucis longus
Peroneus (fibularis) longus
Peroneus (fibularis) brevis
Plantaris
Soleus
Tibialis posterior
Evertors
Peroneus (fibularis) longus
Peroneus (fibularis) brevis
Peroneus (fibularis) tertius
Extensor digitorum longus
Dorsiflexors
Tibialis anterior
Peroneus (fibularis) tertius
Extensor digitorum longus (extensor of the
lesser toes)
Extensor hallucis longus (extensor of the great
toe)
Invertors
Tibialis anterior
Tibialis posterior
Flexor digitorum longus (flexor of the lesser toes)
Flexor hallucis longus (flexor of the great toe)

Ankle and foot muscles by compartment

Anterior compartment
Tibialis anterior
Extensor hallucis longus
Extensor digitorum longus
Peroneus (fibularis) tertius
Lateral compartment
Peroneus (fibularis) longus
Peroneus (fibularis) brevis
Deep posterior compartment
Flexor digitorum longus
Flexor hallucis longus
Tibialis posterior
Superficial posterior compartment
Gastrocnemius (medial head)
Gastrocnemius (lateral head)
Soleus

While viewing the muscles in Figs. 11.8 and 11.9
correlate them with Table 11.1.

Chapter

11

TABLE 11.1 • Agonist muscles of the ankle and foot joints

	Muscle	Origin	Insertion	Action	Plane of motion	Palpation	Innervation
Superficial posterior compartment	Gastrocnemius	Medial head: posterior surface of the medial femoral condyle Lateral head: posterior surface of the lateral femoral condyle	Posterior surface of the calcaneus (Achilles tendon)	Plantar flexion of the ankle	Sagittal	Upper 1/2 of posterior aspect of the lower leg	Tibial nerve (S1, S2)
				Flexion of the knee			
	Soleus	Posterior surface of the proximal fibula and proximal 2/3 of the posterior tibial surface	Posterior surface of the calcaneus (Achilles tendon)	Plantar flexion of the ankle	Sagittal	Posteriorly under the gastrocnemius muscle on the medial and lateral side of the lower leg, particularly while prone with knee flexed approximately 90 degrees and actively plantarflexing ankle	Tibial nerve (S1, S2)
Deep posterior compartment	Tibialis posterior	Posterior surface of the upper 1/2 of the interosseus membrane and the adjacent surfaces of the tibia and fibula	Lower inner surfaces of the navicular and cuneiform bones and bases of the 2nd, 3rd, 4th, and 5th metatarsal bones	Inversion of the foot	Frontal	The tendon may be palpated both proximally and distally immediately behind medial malleolus with inversion and plantar flexion and is better distinguished from flexor digitorum longus and flexor hallucis longus if toes can be maintained in slight extension	Tibial nerve (L5, S1)
				Plantar flexion of the ankle	Sagittal		
	Flexor digitorum longus	Middle 1/3 of the posterior surface of the tibia	Base of the distal phalanx of each of the four lesser toes	Flexion of the four lesser toes at the metatarso-phalangeal, proximal, and distal inter-phalangeal joints	Sagittal	The tendon may be palpated immediately posterior to the medial malleolus and tibialis posterior and immediately anterior to the flexor hallucis longus with flexion of the lesser toes while maintaining great toe extension, ankle dorsiflexion, and foot eversion	Tibial nerve (L5, S1)
				Plantar flexion of the ankle			
				Inversion of the foot	Frontal		
	Flexor hallucis longus	Middle 2/3 of the posterior surface of the fibula	Base of the distal phalanx of the big toe; plantar surface	Flexion of the great toe at the metatarso-phalangeal and interphalangeal joints	Sagittal	Most posterior of three tendons immediately behind medial malleolus; between medial soleus and tibia with active great toe flexion while maintaining extension of four lesser toes, ankle dorsiflexion, and foot eversion	Tibial nerve (L5, S1, S2)
				Plantar flexion of the ankle			
				Inversion of the foot	Frontal		

Chapter

11

TABLE 11.1 (continued) • Agonist muscles of the ankle and foot joints

	Muscle	Origin	Insertion	Action	Plane of motion	Palpation	Innervation
Lateral compartment	Peroneus (fibularis) longus	Head and upper 2/3 of the lateral surface of the fibula	Under surfaces of the medial cuneiform and 1st metatarsal bone	Eversion of the foot	Frontal	Upper lateral side of tibia just distal to fibular head and down to immediately posterior to lateral malleolus; just posterolateral from tibialis anterior and extensor digitorum longus with active eversion	Superficial peroneal nerve (L4, L5, S1)
				Plantar flexion of the ankle	Sagittal		
	Peroneus (fibularis) brevis	Lower 2/3 of the lateral surface of the fibula	Tuberosity of the 5th metatarsal bone	Eversion of the foot	Frontal	Tendon of muscle at proximal end of 5th metatarsal; just proximal and posterior to lateral malleolus; immediately deep anteriorly and posteriorly to peroneus longus with active eversion	Superficial peroneal nerve (L4, L5, S1)
				Plantar flexion of the ankle	Sagittal		
Anterior compartment	Peroneus (fibularis) tertius	Distal 1/3 of the anterior fibula	Base of the 5th metatarsal	Dorsiflexion of the ankle	Sagittal	Just medial to distal fibula; lateral to extensor digitorum longus tendon on anterolateral aspect of foot, down to medial side of base of 5th metatarsal with dorsiflexion and eversion	Deep peroneal nerve (L4, L5, S1)
				Eversion of the foot	Frontal		
	Extensor digitorum longus	Lateral condyle of the tibia, head of the fibula, and upper 2/3 of the anterior surface of the fibula	Tops of the middle and distal phalanges of the lesser four toes	Extension of the lesser toes at the metatarso-phalangeal, proximal and distal interphalangeal joints	Sagittal	Second muscle to lateral side of anterior tibial border; upper lateral side of tibia between tibialis anterior medially and fibula laterally; divides into four tendons just distal to anterior ankle with active toe extension	Deep peroneal nerve (L4, L5, S1)
				Dorsiflexion of the ankle			
				Eversion of the foot	Frontal		
	Extensor hallucis longus	Middle 2/3 of the medial surface of the anterior fibula	Base of the distal phalanx of the great toe	Extension of great toe at the metatarso-phalangeal and interphalangeal joints	Sagittal	From dorsal aspect of great toe to just lateral of tibialis anterior and medial to extensor digitorum longus at anterior ankle joint	Deep peroneal nerve (L4, L5, S1)
				Dorsiflexion of the ankle			
				Weak inversion of the foot	Frontal		
	Tibialis anterior	Upper 2/3 of the lateral surface of the tibia	Inner surface of the medial cuneiform and the base of the 1st metatarsal bone	Dorsiflexion of the ankle	Sagittal	First muscle to the lateral side of the anterior tibial border, particularly palpable with fully active ankle dorsiflexion; most prominent tendon crossing the ankle anteromedially	Deep peroneal nerve (L4, L5, S1)
				Inversion of the foot	Frontal		

Chapter **11**

Nerves

As described in Chapter 9, the sciatic nerve originates from the sacral plexus and becomes the tibial nerve and peroneal nerve. The tibial division of the sciatic nerve (Fig. 9.20) continues down to the posterior aspect of the lower leg to innervate the gastrocnemius (medial head), soleus, tibialis posterior, flexor digitorum longus, and flexor hallucis longus. Just before reaching the ankle the tibial nerve branches to become the medial and lateral plantar nerves, which innervate the intrinsic muscles of the foot. The medial plantar nerve innervates the abductor hallucis, flexor hallucis brevis, first lumbricale, and the flexor digitorum brevis. The lateral plantar nerve supplies the adductor hallucis, quadratus plantae, lumbricales (2, 4, and 4), dorsal interossei, plantar interossei, abductor digiti minimi, and the flexor digiti minimi.

The peroneal or fibular nerve (Fig. 11.10) divides just below the head of the fibula to become the superficial and deep peroneal nerve. The superficial branch innervates the peroneus longus and peroneus brevis, while the deep branch innervates the tibialis anterior, extensor digitorum longus, extensor hallucis longus, peroneus tertius, and extensor digitorum brevis.

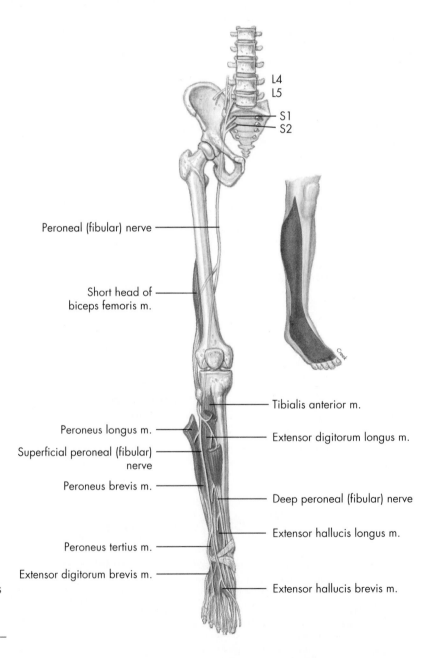

FIG. 11.10 ● Muscular and cutaneous distribution of the peroneal (fibular) nerve.

Gastrocnemius muscle   FIG. 11.11

(gas-trok-ne´mi-us)

Origin

Medial head: posterior surface of the medial femoral condyle

Lateral head: posterior surface of the lateral femoral condyle

Insertion

Posterior surface of the calcaneus (Achilles tendon)

Action

Plantar flexion of the ankle

Flexion of the knee

Palpation

Easiest muscle in the lower extremity to palpate; upper one-half posterior aspect of the lower leg

Innervation

Tibial nerve (S1, S2)

Application, strengthening, and flexibility

The gastrocnemius and soleus together are known as the **triceps surae** with triceps referring to the heads of the medial and lateral gastrocnemius and the soleus and surae referring to the calf. Because the gastrocnemius is a biarticular muscle, it is more effective as a knee flexor if the ankle is dorsiflexed and more effective as a plantar flexor of the foot if the knee is held in extension. This is observed when one sits too close to the wheel in driving a car, which significantly shortens the entire muscle, reducing its effectiveness. When the knees are bent, the muscle becomes an ineffective plantar flexor, and it is more difficult to depress the brakes. Running, jumping, hopping, and skipping exercises all depend significantly on the gastrocnemius and soleus to propel the body upward and forward. Heel-raising exercises with the knees in full extension and the toes resting on a block of wood are an excellent way to strengthen the muscle through the full range of motion. By holding a barbell on the shoulders, the resistance may be increased.

The gastrocnemius may be stretched by standing and placing both palms on a wall about 3 feet away and leaning into the wall. The feet should be pointed straight ahead, and the heels should remain on the floor. The knees should remain fully extended throughout the exercise to accentuate the stretch on the gastrocnemius.

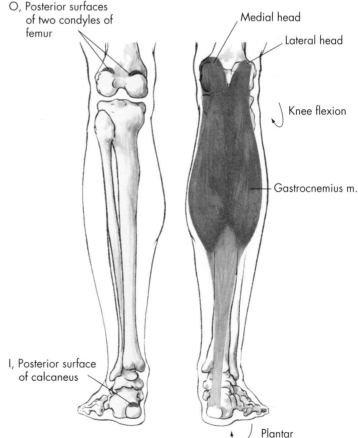

O, Posterior surfaces of two condyles of femur

Medial head

Lateral head

Knee flexion

Gastrocnemius m.

I, Posterior surface of calcaneus

Plantar flexion

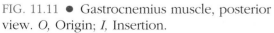

FIG. 11.11 ● Gastrocnemius muscle, posterior view. *O,* Origin; *I,* Insertion.

Chapter
11

Soleus muscle   FIG. 11.12

(so´le-us)

Origin

Posterior surface of the proximal fibula and proximal two-thirds of the posterior tibial surface

Insertion

Posterior surface of the calcaneus (Achilles tendon)

Action

Plantar flexion of the ankle

Palpation

Posteriorly under the gastrocnemius muscle on the medial and lateral side of the lower leg, particularly prone with knee flexed approximately 90 degrees and actively plantarflexing ankle

Innervation

Tibial nerve (S1, S2)

Application, strengthening, and flexibility

The soleus muscle is one of the most important plantar flexors of the ankle. Some anatomists believe that it is nearly as important in this movement as the gastrocnemius. This is especially true when the knee is flexed. When one rises up on the toes, the soleus muscle can plainly be seen on the outside of the lower leg if one has exercised the legs extensively, as in running and walking.

The soleus muscle is used whenever the ankle plantar flexes. Any movement with body weight on the foot with the knee flexed or extended calls it into action. When the knee is flexed slightly, the effect of the gastrocnemius is reduced, thereby placing more work on the soleus. Running, jumping, hopping, skipping, and dancing on the toes are all exercises that depend heavily on the soleus. It may be strengthened through any plantar flexion exercise against resistance, particularly if the knee is flexed slightly to deemphasize the gastrocnemius. Heel-raising exercises as described for the gastrocnemius, except with the knees flexed slightly, are one way to isolate this muscle for strengthening. Resistance may be increased by holding a barbell on the shoulders.

The soleus is stretched in the same manner as is the gastrocnemius, except that the knees must be flexed slightly, which releases the stretch on the gastrocnemius and places it on the soleus. Again, it is important to attempt to keep the heels on the floor.

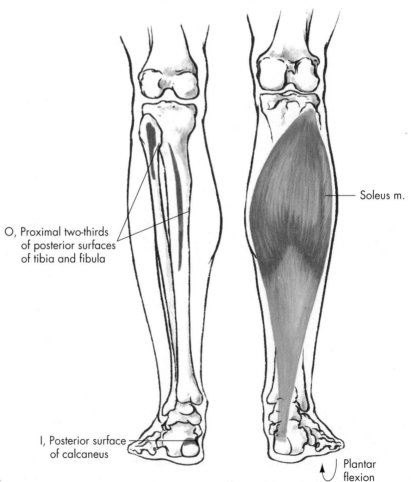

O, Proximal two-thirds of posterior surfaces of tibia and fibula

I, Posterior surface of calcaneus

Soleus m.

Plantar flexion

FIG. 11.12 ● Soleus muscle, posterior view. *O,* Origin; *I,* Insertion.

Peroneus (fibularis) longus muscle FIG. 11.13

(per-o-ne´us long´gus)

Origin

Head and upper two-thirds of the lateral surface of the fibula

Insertion

Undersurfaces of the medial cuneiform and first metatarsal bones

Action

Eversion of the foot
Plantar flexion of the ankle

Palpation

Upper lateral side of the tibia; just distal to fibular head and down to immediately posterior to lateral malleolus; just posterolateral from the tibialis anterior and extensor digitorum longus with active eversion

Innervation

Superficial peroneal nerve (L4–L5, S1)

Application, strengthening, and flexibility

The peroneus longus muscle passes posteroinferiorly to the lateral malleolus and under the foot from the outside to under the inner surface. Because of its line of pull, it is a strong evertor and assists in plantar flexion.

When the peroneus longus muscle is used effectively with the other ankle flexors, it helps bind the transverse arch as it flexes. Developed without the other plantar flexors, it would produce a weak, everted foot. In running, jumping, hopping, and skipping, the foot should be placed so that it is pointing forward to ensure proper development of the group. Walking barefoot or in stocking feet on the inside of the foot (everted position) is the best exercise for this muscle.

Eversion exercises to strengthen this muscle may be performed by turning the sole of the foot outward while resistance is applied in the opposite direction.

The peroneus longus may be stretched by passively taking the foot into extreme inversion and dorsiflexion while the knee is flexed.

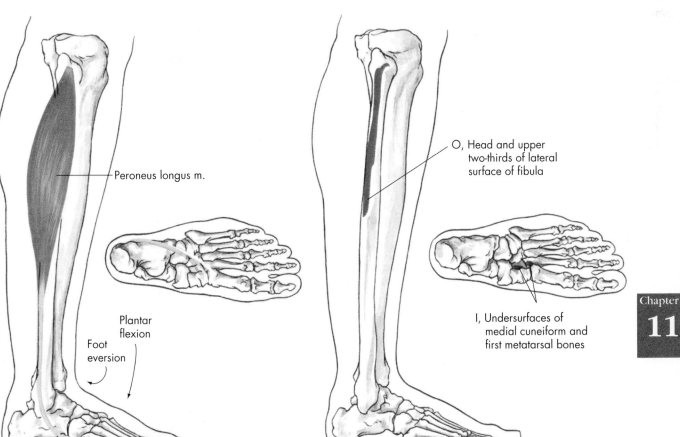

Peroneus longus m.

Plantar flexion

Foot eversion

O, Head and upper two-thirds of lateral surface of fibula

I, Undersurfaces of medial cuneiform and first metatarsal bones

Chapter 11

FIG. 11.13 ● Peroneus longus muscle, lateral and plantar views. *O,* Origin; *I,* Insertion.

Peroneus (fibularis) brevis muscle   FIG. 11.14

(per-o-ne´us bre´vis)

Origin

Lower two-thirds of the lateral surface of the fibula

Insertion

Tuberosity of the fifth metatarsal bone

Action

Eversion of the foot
Plantar flexion of the ankle

Palpation

Tendon of the muscle at the proximal end of the fifth metatarsal just proximal and posterior to the lateral malleolus; immediately deep anteriorly and posteriorly to the peroneus longus with active eversion

Innervation

Superficial peroneal nerve (L4–L5, S1)

Application, strengthening, and flexibility

The peroneus brevis muscle passes posteroinferiorly to the lateral malleolus to pull on the base of the fifth metatarsal. It is a primary evertor of the foot and assists in plantar flexion. In addition, it aids in maintaining the lateral longitudinal arch as it depresses the foot.

The peroneus brevis muscle is exercised with other plantar flexors in the powerful movements of running, jumping, hopping, and skipping. It may be strengthened in a fashion similar to that for the peroneus longus by performing eversion exercises, such as turning the sole of the foot outward against resistance.

The peroneus brevis is stretched in the same manner as the peroneus longus.

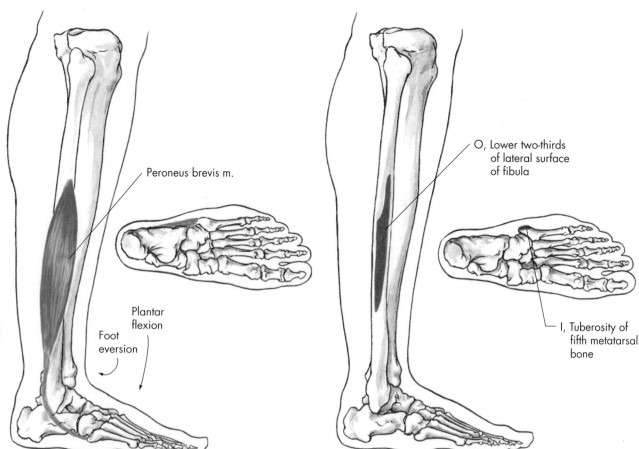

FIG. 11.14 ● Peroneus brevis muscle, lateral and plantar views. *O,* Origin; *I,* Insertion.

Peroneus (fibularis) tertius muscle FIG. 11.15

(per-o-ne´us ter´shi-us)

Origin
Distal third of the anterior fibula

Insertion
Base of the fifth metatarsal

Action
Eversion of the foot
Dorsiflexion of the ankle

Palpation
Just medial to distal fibula; lateral to the extensor digitorum longus tendon on the anterolateral aspect of the foot, down to the medial side of the base of the fifth metatarsal with dorsiflexion and eversion

Innervation
Deep peroneal nerve (L4–L5, S1)

Application, strengthening, and flexibility

The peroneus tertius, absent in some humans, is a small muscle that assists in dorsal flexion and eversion. Some authorities refer to it as the fifth tendon of the extensor digitorum longus. It may be strengthened by pulling the foot up toward the shin against a weight or resistance. Everting the foot against resistance, such as weighted eversion towel drags, can also be used for strength development.

The peroneus tertius may be stretched by passively taking the foot into extreme inversion and plantar flexion.

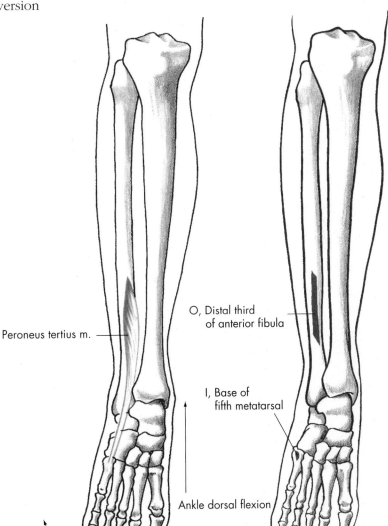

Peroneus tertius m.

O, Distal third of anterior fibula

I, Base of fifth metatarsal

Ankle dorsal flexion

Foot eversion

FIG. 11.15 ● Peroneus tertius muscle, anterior view. *O,* Origin; *I,* Insertion.

Chapter
11

Extensor digitorum longus muscle   FIG. 11.16

(eks-ten´sor dij-i-to´rum long´gus)

Origin

Lateral condyle of the tibia, head of the fibula, and upper two-thirds of the anterior surface of the fibula

Insertion

Tops of the middle and distal phalanges of the four lesser toes

Action

Extension of the four lesser toes at the metatarsophalangeal, proximal and distal interphalangeal joints
Dorsiflexion of the ankle
Eversion of the foot

Palpation

Second muscle to the lateral side of the anterior tibial border; upper lateral side of the tibia between the tibialis anterior medially and the fibula laterally; divides into four tendons just distal to anterior ankle with active toe extension

Innervation

Deep peroneal nerve (L4–L5, S1)

Application, strengthening, and flexibility

Strength is necessary in the extensor digitorum longus muscle to maintain balance between the plantar and the dorsal flexors.

Action that involves dorsal flexion of the ankle and extension of the toes against resistance strengthens both the extensor digitorum longus and the extensor hallucis longus muscles. This may be accomplished by manually applying a downward force on the toes while attempting to extend them up.

The extensor digitorum longus may be stretched by passively taking the four lesser toes into full flexion while the foot is inverted and plantar flexed.

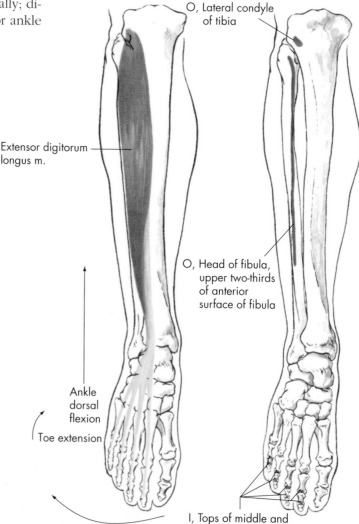

O, Lateral condyle of tibia

Extensor digitorum longus m.

O, Head of fibula, upper two-thirds of anterior surface of fibula

Ankle dorsal flexion

Toe extension

Eversion

I, Tops of middle and distal phalanges of four lesser toes

FIG. 11.16 ● Extensor digitorum longus muscle, anterior view. *O*, Origin; *I*, Insertion.

Extensor hallucis longus muscle
FIG. 11.17

(eks-ten´sor hal-u´sis long´gus)

Origin
Middle two-thirds of the medial surface of the anterior fibula

Insertion
Base of the distal phalanx of the great toe

Action
Dorsiflexion of the ankle

Extension of the great toe at the metatarsophalangeal and interphalangeal joints

Weak inversion of the foot

Palpation
From the dorsal aspect of the great toe to just lateral of the tibialis anterior and medial to the extensor digitorum longus at the anterior ankle joint

Innervation
Deep peroneal nerve (L4–L5, S1)

Application, strengthening, and flexibility
The four dorsiflexors of the foot—tibialis anterior, extensor digitorum longus, extensor hallucis longus, and peroneus tertius—may be exercised by attempting to walk on the heels with the ankle flexed dorsally and toes extended. Extension of the great toe, as well as ankle dorsiflexion against resistance, will provide strengthening for this muscle.

The extensor hallucis longus may be stretched by passively taking the great toe into full flexion while the foot is everted and plantar flexed.

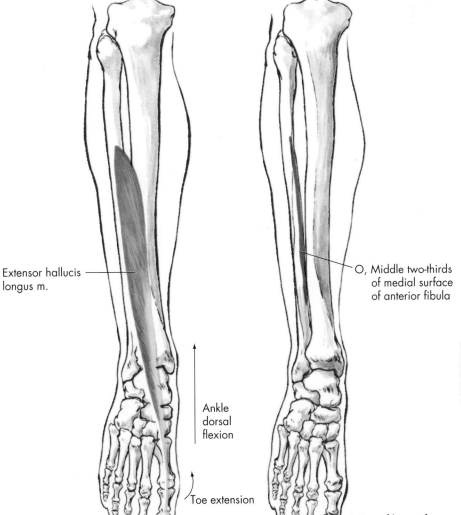

Extensor hallucis longus m.

O, Middle two-thirds of medial surface of anterior fibula

Ankle dorsal flexion

Toe extension

I, Top of base of distal phalanx of great toe

FIG. 11.17 ● Extensor hallucis longus muscle, anterior view. *O*, Origin; *I*, Insertion.

Chapter
11

Tibialis anterior muscle FIG. 11.18

(tib-i-aʹlis ant-teʹri-or)

Origin

Upper two-thirds of the lateral surface of the tibia

Insertion

Inner surface of the medial cuneiform and the base of the first metatarsal bone

Action

Dorsiflexion of the ankle
Inversion of the foot

Palpation

First muscle to the lateral side of the anterior tibial border, particularly palpable with fully active ankle dorsiflexion; most prominent tendon crossing the ankle anteromedially

Innervation

Deep peroneal nerve (L4–L5, S1)

Application, strengthening, and flexibility

By its insertion, the tibialis anterior muscle is in a fine position to hold up the inner margin of the foot. However, as it contracts concentrically, it dorsiflexes the ankle and is used as an antagonist to the plantar flexors of the ankle. The tibialis anterior is forced to contract strongly when a person ice skates or walks on the outside of the foot. It strongly supports the medial longitudinal arch in inversion.

Walking barefoot or in stocking feet on the outside of the foot (inversion) is an excellent exercise for the tibialis anterior muscle.

Turning the sole of the foot to the inside against resistance to perform inversion exercises is one way to strengthen this muscle. Dorsal flexion exercises against resistance may also be used for this purpose.

The tibialis anterior may be stretched by passively taking the foot into extreme eversion and plantar flexion.

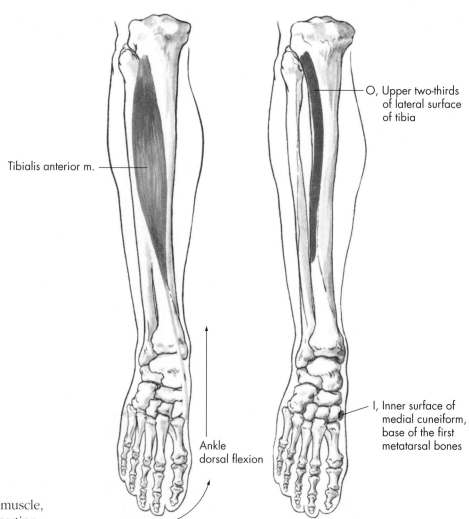

Tibialis anterior m.

O, Upper two-thirds of lateral surface of tibia

I, Inner surface of medial cuneiform, base of the first metatarsal bones

Ankle dorsal flexion

Foot inversion

FIG. 11.18 ● Tibialis anterior muscle, anterior view. *O,* Origin; *I,* Insertion.

Tibialis posterior muscle   FIG. 11.19

(tib-i-a´lis pos-te´ri-or)

Origin

Posterior surface of the upper half of the interosseus membrane and adjacent surfaces of the tibia and fibula

Insertion

Lower inner surfaces of the navicular and cuneiform bones and bases of the second, third, fourth, and fifth metatarsal bones

Action

Plantar flexion of the ankle
Inversion of the foot

Palpation

The tendon may be palpated both proximally and distally immediately behind the medial malleolus with inversion and plantar flexion and is better distinguished from the flexor digitorum longus and flexor hallucis longus if the toes can be maintained in slight extension

Innervation

Tibial nerve (L5, S1)

Application, strengthening, and flexibility

Passing down the back of the leg, under the medial malleolus, then forward to the navicular and medial cuneiform bones, the tibialis posterior muscle pulls down from the underside and, when contracted concentrically, inverts and plantarflexes the foot. As a result it is in position to support the medial longitudinal arch. **Shin splints** is a slang term frequently used to describe an often chronic condition in which the tibialis posterior, tibialis anterior, and extensor digitorum longus muscles are inflamed. This inflammation is usually a tendinitis of one or more of these structures but may be a result of stress fracture, periostitis, tibial stress syndrome, or compartment syndrome. Sprints and long-distance running are common causes, particularly if the athlete has not developed appropriate strength, flexibility, and endurance in the lower-leg musculature.

Use of the tibialis posterior muscle in plantar flexion and inversion gives support to the longitudinal arch of the foot. This muscle is generally strengthened by performing heel raises, as described for the gastrocnemius and soleus, as well as inversion exercises against resistance.

The tibialis posterior may be stretched by passively taking the foot into extreme eversion and dorsiflexion while the knee is flexed.

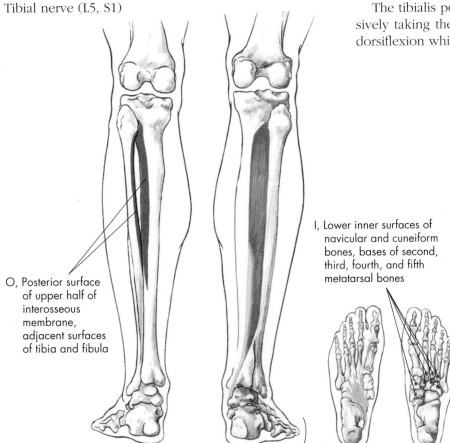

O, Posterior surface of upper half of interosseous membrane, adjacent surfaces of tibia and fibula

I, Lower inner surfaces of navicular and cuneiform bones, bases of second, third, fourth, and fifth metatarsal bones

Foot inversion

Plantar flexion

FIG. 11.19 ● Tibialis posterior muscle, posterior and plantar views. *O,* Origin; *I,* Insertion.

Chapter
11

Flexor digitorum longus muscle

FIG. 11.20

(fleks´or dij-i-to´rum long´gus)

Origin

Middle third of the posterior surface of the tibia

Insertion

Base of the distal phalanx of each of the four lesser toes

Action

Flexion of the four lesser toes at the metatarsophalangeal, proximal and distal interphalangeal joints
Inversion of the foot
Plantar flexion of the ankle

Palpation

The tendon may be palpated immediately posterior to the medial malleolus and tibialis posterior and immediately anterior to the flexor hallucis longus with flexion of the lesser toes while maintaining great toe extension, ankle dorsiflexion, and foot eversion

Innervation

Tibial nerve (L5, S1)

Application, strengthening, and flexibility

Passing down the back of the lower leg under the medial malleolus and then forward, the flexor digitorum longus muscle draws the four lesser toes down into flexion toward the heel as the plantar flexes the ankle. It is very important in helping other foot muscles maintain the longitudinal arch. Walking, running, and jumping do not necessarily call the flexor digitorum longus muscle into action. Some of the weak foot and ankle conditions result from ineffective use of the flexor digitorum longus. Walking barefoot with the toes curled downward toward the heels and with the foot inverted will exercise this muscle. It may be strengthened by performing towel grabs against resistance in which the heel rests on the floor while the toes extend to grab a flat towel and then flex to pull the towel under the foot. This may be repeated numerous times, with a small weight placed on the opposite end of the towel for added resistance.

The flexor digitorum longus may be stretched by passively taking the four lesser toes into extreme extension while the foot is everted and dorsiflexed. The knee should be flexed.

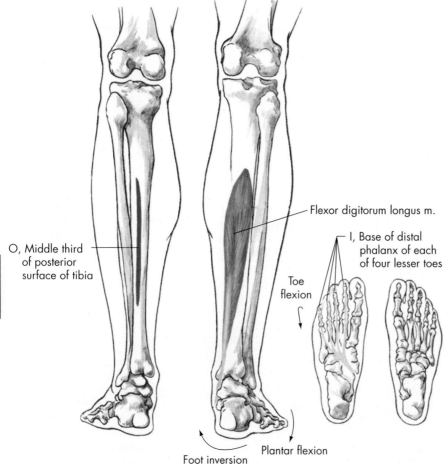

FIG. 11.20 ● Flexor digitorum longus muscle, posterior and plantar views. *O,* Origin; *I,* Insertion.

Chapter

11

Flexor hallucis longus muscle FIG. 11.21

(fleks´or hal-u´sis long´gus)

Origin

Middle two-thirds of the posterior surface of the fibula

Insertion

Base of the distal phalanx of the big toe, plantar
surface

Action

Flexion of the great toe at the metatarsophalangeal
and interphalangeal joints
Inversion of the foot
Plantar flexion of the ankle

Palpation

Most posterior of the three tendons immediately behind the medial malleolus; between the medial
soleus and the tibia with active great toe flexion
while maintaining extension of four lesser toes,
ankle dorsiflexion, and foot eversion

Innervation

Tibial nerve (L5, S1–S2)

Application, strengthening, and flexibility

Pulling from the underside of the great toe,
the flexor hallucis longus muscle may work

independently of the flexor digitorum longus
muscle or with it. If these two muscles are poorly
developed, they cramp easily when they are
called on to do activities to which they are unac-
customed.

These muscles are used effectively in walking if
the toes are used (as they should be) in maintaining balance as each step is taken. Walking "with
the toes" rather than "over" them is an important
action for them.

When the gastrocnemius, soleus, tibialis posterior, peroneus longus, peroneus brevis, flexor digitorum longus, flexor digitorum brevis, and flexor
hallucis longus muscles are all used effectively in
walking, the strength of the ankle is evident. If
an ankle and a foot are weak, in most cases it is
because of lack of use of all the muscles just mentioned. Running, walking, jumping, hopping, and
skipping provide exercise for this muscle group.
The flexor hallucis longus muscle may be specifically strengthened by performing towel grabs as
described for the flexor digitorum longus.

The flexor hallucis longus may be stretched by
passively taking the great toe into extreme extension while the foot is everted and dorsiflexed. The
knee should be flexed.

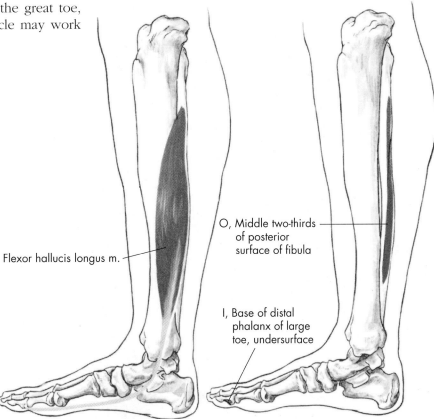

FIG. 11.21 ● Flexor hallucis
longus muscle, medial view.
O, Origin; *I,* Insertion.

O, Middle two-thirds
of posterior
surface of fibula

I, Base of distal
phalanx of large
toe, undersurface

Flexor hallucis longus m.

Toe flexion

Plantar flexion

Foot inversion

Intrinsic muscles of the foot FIGS. 11.22, 11.23

The intrinsic muscles of the foot have their origins and insertions on the bones within the foot (Figs. 11.22 and 11.23). One of these muscles, the extensor digitorum brevis, is found on the dorsum of the foot. The remainder are found in a plantar compartment in four layers on the plantar surface of the foot. The following muscles are found in the four layers:

First (superficial) layer: abductor hallucis, flexor digitorum brevis, abductor digiti minimi (quinti)
Second layer: quadratus plantae, lumbricales (four)
Third layer: flexor hallucis brevis, adductor hallucis, flexor digiti minimi (quinti) brevis
Fourth (deep) layer: dorsal interossei (four), plantar interossei (three)

The intrinsic foot muscles may be grouped by location as well as by the parts of the foot upon which they act. The abductor hallucis, flexor hallucis brevis, and adductor hallucis all insert either medially or laterally on the proximal phalanx of the great toe. The abductor hallucis and flexor hallucis brevis are located somewhat medially, whereas the adductor hallucis is more centrally located beneath the metatarsals.

The quadratus plantae, four lumbricales, four dorsal interossei, three plantar interossei, flexor digitorum brevis, and extensor digitorum brevis are all located somewhat centrally. All are beneath the foot except the extensor digitorum brevis, which is the only intrinsic muscle in the foot located in the dorsal compartment. Although the entire extensor digitorum brevis has its origin on the anterior and lateral calcaneus, some anatomists refer to its first tendon as the extensor hallucis brevis in order to maintain consistency in naming according to function and location.

Located laterally beneath the foot are the abductor digiti minimi and the flexor digiti minimi brevis, which both insert on the lateral aspect of the base of the proximal phalanx of the fifth phalange. Because of these two muscles' insertion and action on the fifth toe, the name "quinti" is sometimes used instead of minimi.

Four muscles act on the great toe. The abductor hallucis is solely responsible for abduction of the great toe, but assists the flexor hallucis brevis in flexing the great toe at the metatarsophalangeal joint. The adductor hallucis is the sole adductor of the great toe, while the extensor digitorum brevis is the only intrinsic extensor of the great toe at the metatarsophalangeal joint.

The four lumbricales are flexors of the second, third, fourth, and fifth phalanges at their metatarsophalangeal joints, while the quadratus plantae are flexors of these phalanges at their distal interphalangeal joints. The three plantar interossei are adductors and flexors of the proximal phalanxes of the third, fourth, and fifth phalanges, while the four dorsal interossei are abductors and flexors of the second, third, and fourth phalanges, also at their metatarsophalangeal joints. The flexor digitorum brevis flexes the middle phalanxes of the second, third, fourth, and fifth phalanges. The extensor digitorum brevis, as previously mentioned, is an extensor of the great toe but also extends the second, third, and fourth phalanges at their metatarsophalangeal joints.

There are two muscles that act solely on the fifth toe. The proximal phalanx of the fifth phalange is abducted by the abductor digiti minimi and is flexed by the flexor digiti minimi brevis.

Refer to Table 11.2 for further details regarding the intrinsic muscles of the foot.

Muscles are developed and maintain their strength only when they are used. One factor in the great increase in weak foot conditions is the lack of exercise to develop these muscles. Walking is one of the best activities for maintaining and developing the many small muscles that help support the arch of the foot. Some authorities advocate walking without shoes or with some of the newer shoes designed to enhance proper mechanics. Additionally, towel exercises such as those described for the flexor digitorum longus and flexor hallucis longus are helpful in strengthening the intrinsic muscles of the foot.

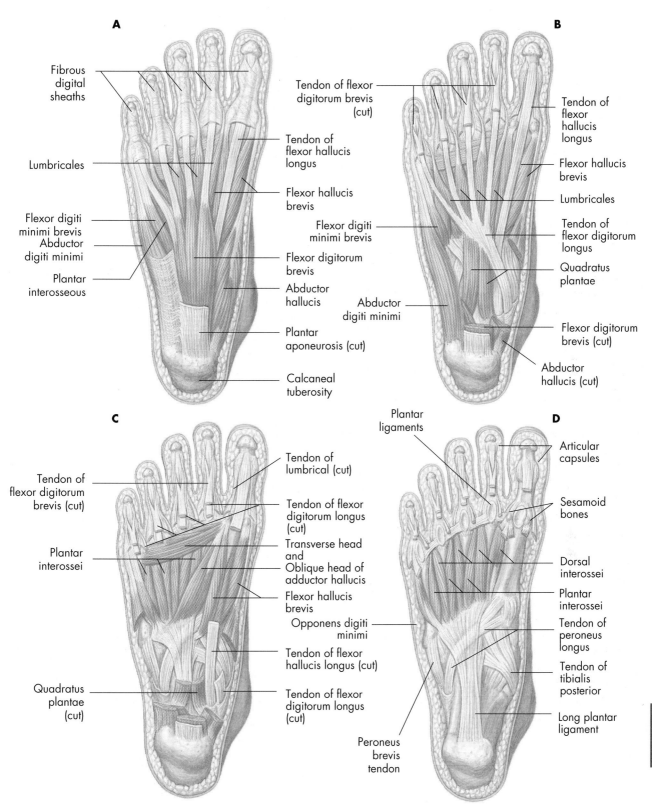

A

Fibrous digital sheaths

Lumbricales

Flexor digiti minimi brevis
Abductor digiti minimi

Plantar interosseous

Tendon of flexor hallucis longus

Flexor hallucis brevis

Flexor digitorum brevis

Abductor hallucis

Plantar aponeurosis (cut)

Calcaneal tuberosity

B

Tendon of flexor digitorum brevis (cut)

Tendon of flexor hallucis longus

Flexor hallucis brevis

Lumbricales

Tendon of flexor digitorum longus

Quadratus plantae

Flexor digiti minimi brevis

Abductor digiti minimi

Flexor digitorum brevis (cut)

Abductor hallucis (cut)

C

Tendon of flexor digitorum brevis (cut)

Plantar interossei

Quadratus plantae (cut)

Tendon of lumbrical (cut)

Tendon of flexor digitorum longus (cut)

Transverse head and

Oblique head of adductor hallucis

Flexor hallucis brevis

Opponens digiti minimi

Tendon of flexor hallucis longus (cut)

Tendon of flexor digitorum longus (cut)

D

Plantar ligaments

Articular capsules

Sesamoid bones

Dorsal interossei

Plantar interossei

Tendon of peroneus longus

Tendon of tibialis posterior

Long plantar ligament

Peroneus brevis tendon

FIG. 11.22 ● The four musculotendinous layers of the plantar aspect of the foot, detailing the intrinsic muscles. **A,** Superficial layer; **B,** Second layer; **C,** Third layer; **D,** Deep layer.

Modified from Van De Graaff KM: *Human anatomy,* ed 4, New York, 1995, McGraw-Hill.

Chapter

11

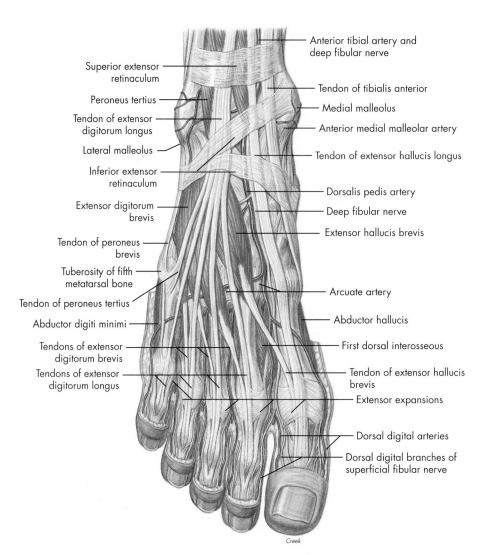

FIG. 11.23 ● Anterior view of dorsum of foot.

TABLE 11.2 • **Intrinsic muscles of the foot**

	Muscle	Origin	Insertion	Action	Palpation	Innervation
Superficial layer	Flexor digitorum brevis	Tuberosity of calcaneus, plantar aponeurosis	Medial and lateral aspects of 2nd, 3rd, 4th, and 5th middle phalanxes	MP and PIP flexion of 2nd, 3rd, 4th, and 5th phalanges	Cannot be palpated	Medial plantar nerve (L4, L5)
	Abductor digiti minimi (quinti)	Tuberosity of calcaneus, plantar aponeurosis	Lateral aspect of 5th proximal phalanx	MP abduction of 5th phalange	Cannot be palpated	Lateral plantar nerve (S1, S2)
	Flexor hallucis brevis	Cuboid, lateral cuneiform	Medial head: medial aspect of 1st proximal phalanx Lateral head: lateral aspect of 1st proximal phalanx	MP flexion of 1st phalange	Cannot be palpated	Medial plantar nerve (L4, L5, S1)

TABLE 11.2 (continued) • Intrinsic muscles of the foot

	Muscle	Origin	Insertion	Action	Palpation	Innervation
Superficial layer	Abductor hallucis	Tuberosity of calcaneus, flexor retinaculum, plantar aponeurosis	Medial aspect of base of 1st proximal phalanx	MP flexion, abduction of 1st phalanx	On the plantar aspect of the foot from medial tubercle of calcaneus to medial side of great toe proximal phalanx with great toe abduction	Medial plantar nerve (L4, L5)
Second layer	Flexor digiti minimi (quinti) brevis	Base of 5th metatarsal, sheath of peroneus longus tendon	Lateral aspect of base of 5th proximal phalanx	MP flexion of 5th phalange	Cannot be palpated	Lateral plantar nerve (S2, S3)
	Quadratus plantae	Medial head: medial surface of calcaneus Lateral head: lateral border of inferior surface of calcaneus	Lateral margin of flexor digitorum longus tendon	DIP flexion of 2nd, 3rd, 4th, and 5th phalanges	Cannot be palpated	Lateral plantar nerve (S1, S2)
	Lumbricales (4)	Tendons of flexor digitorum longus	Dorsal surface of 2nd, 3rd, 4th, and 5th proximal phalanxes	MP flexion of 2nd, 3rd, 4th, and 5th phalanges	Cannot be palpated	1st lumbricales: medial plantar nerve (L4, L5) 2nd, 3rd, 4th lumbricales: lateral plantar nerve (S1, S2)
Third layer	Adductor hallucis	Oblique head: 2nd, 3rd, and 4th metatarsals and sheath of peroneus longus tendon Transverse head: plantar metatarsophalangeal ligaments of 3rd, 4th, and 5th phalanges and transverse metatarsal ligaments	Lateral aspect of base of 1st proximal phalanx	MP adduction of 1st phalange	Cannot be palpated	Lateral plantar nerve (S1, S2)
	Plantar interossei (3)	Bases and medial shafts of 3rd, 4th, and 5th metatarsals	Medial aspects of bases of 3rd, 4th, and 5th proximal phalanxes	MP adduction and flexion of 3rd, 4th, and 5th phalanges	Cannot be palpated	Lateral plantar nerve (S1, S2)

Chapter

11

TABLE 11.2 (continued) • Intrinsic muscles of the foot

	Muscle	Origin	Insertion	Action	Palpation	Innervation
Dorsal layer	Dorsal interossei (4)	Two heads on shafts of adjacent metatarsals	1st interossei: medial aspect of 2nd proximal phalanx 2nd, 3rd, and 4th interossei: lateral aspect of 2nd, 3rd, and 4th proximal phalanxes	MP abduction and flexion of 2nd, 3rd, and 4th phalanges	Cannot be palpated	Lateral plantar nerve (S1, S2)
	Extensor digitorum brevis	Anterior and lateral calcaneus, lateral talocalcaneal ligament, inferior extensor retinaculum	Base of proximal phalanx of 1st phalange, lateral sides of extensor digitorum longus tendons of 2nd, 3rd, and 4th phalanges	Assists in MP extension of 1st phalanx and extension of middle three phalanges	As a mass anterior to and slightly below lateral malleolus on dorsum of foot	Deep peroneal nerve (L5, S1)

Web sites

Radiologic Anatomy Browser

http://radlinux1.usuf1.usuhs.mil/rad/iong

Numerous radiological views of the musculoskeletal system.

University of Arkansas Medical School Gross

Anatomy for Medical Students

http://anatomy.uams.edu/anatomyhtml/gross.html

Dissections, anatomy tables, atlas images, links, etc.

Loyola University Medical Center: Structure of the

Human Body

www.meddean.luc.edu/lumen/meded/grossanatomy/index.html

Excellent site with many slides, dissections, tutorials, etc., for the study of human anatomy.

Wheeless' Textbook of Orthopaedics

www.wheelessonline.com

An extensive index of links to the fractures, joints, muscles, nerves, trauma, medications, medical topics, and lab tests, as well as links to orthopaedic journals and other orthopaedic and medical news.

Foot and Ankle Web Index

www.footandankle.com

The foot and ankle link library is very helpful.

American College of Foot and Ankle Surgeons

www.acfas.org

Sponsored by podiatric surgeons and doctors of podiatric medicine (DPM), has information on topics relating to foot health—foot and ankle deformities and injuries; care of the diabetic foot; foot and ankle disorders caused by arthritis, aging, trauma and sports injuries; and congenital deformities and disease.

The University of Texas MD Anderson Cancer

Center Multimedia and Learning Resources

www.mdacc.tmc.edu/mmlearn/anatomy.html

Numerous cadaveric cuts of the foot, knee, hand, arm, and elbow; an interactive ankle; and a rotating foot and ankle.

American Orthopaedic Foot and Ankle Society

www.aofas.org

Numerous patient education brochures regarding foot and ankle problems are found here.

Premiere Medical Search Engine

www.medsite.com

Allows the reader to enter any medical condition and it will search the net to find relevant articles.

Virtual Hospital

www.vh.org

Numerous slides, patient information, etc.

Arthroscopy.com

www.arthroscopy.com/sports.htm

Patient information on various musculoskeletal problems of the lower extremity.

Human Anatomy Online

www.innerbody.com/image/musc08.html

Interactive musculoskeletal anatomy.

Agility total ankle system

www.agilityankle.com

Text and graphics on ankle anatomy with information on an ankle replacement system.

American Academy of Orthopaedic Surgeons

http://orthoinfo.aaos.org/category.cfm? topcategory=Foot

Patient education library on the foot and ankle.

Foot and Ankle Link Library

www.footandankle.com/podmed

A list of links to professional organizations and educational sites.

The Journal of Foot and Ankle Surgery

www2.jfas.org/scripts/om.dll/serve?action= searchDB&searchDBfor=hom&id=es

Allows you to search for articles and has a link to the American College of Foot and Ankle Surgeons.

Joint Healing.com

www.jointhealing.com/pages/foot/foot_ anatomy.html

Foot and ankle anatomy.

Images of the Foot and Ankle

www.thefoot.org/imagebank.htm

Different views of the bony anatomy and conditions of the foot and ankle.

Dr. Foot's Health Information Reference

www.doctorfoot.com/diagrams.html

Illustrations of the foot and ankle (muscles, nerves, bones, and deformities).

Worksheet exercise

As an aid to learning, for in-class or out-of-class assignments, or for testing, tear-out worksheets are found at the end of the text (p. 386).

Anterior and posterior skeletal worksheet (no. 1)

Draw and label on the worksheet the following muscles of the ankle and foot:

a. Tibialis anterior
b. Extensor digitorum longus
c. Peroneus longus
d. Peroneus brevis
e. Peroneus tertius
f. Soleus
g. Gastrocnemius
h. Extensor hallucis longus
i. Tibialis posterior
j. Flexor digitorum longus
k. Flexor hallucis longus

LABORATORY AND REVIEW EXERCISES

1. Locate the following parts of the ankle and foot on a human skeleton and on a subject:
   a. Lateral malleolus
   b. Medial malleolus
   c. Calcaneus
   d. Navicular
   e. Three cuneiform bones
   f. Metatarsal bones
   g. Phalanges
2. How and where can the following muscles be palpated on a human subject?
   a. Tibialis anterior
   b. Extensor digitorum longus
   c. Peroneus longus
   d. Peroneus brevis
   e. Soleus
   f. Gastrocnemius
   g. Extensor hallucis longus
   h. Flexor digitorum longus
   i. Flexor hallucis longus
3. Demonstrate and palpate the following movements:
   a. Plantar flexion
   b. Dorsal flexion
   c. Inversion
   d. Eversion
   e. Flexion of the toes
   f. Extension of the toes
4. List the planes in which each of the following movements occurs. List the respective axis of rotation for each movement in each plane.
   a. Plantar flexion
   b. Dorsal flexion
   c. Inversion
   d. Eversion
   e. Flexion of the toes
   f. Extension of the toes
5. Why are "low arches" and "flat feet" not synonymous terms?

Chapter
11

6. Discuss the value of proper footwear in various sports and activities.
7. What are orthotics and how do they function?
8. Research common foot and ankle disorders, such as flat feet, lateral ankle sprains, high ankle sprains, bunions, plantar fasciitis, and hammertoes. Report your findings in class.
9. Research the anatomical factors relating to the prevalence of inversion versus eversion ankle sprains and report your findings in class.
10. Report orally or in writing on magazine articles that rate running and walking shoes.
11. Have a laboratory partner raise up on the toes (heel raise) with the knees fully extended and then repeat with the knees flexed approximately 20 degrees. Which exercise position appears to be more difficult to maintain for an extended period of time and why? What are the implications for strengthening these muscles? For stretching these muscles?
12. Fill in the muscle analysis chart below by listing the muscles primarily involved in each joint movement.
13. Fill in the antagonistic muscle action chart (p. 311) by listing the muscle(s) or parts of muscles that are antagonistic in their actions to the muscles in the left column.
14. After analyzing each of the exercises in the ankle and foot joint movement analysis chart (p. 312), break each into two primary movement phases such as a lifting phase and lowering phase. For each phase, determine the ankle and foot joint movements occurring, and then list the ankle and foot joint muscles primarily responsible for causing/controlling those movements. Beside each muscle in each movement indicate the type of contraction as follows: I-isometric; C-concentric; E-eccentric.
15. Analyze each skill in the ankle and foot joint sport skill analysis chart (p. 313) and list the movement of the right and left ankle and foot joints in each phase of the skill. You may prefer to list the initial positions that the ankle and foot joints are in for the stance phase. After each movement, list the ankle and foot joint muscle(s) primarily responsible for causing/controlling those movements. Beside each muscle in each movement indicate the type of contraction as follows: I-isometric; C-concentric; E-eccentric. It may be desirable to review the concepts for analysis in Chapter 8 for the various phases.

Muscle analysis chart • Ankle, transverse tarsal and subtalar joints, and toes

Ankle	
Dorsiflexion	Plantar flexion
Transverse tarsal and subtalar joints	
Eversion	Inversion
Toes	
Flexion	Extension

Antagonistic muscle action chart • Ankle, transverse tarsal and subtalar joints, and toes

Agonist	Antagonist
Gastrocnemius	
Soleus	
Tibialis posterior	
Flexor digitorum longus	
Flexor hallucis longus	
Peroneus longus/ peroneus brevis	
Peroneus tertius	
Tibialis anterior	
Extensor digitorum longus	
Extensor hallucis longus	

Ankle and foot joint movement analysis chart

Exercise	Initial movement phase		Secondary movement phase	
	Movement(s)	Agonist(s)–(contraction type)	Movement(s)	Agonist(s)–(contraction type)
Push-up				
Squat				
Dead lift				
Hip sled				
Forward lunge				
Rowing exercise				
Stair machine				

Chapter

11

Ankle and foot joint sport skill analysis chart

Exercise		Stance phase	Preparatory phase	Movement phase	Follow-through phase
Baseball pitch	(R)				
	(L)				
Football punting	(R)				
	(L)				
Walking	(R)				
	(L)				
Softball pitch	(R)				
	(L)				
Soccer pass	(R)				
	(L)				
Batting	(R)				
	(L)				
Bowling	(R)				
	(L)				
Basketball jump shot	(R)				
	(L)				

References

Astrom M, Arvidson T: Alignment and joint motion in the normal foot, *Journal of Orthopaedic and Sports Physical Therapy* 22:5, November 1995.

Booher JM, Thibodeau GA: *Athletic injury assessment,* ed 4, New York, 2000, McGraw-Hill.

Field D: *Anatomy: palpation and surface markings,* ed 3, Oxford, 2001, Butterworth-Heinemann.

Franco AH: Pes cavus and pes planus—analysis and treatment, *Physical Therapy* 67:688, May 1987.

Gench BE, Hinson MM, Harvey PT: *Anatomical kinesiology,* Dubuque, IA, 1995, Eddie Bowers.

Hamilton N, Luttgens K: *Kinesiology: scientific basis of human motion,* ed 10, Boston, 2002, McGraw-Hill.

Henderson J: Baring the soles, *Runners World* 22:14, November 1987.

Hislop HJ, Montgomery J: *Daniels and Worthingham's muscle testing: techniques of manual examination,* ed. 7, Philadelphia, 2002, Saunders.

Lindsay DT: *Functional human anatomy,* St. Louis, 1996, Mosby.

Magee DJ: *Orthopedic physical assessment,* ed 4, Philadelphia, 2002, Saunders.

Muscolino JE: *The muscular system manual: the skeletal muscles of the human body,* ed 2, St. Louis, 2003, Elsevier Mosby.

Oatis CA: *Kinesiology: the mechanics and pathomechanics of human movement,* Philadelphia, 2004, Lippincott Williams & Wilkins.

Prentice WE: *Arnheim's principles of athletic training,* ed 12, New York, 2006, McGraw-Hill.

Robinson M: Feet first, *Coach and Athlete* 44:30, August–September 1981.

Rockar PA: The subtalar joint: anatomy and joint motion, *Journal of Orthopaedic and Sports Physical Therapy* 21:6, June 1995.

Sammarco GJ: Foot and ankle injuries in sports, *American Journal of Sports Medicine* 14:6, November–December 1986.

Seeley RR, Stephens TD, Tate P: *Anatomy & physiology,* ed 7, New York, 2006, McGraw-Hill.

Sieg KW, Adams SP: *Illustrated essentials of musculoskeletal anatomy,* ed 2, Gainesville, FL, 1985, Megabooks.

Stone RJ, Stone JA: *Atlas of the skeletal muscles,* New York, 1990, McGraw-Hill.

Thibodeau GA, Patton KT: *Anatomy & physiology,* ed 9, St. Louis, 1993, Mosby.

Van De Graaff KM: *Human anatomy,* ed 6, Dubuque, IA, 2002, McGraw-Hill.

Chapter
11

The Trunk and Spinal Column

Objectives

- To identify and differentiate the different types of vertebrae in the spinal column

- To label on a skeletal chart the types of vertebrae and their important features

- To draw and label on a skeletal chart some of the muscles of the trunk and the spinal column

- To demonstrate and palpate with a fellow student the movements of the spine and trunk and list their respective planes of motion and axes of rotation

- To palpate on a human subject some of the muscles of the trunk and spinal column

- To list and organize the muscles that produce the primary movements of the trunk and spinal column and their antagonists

---

Online Learning Center Resources

Visit *Manual of Structural Kinesiology's* **Online Learning Center** at **www.mhhe.com/floyd16e** for additional information and study material for this chapter, including:

- *Self-grading quizzes*
- *Anatomy flashcards*
- *Animations*

The truck and spinal column present problems in kinesiology that are not found in the study of other parts of the body. The vertebral column is quite elaborate, consisting of 24 intricate and complex articulating vertebrae with an additional nine nonmovable vertebrae. These vertebrae contain the spinal column, with its 31 pairs of spinal nerves. Most would agree that it is one of the more complex parts of the human body.

The anterior portion of the trunk contains the abdominal muscles, which are somewhat different from other muscles in that some sections are linked by fascia and tendinous bands and thus do not attach from bone to bone. In addition, there are many small intrinsic muscles acting on the head, vertebral column, and thorax that assist in spinal stabilization or respiration, depending on their location. These muscles are generally too deep to palpate and consequently will not be given the full attention that the larger superficial muscles will receive in this chapter.

Bones

Vertebral column

The intricate and complex bony structure of the vertebral column consists of 24 articulating vertebrae and nine that are fused together (Fig. 12.1). The column is further divided into the seven cervical (neck) vertebrae, 12 thoracic (chest) vertebrae, and five lumbar (lower back) vertebrae. The sacrum (posterior pelvic girdle) and the coccyx

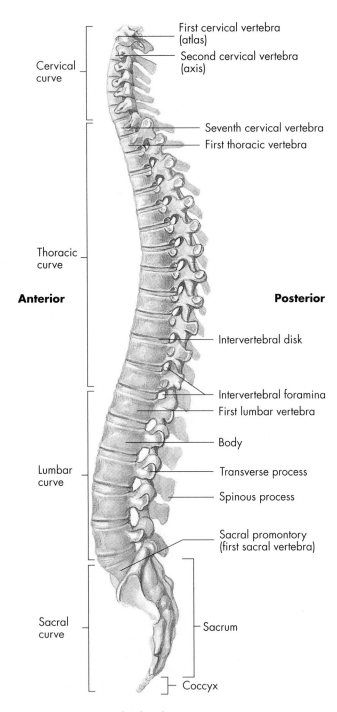

FIG. 12.1 ● Vertebral column.

From Seeley RR, et al: *Anatomy & physiology,* ed 3, St. Louis, 1995, Mosby.

curves are concave posteriorly and convex anteriorly. The normal curves of the spine enable it to absorb blows and shocks.

The bones in each region of the spine have slightly different sizes and shapes to allow for various functions (Fig. 12.2). The vertebrae increase in size from the cervical region to the lumbar region, primarily because they have to support more weight in the lower back than in the neck. The first two cervical vertebrae are known as the atlas and axis, respectively. The vertebrae C2 through L5 have similar architecture: Each has a bony block anteriorly, known as the body, a vertebral foramen centrally for the spinal cord to pass through, a transverse process projecting out laterally to each side, and a spinous process projecting posteriorly that is easily palpable.

Undesirable deviations from the normal curvatures occur due to a number of factors. Increased posterior concavity of the lumbar and cervical curves is known as **lordosis**, while increased anterior concavity of the normal thoracic curve is known as **kyphosis**. The lumbar spine may have a reduction of its normal lordotic curve, resulting in a flat-back appearance referred to as **lumbar kyphosis**. **Scoliosis** refers to lateral curvatures or sideward deviations of the spine.

Thorax

The skeletal foundation of the thorax is formed by 12 pairs of ribs (Fig. 12.3). Seven pairs are true ribs, in that they attach directly to the sternum. Five pairs are considered false ribs. Of these, three pairs attach indirectly to the sternum and two pairs are floating ribs, in that their ends are free. The manubrium, the body of the sternum, and the xiphoid process are the other bones of the thorax. All of the ribs are attached posteriorly to the thoracic vertebrae.

Key bony landmarks for locating the muscles of the neck include the mastoid process, transverse and spinous processes of the cervical spine, spinous processes of the upper four thoracic vertebrae, manubrium of the sternum, and medial clavicle. The spinous and transverse processes of the thoracic spine and the posterior ribs are key areas of attachment for the posterior muscles of the spine. The anterior trunk muscles have attachments on the borders of the lower eight ribs, the costal cartilages of the ribs, the iliac crest, and the pubic crest. The transverse processes of the upper four lumbar vertebrae also serve as points of insertion for the quadratus lumborum, along with the lower border of the twelfth rib.

(tail bone) consist of five and four fused vertebrae, respectively. The first two cervical vertebrae are unique in that their shapes allow for extensive rotary movements of the head to the sides, as well as forward and backward. The spine has three normal curves within its movable vertebrae. The thoracic spine curve is concave anteriorly and convex posteriorly, while the cervical and lumbar

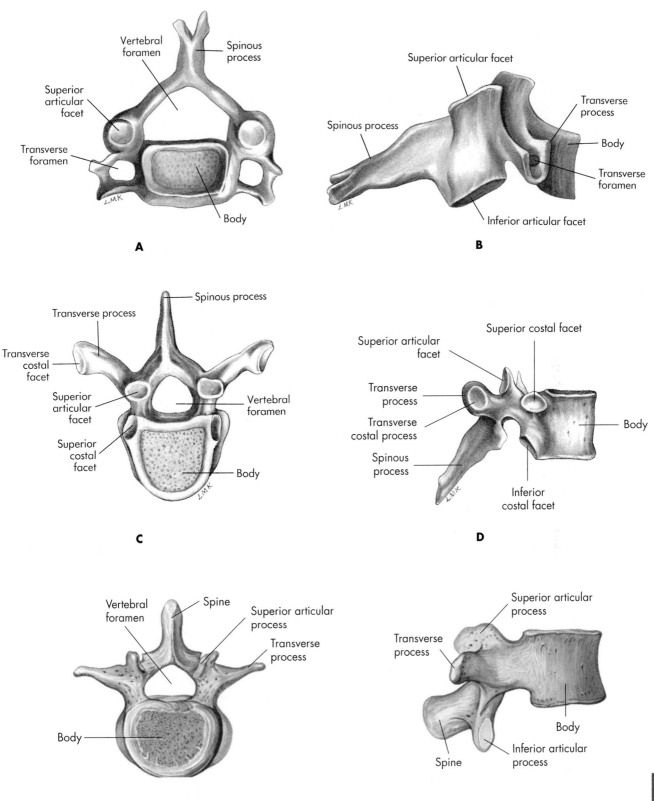

FIG. 12.2 ● Vertebral column. **A,** Typical cervical vertebra viewed from above; **B,** Typical cervical vertebra viewed from the side; **C,** Typical thoracic vertebra viewed from above; **D,** Typical thoracic vertebra viewed from the side; **E,** Third lumbar vertebra viewed from above; **F,** Third lumbar vertebra viewed from the side.

From Anthony CP, Kolthoff NJ: *Textbook of anatomy and physiology,* ed 9, St. Louis, 1975, Mosby.

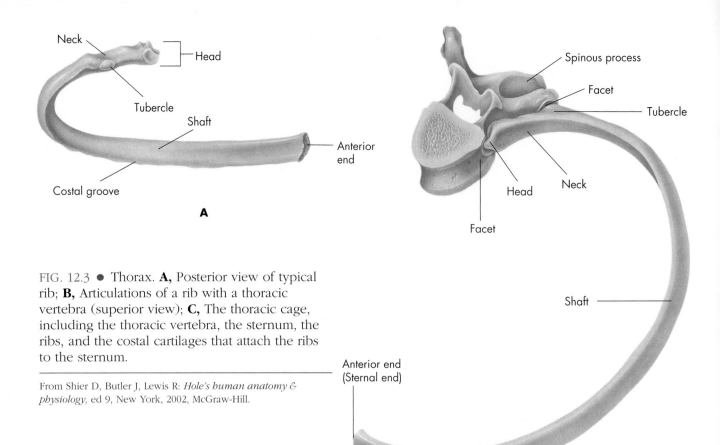

A

B

FIG. 12.3 ● Thorax. **A,** Posterior view of typical rib; **B,** Articulations of a rib with a thoracic vertebra (superior view); **C,** The thoracic cage, including the thoracic vertebra, the sternum, the ribs, and the costal cartilages that attach the ribs to the sternum.

From Shier D, Butler J, Lewis R: *Hole's human anatomy & physiology,* ed 9, New York, 2002, McGraw-Hill.

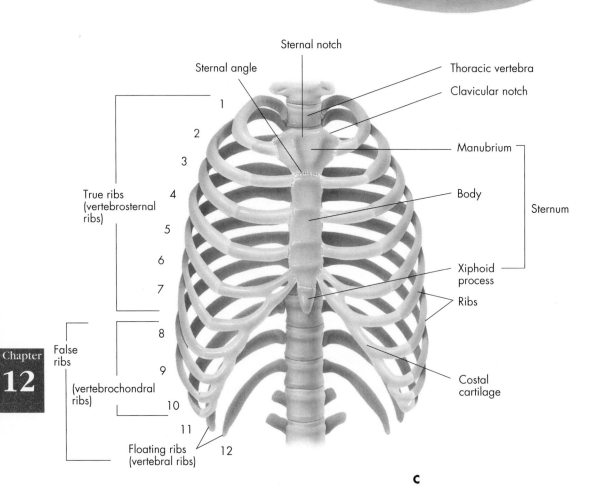

C

Joints

The first joint in the axial skeleton is the atlanto-occipital joint, formed by the occipital condyles of the skull sitting on the articular fossa of the first vertebra, which allows flexion and extension. The atlas (C1) in turn sits on the axis (C2) to form the atlantoaxial joint (Fig. 12.4, *A*). Except for the atlantoaxial joint, there is not a great deal of movement possible between any two vertebrae. However, the cumulative effect of combining the movement from several adjacent vertebrae allows for substantial movements within a given area. Most of the rotation within the cervical region occurs in the atlantoaxial joint, which is classified as a trochoid or pivot-type joint. The remainder of the vertebral articulations are classified as arthrodial or gliding-type joints because of their limited gliding movements.

Gliding movement occurs between the superior and inferior articular processes that form the facet joints of the vertebrae, as depicted in Figs. 12.2 and 12.4, *B*. Located in between and adhering to the articular cartilage of the vertebral bodies are the intervertebral disks (Fig. 12.4, *C*). These disks are composed of an outer rim of dense fibrocartilage known as the annulus fibrosus and a central gelatinous, pulpy substance known as the nucleus pulposus. This arrangement of compressed elastic material allows compression in all directions, along with torsion. With age, injury, or improper use of the spine, the intervertebral disks become less resilient, resulting in a weakened annulus fibrosus. Substantial weakening combined with compression can result in the nucleus protruding through the annulus, which is known as a herniated nucleus pulposus. Commonly referred to as a herniated or "slipped" disk, this protrusion puts pressure on the spinal nerve root, causing a variety of symptoms, including radiating pain, tingling, numbness, and weakness in the dermatomes and myotomes of the extremity supplied by the spinal nerve (Fig. 12.5).

A substantial number of low back problems are caused by improper use of the back over time. These improper mechanics often result in acute strains and muscle spasm of the lumbar extensors and chronic mechanical changes leading to disk herniation. Most problems occur from using the relatively small back muscles to lift objects from a lumbar spine flexed position instead of keeping the lumbar spine in a neutral position

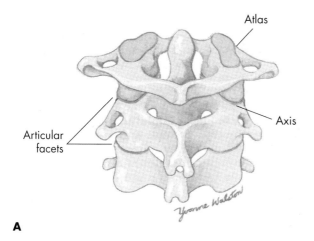

A

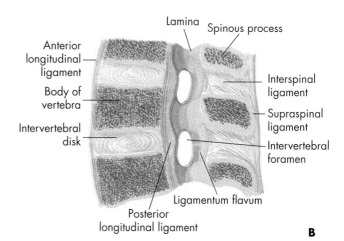

B

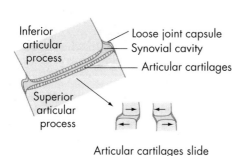

C

FIG. 12.4 ● Articular facets of vertebrae. **A,** The facets of superior and inferior articular processes articulate between adjacent cervical vertebrae; **B,** Ligaments limit motion between vertebrae, shown in sagittal section through three lumbar vertebrae; **C,** Articular cartilages slide back and forth on each other, and the loose articular capsule allows this motion.

Chapter
12

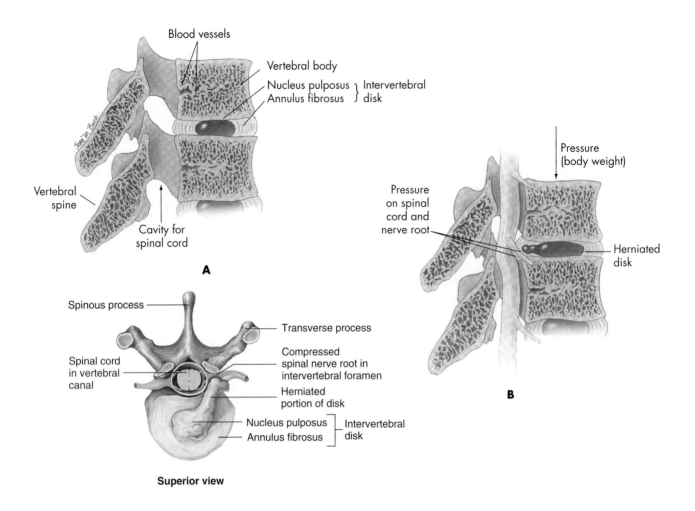

Blood vessels

Vertebral body

Nucleus pulposus } Intervertebral
Annulus fibrosus } disk

Vertebral spine

Cavity for spinal cord

A

Pressure (body weight)

Pressure on spinal cord and nerve root

Herniated disk

B

Spinous process

Transverse process

Spinal cord in vertebral canal

Compressed spinal nerve root in intervertebral foramen

Herniated portion of disk

Nucleus pulposus } Intervertebral
Annulus fibrosus } disk

Superior view

FIG. 12.5 ● Intervertebral disks. **A,** Sagittal section of normal disks; **B,** Sagittal section of herniated disks; **C,** Herniated disk, superior view.

From Thibodeau GA, Paton KT: *Anatomy & physiology,* ed 9, St. Louis, 1993, Mosby; Seeley RR, Stephens TD, Tate P: *Anatomy & physiology,* ed 7, New York, 2006, McGraw-Hill.

while squatting and using the larger, more powerful muscles of the lower extremity. Additionally, our lifestyles chronically place us in lumbar flexion, which over time leads to a gradual loss of lumbar lordosis. This "flat-back syndrome" results in increased pressure on the lumbar disk and intermittent or chronic low back pain.

Most of the spinal column movement occurs in the cervical and lumbar regions. There is, of course, some thoracic movement, but it is slight in comparison with that of the neck and low back. In discussing movements of the head, it must be remembered that this movement occurs between the cranium and the first cervical vertebra, as well as within the cervical vertebrae. With the understanding that these motions usually occur together, for simplification purposes this text re-

fers to all movements of the head and neck as cervical movements. Similarly, in discussing trunk movements, lumbar motion terminology is used to describe the combined motion that occurs in both the thoracic and lumbar regions. A closer investigation of specific motion between any two vertebrae is beyond the scope of this text.

The cervical region (Fig. 12.6) can flex 45 degrees and extend 45 degrees. The cervical area laterally flexes 45 degrees and can rotate approximately 60 degrees. The lumbar spine, accounting for most of the trunk movement, (Fig. 12.7) flexes approximately 80 degrees and extends 20 to 30 degrees. Lumbar lateral flexion to each side is usually within 35 degrees, and approximately 45 degrees of rotation occurs to the left and right.

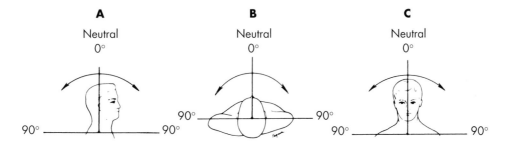

FIG. 12.6 ● Active ROM of the cervical spine. **A,** Flexion and extension. These motions can be estimated in degrees or indicated by the distance the chin lacks from touching the chest; **B,** Rotation can be estimated in degrees or in percentages of motion compared in each direction; **C,** Lateral flexion can be estimated in degrees or indicated by the number of inches the ear lacks from reaching the shoulders.

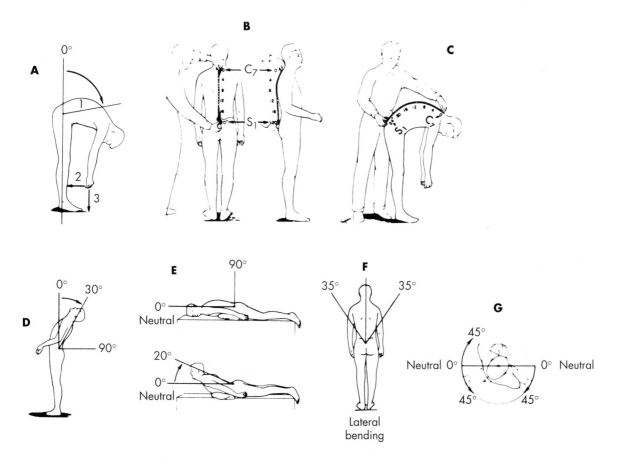

FIG. 12.7 ● Active ROM of the thoracic and lumbar spine. **A,** Forward flexion. Motion can be estimated in degrees as indicated by 1 or measurement of fingertips to leg or from floor as indicated by 2 and 3, respectively; **B** and **C,** Flexion tape measuring method; **D,** Hyperextension with the athlete standing; **E,** Hyperextension with the athlete lying prone; **F,** Lateral bending; **G,** Rotation of the spine.

Movements   FIG. 12.8

Spinal movements are often preceded by the name given to the region of movement. For example, flexion of the trunk at the lumbar spine is known as lumbar flexion, and extension of the neck is often referred to as cervical extension. Additionally, as was discussed in Chapter 9, the pelvic girdle rotates as a unit due to movement occurring in the hip joints and the lumbar spine. Refer to Table 9.1.

Spinal flexion: anterior movement of the spine in the sagittal plane; in the cervical region, the head moves toward the chest; in the lumbar region, the thorax moves toward the pelvis

Spinal extension: return from flexion; posterior movement of the spine in the sagittal plane; in the cervical spine, the head moves away from the chest; in the lumbar spine, the thorax moves away from the pelvis; sometimes referred to as hyperextension

Lateral flexion (left or right): sometimes referred to as side bending; the head moves laterally toward the shoulder, and the thorax moves laterally toward the pelvis; both in the frontal plane

Spinal rotation (left or right): rotary movement of the spine in the transverse plane; the chin rotates from neutral toward the shoulder, and the thorax rotates toward one iliac crest

Reduction: return movement from lateral flexion to neutral in the frontal plane

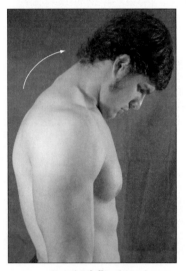

Cervical flexion

A

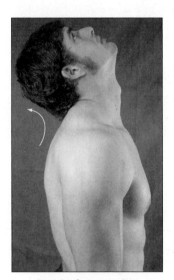

Cervical extension (hyperextension)

B

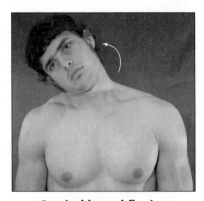

Cervical lateral flexion to the right

C

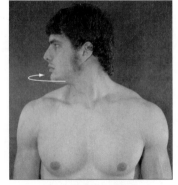

Cervical rotation to the right

D

FIG. 12.8 ● Movements of the spine.

Chapter
12

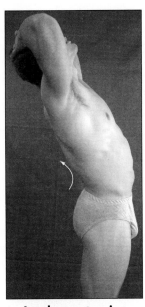
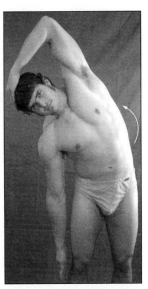
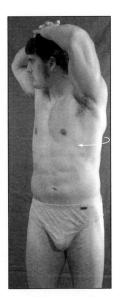

Lumbar flexion	**Lumbar extension (hyperextension)**	**Lumbar lateral flexion to the right**	**Lumbar rotation to the right**
E	F	G	H

FIG. 12.8 (continued) ● Movements of the spine.

Trunk and spinal column muscles

The largest muscle group in this area is the erector spinae (sacrospinalis), which extends on each side of the spinal column from the pelvic region to the cranium. It is divided into three muscles: the spinalis, the longissimus, and the iliocostalis. From the medial to the lateral side, it has attachments in the lumbar, thoracic, and cervical regions. Thus, the erector spinae group is actually made up of nine muscles. Additionally, the sternocleidomastoid and splenius muscles are large muscles involved in cervical and head movements. Large abdominal muscles involved in lumbar movements include the rectus abdominis, external oblique abdominal, internal oblique abdominal, and quadratus lumborum (Table 12.1)

Numerous small muscles are found in the spinal column region. Many of them originate on one vertebra and insert on the next vertebra. They are important in the functioning of the spine, but knowledge of these muscles is of limited value to most people who use this text. Consequently, discussion will concentrate on the larger muscles primarily involved in trunk and spinal column movements (Table 12.1) and will only briefly address the smaller muscles.

So that the muscles of the trunk and spinal column may be better understood, they may be grouped according to both location and function. It should be noted that some muscles have multiple segments. As a result, one segment of a particular muscle may be located and perform movement in one region, while another segment of the same muscle may be located in another region to perform movements in that region. Many of the muscles of the trunk and spinal column function in moving the spine as well as in aiding respiration. All of the muscles of the thorax are primarily involved in respiration. The abdominal wall muscles are different from other muscles that have been studied. They do not go from bone to bone but attach into an aponeurosis (fascia) around the rectus abdominis area. They are the external oblique abdominal, internal oblique abdominal, and transversus abdominis.

Chapter
12

TABLE 12.1 • Agonist muscles of the spine

	Muscle	Origin	Insertion	Action	Plane of motion	Palpation	Innervation
Anterior neck	Sternocleid-omastoid (both sides)	Manubrium of sternum, anterior superior surface medial clavicle	Mastoid process	Extension of head at atlantooccipital joint	Sagittal	Anterolateral neck, diagonally between the origin and insertion, particularly with rotation to contralateral side	Spinal accessory (Cr 11, C2–C3)
				Flexion of the cervical spine	Sagittal		
				Each side: rotation to contralateral side	Transverse		
				Each side: lateral flexion to ipsilateral side	Frontal		
Posterior neck	Splenius cervicis	Spinous processes of the 3rd through the 6th thoracic vertebrae	Transverse processes of the first three cervical vertebrae	Both sides: extension of the cervical spine	Sagittal	Palpate in lower posterior cervical spine just medial to inferior levator scapulae with resisted ipsilateral rotation	Posterior lateral branches of cervical nerves four through eight (C4–C8)
				Each side: rotation to ipsilateral side	Transverse		
				Each side: lateral flexion to ipsilateral side	Frontal		
	Splenius capitis	Lower 1/2 of the ligamentum nuchae and the spinous processes of the 7th cervical and the upper three or four thoracic vertebrae	Mastoid process and occipital bone	Both sides: extension of the head and cervical spine	Sagittal	Deep to trapezius inferiorly and sternocleidomastoid superiorly; with subject seated palpate in posterior triangle of neck between upper trapezius and sternocleidomastoid with resisted rotation to ipsilateral side	Posterior lateral branches of cervical nerves four through eight (C4–C8)
				Each side: rotation to ipsilateral side	Transverse		
				Each side: lateral flexion to ipsilateral side	Frontal		
Posterior spine	Erector spinae: Iliocostalis	Medial iliac crest, thoracolumbar aponeurosis from sacrum, posterior ribs 3–12	Posterior ribs 1–12, cervical 4–7 transverse processes	Extension of the spine	Sagittal	Deep and difficult to distinguish from other muscles in the cervical and thoracic regions; with subject prone, palpate immediately lateral to spinous processes in lumbar region with active extension	Posterior branches of the spinal nerves
				Anterior pelvic rotation	Sagittal		
				Lateral flexion of the spine	Frontal		
				Lateral pelvic rotation to contralateral side	Frontal		
				Ipsilateral rotation of the spine and head	Transverse		
	Erector spinae: Longissimus	Medial iliac crest, thoracolumbar aponeurosis from sacrum, lumbar 1–5 transverse processes, and thoracic 1–5 transverse processes, cervical 5–7 articular processes	Cervical 2–6 spinous processes, thoracic 1–12 transverse processes, lower 9 ribs, mastoid process	Extension of the spine and head	Sagittal		Posterior branches of the spinal nerves
				Anterior pelvic rotation	Sagittal		
				Lateral flexion of the spine and head	Frontal		
				Lateral pelvic rotation to contralateral side	Frontal		
				Ipsilateral rotation of the spine and head	Transverse		
	Erector spinae: Spinalis	Ligamentum nuchae, 7th cervical spinous process, thoracic 11–12 spinous processes, and lumbar 1–2 spinous processes	2nd cervical spinous process, thoracic 5–12 spinous processes, occipital bone	Extension of the spine	Sagittal		Posterior branches of the spinal nerves
				Anterior pelvic rotation	Sagittal		
				Lateral flexion of the spine and head	Frontal		
				Lateral pelvic rotation to contralateral side	Frontal		
				Ipsilateral rotation of the spine and head	Transverse		

Chapter **12**

TABLE 12.1 (continued) • Agonist muscles of the spine

	Muscle	Origin	Insertion	Action	Plane of motion	Palpation	Innervation
Anterior Trunk	Rectus abdominis	Crest of pubis	Cartilage of 5th, 6th, and 7th ribs and xiphoid process	Both sides: lumbar flexion	Sagittal	Anteromedial surface of abdomen, between rib cage and pubic bone with isometric trunk flexion	Intercostal nerves (T7–T12)
				Both sides: posterior pelvic rotation			
				Each side: weak lateral flexion to ipsilateral side	Frontal		
	External oblique abdominal	Borders of lower eight ribs at side of chest dove-tailing with serratus anterior	Anterior half of crest of ilium, inguinal ligament, crest of pubis, and fascia of rectus abdominis at lower front	Both sides: lumbar flexion	Sagittal	With subject supine palpate lateral to the rectus abdominis between iliac crest and lower ribs with active rotation to the contralateral side	Intercostal nerves (T8–T12), iliohypogastric nerve (T12, L1), and ilioinguinal nerve (L1)
				Both sides: posterior pelvic rotation			
				Each side: lumbar lateral flexion to ipsilateral side	Frontal		
				Each side: lateral pelvic rotation to contralateral side			
				Each side: lumbar rotation to contralateral side	Transverse		
	Internal oblique abdominal	Upper 1/2 of inguinal ligament, anterior 2/3 of crest of ilium, and lumbar fascia	Costal cartilage of 8th, 9th, and 10th ribs and linea alba	Both sides: lumbar flexion	Sagittal	With subject supine palpate anterolateral abdomen between iliac crest and lower ribs with active rotation to the ipsilateral side	Intercostal nerves (T8–T12), iliohypogastric nerve (T12, L1), and ilioinguinal nerve (L1)
				Both sides: posterior pelvic rotation			
				Each side: lumbar lateral flexion to ipsilateral side	Frontal		
				Each side: lateral pelvic rotation to contralateral side			
				Each side: lumbar rotation to ipsilateral side	Transverse		
	Transversus abdominis	Lateral 1/3 of inguinal ligament, inner rim of iliac crest, inner surface of costal cartilages of lower six ribs, lumbar fascia	Crest of the pubis and the iliopectineal line, abdominal aponeurosis to the linea alba	Forced expiration by pulling the abdominal wall inward	Transverse	With subject supine; anterolateral abdomen between iliac crest and lower ribs during forceful exhalation; very difficult to distinguish from abdominal obliques	Intercostal nerves (T7–T12), iliohypogastric nerve (T12, L1), and ilioinguinal nerve (L1)
Lateral lumbar	Quadratus lumborum	Posterior inner lip of the iliac crest	Approximately one half the length of the lower border of the 12th rib and the transverse process of the upper four lumbar vertebrae	Lateral flexion to the ipsilateral side	Frontal	With subject prone just superior to iliac crest and lateral to lumbar erector spinae with isometric lateral flexion	Branches of T12, L1 nerves
				Lateral pelvic rotation to contralateral side			
				Lumbar extension	Sagittal		
				Anterior pelvic rotation			
				Stabilizes the pelvis and lumbar spine	All planes		

Chapter 12

Muscles that move the head

Anterior
    Rectus capitis anterior
    Longus capitis
Posterior
    Longissimus capitis
    Obliquus capitis superior
    Obliquus capitis inferior
    Rectus capitis posterior—major and minor
    Trapezius, superior fibers
    Splenius capitis
    Semispinalis capitis
Lateral
    Rectus capitis lateralis
    Sternocleidomastoid

Muscles of the vertebral column

Superficial
    Erector spinae (sacrospinalis)
        Spinalis—capitis, cervicis, thoracis
        Longissimus—capitis, cervicis, thoracis
        Iliocostalis—cervicis, thoracis, lumborum
    Splenius cervicis
Deep
    Longus colli—superior oblique, inferior oblique, vertical
    Interspinales—entire spinal column
    Intertransversales—entire spinal column
    Multifidus—entire spinal column
    Psoas minor
    Rotatores—entire spinal column
    Semispinalis—cervicis, thoracis

Muscles of the thorax

Diaphragm
Intercostalis—external, internal
Levator costarum
Subcostales
Scalenus—anterior, medius, posterior
Serratus posterior—superior, inferior
Transversus thoracis

Muscles of the abdominal wall

Rectus abdominis
External oblique abdominal (obliquus externus abdominis)
Internal oblique abdominal (obliquus internus abdominis)
Transverse abdominis (transversus abdominis)
Quadratus lumborum

Nerves

Cranial nerve 11 and the spinal nerves of C2 and C3 innervate the sternocleidomastoid. The posterior lateral branches of C4 through C8 innervate the splenius muscles. The entire erector spinae group is supplied by the posterior branches of the spinal nerves, whereas the intercostal nerves of T7 through T12 innervate the rectus abdominis. Both the internal and external oblique abdominal muscles receive innervation from the intercostal nerves (T8–T12), the iliohypogastric nerve (T12, L1), and the ilioinguinal nerve (L1). The same innervation is provided to the transverse abdominis, except that innervation begins with the T7 intercostal nerve. Branches from T12 and L1 supply the quadratus lumborum. Review Figures 2.6, 4.6, 4.7, and 9.16.

Chapter
12

Muscles that move the head

All muscles featured here originate on the cervical vertebrae and insert on the occipital bone of the skull, as implied by their "capitis" name (Figs. 12.9 and 12.10; Table 12.2). Three muscles make

FIG. 12.9 ● Anterior muscles of the neck.

Modified from Lindsay DT: *Functional human anatomy*, St. Louis, 1996, Mosby.

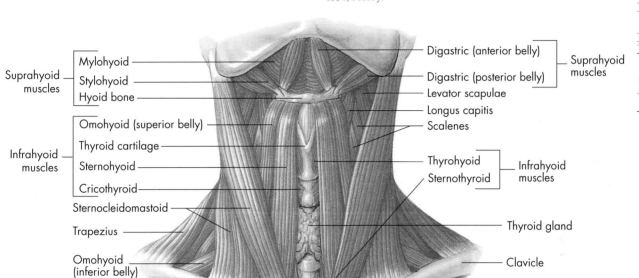

FIG. 12.10 ● Deep muscles of the posterior neck and upper back regions.

Modified from Van De Graaff KM: *Human anatomy*, ed 4, New York, 1995, McGraw-Hill.

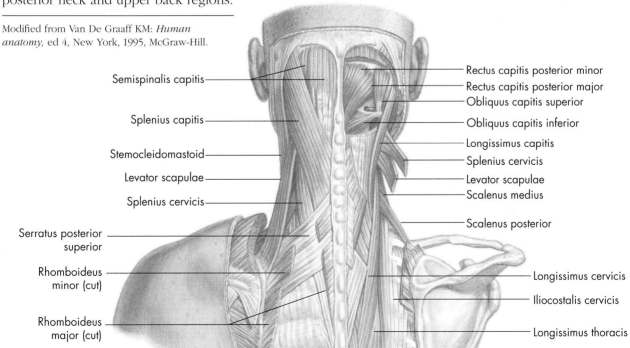

TABLE 12.2 • Muscles that move the head

Muscle	Origin	Insertion	Action	Innervation
Rectus capitis anterior	Anterior surface of lateral mass of atlas	Basilar part of occipital bone anterior to foramen magnum	Flexion of head and stabilization of atlantooccipital joint	C1–C3
Rectus capitis lateralis	Superior surface of transverse processes of atlas	Jugular process of occipital bone	Lateral flexion of head and stabilization of atlantooccipital joint	C1–C3
Rectus capitis posterior (major)	Spinous process of axis	Lateral portion of inferior nuchal line of occipital bone	Extension and rotation of head to ipsilateral side	Posterior rami of C1
Rectus capitis posterior (minor)	Spinous process of atlas	Medial portion of inferior nuchal line of occipital bone	Extension of head	Posterior rami of C1
Longus capitis	Transverse processes of C3–C6	Basilar part of occipital bone	Flexes head and cervical spine	C1–C3
Obliquus capitis superior	Transverse process of atlas	Occipital bone between inferior and superior nuchal line	Extension and lateral flexion of head	Posterior rami of C1
Obliquus capitis inferior	Spinous process of axis	Transverse process of atlas	Rotation of atlas	Posterior rami of C1
Semispinalis capitis	Transverse processes of C4–T7	Occipital bone, between superior and inferior nuchal lines	Extension and contralateral rotation of head	Posterior primary divisions on spinal nerves

up the anterior vertebral muscles—the longus capitis, the rectus capitis anterior, and the rectus capitis lateralis. All are flexors of the head and upper cervical spine. The rectus capitis lateralis laterally flexes the head, in addition to assisting the rectus capitis anterior in stabilizing the atlantooccipital joint.

The rectus capitis posterior major and minor, obliquus capitis superior and inferior, and semispinalis capitis are located posteriorly. All are extensors of the head, except the obliquus capitis inferior, which rotates the atlas. The obliquus

capitis superior assists the rectus capitis lateralis in lateral flexion of the head. In addition to extension, the rectus capitis posterior major is responsible for rotation of the head to the ipsilateral side. It is assisted by the semispinalis capitis, which rotates the head to the contralateral side. The splenius capitis and the sternocleidomastoid (Table 12.1) are much larger and more powerful in moving the head and cervical spine, and will be covered in detail on the following pages. The remaining muscles that act on the cervical spine are addressed with the muscles of the vertebral column.

Chapter
12

Sternocleidomastoid muscle   FIG. 12.11

(ster´no-kli-do-mas-toyd)

Origin

Manubrium of the sternum
Anterior superior surface of medial clavicle

Insertion

Mastoid process

Action

Extension of the head at the atlantooccipital joint
Flexion of the cervical spine

Right side: rotation to the left and lateral flexion to the right

Left side: rotation to the right and lateral flexion to the left

Palpation

Anterolateral side of the neck, diagonally between the origin and insertion, particularly with rotation to contralateral side

Innervation

Spinal accessory nerve (Cr11, C2–C3)

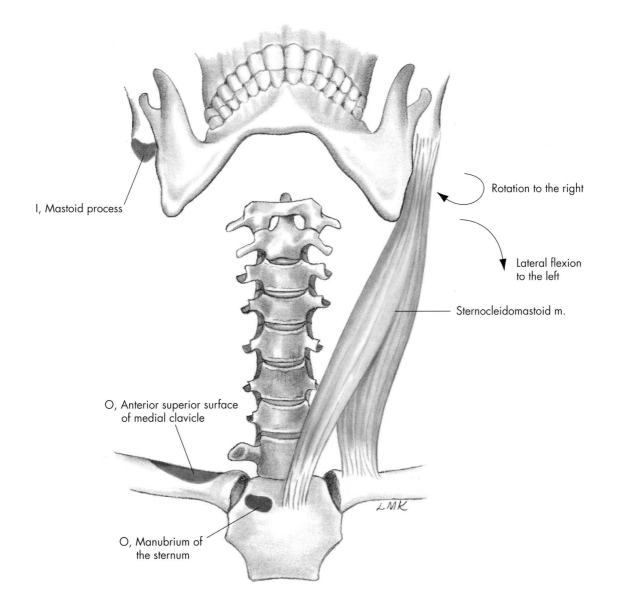

I, Mastoid process

Rotation to the right

Lateral flexion to the left

Sternocleidomastoid m.

O, Anterior superior surface of medial clavicle

O, Manubrium of the sternum

FIG. 12.11 ● Sternocleidomastoid muscle, anterior view. *O,* Origin; *I,* Insertion (continued on next page).

Chapter
12

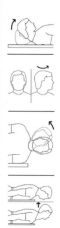

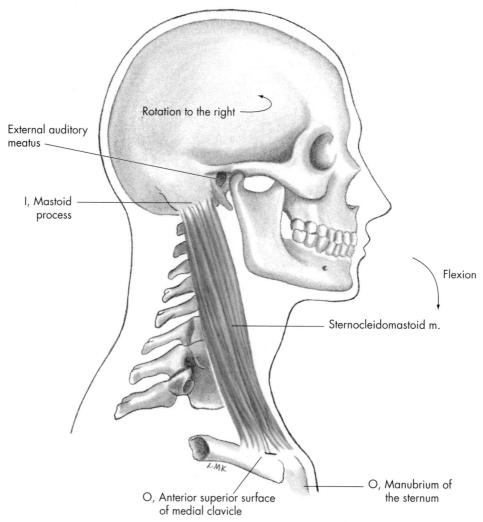

Rotation to the right

External auditory meatus

I, Mastoid process

Flexion

Sternocleidomastoid m.

L.MK

O, Anterior superior surface of medial clavicle

O, Manubrium of the sternum

FIG. 12.11 continued ● Sternocleidomastoid muscle, lateral view. *O,* Origin; *I,* Insertion.

Application, strengthening, and flexibility

The sternocleidomastoid is primarily responsible for flexion and rotation of the head and neck. One side of this muscle may be easily visualized and palpated when rotating the head to the opposite side.

The sternocleidomastoid is easily worked for strength development by placing the hands on the forehead to apply force posteriorly while using these muscles to pull the head forward into flexion. The hand may also be used on one side of the jaw to apply rotary force in the opposite direction while the sternocleidomastoid is contracting concentrically to rotate the head in the direction of the hand.

Cervical hyperextension provides some bilateral stretching of the sternocleidomastoid. Each side may be stretched individually. The right side is stretched by moving into left lateral flexion and right cervical rotation combined with extension. The opposite movements in extension stretch the left side.

Splenius muscles (cervicis, capitis)

FIG. 12.12

(sple´ni-us) (ser´vi-sis) (kap´i-tis)

Origin

Splenius cervicis: spinous processes of the third through the sixth thoracic vertebrae

Splenius capitis: lower half of the ligamentum nuchae and spinous processes of the seventh cervical and the upper three or four thoracic vertebrae

Insertion

Splenius cervicis: transverse processes of the first three cervical vertebrae

Splenius capitis: mastoid process and occipital bone

Action

Both sides: extension of the head (splenius capitis) and neck (splenius capitis and capitis)

Right side: rotation and lateral flexion to the right

Left side: rotation and lateral flexion to the left

Palpation

Splenius capitis: deep to the trapezius inferiorly and the sternocleidomastoid superiorly; with subject seated palpate in the posterior triangle of the neck between the upper trapezius and sternocleidomastoid with resisted rotation to ipsilateral side

Splenius cervicis; palpate in lower posterior cervical spine just medial to inferior levator scapulae with resisted ipsilateral rotation

Innervation

Posterior lateral branches of cervical nerves four through eight (C4–C8)

Application, strengthening, and flexibility

Any movement of the head and neck into extension, particularly extension and rotation, would bring the splenius muscle strongly into play, together with the erector spinae and the upper trapezius muscles. Tone in the splenius muscle tends to hold the head and neck in proper posture position.

A good exercise for the splenius muscle is to lace the fingers behind the head with it in flexion and then to slowly contract the posterior head and neck muscles to move the head and neck into full extension. This exercise may also be performed by using a towel or a partner for resistance.

The entire splenius may be stretched with maximal flexion of the head and cervical spine. The right side can be stretched through combined movements of left rotation, left lateral flexion, and flexion. The same movements to the right side apply stretch to the left side.

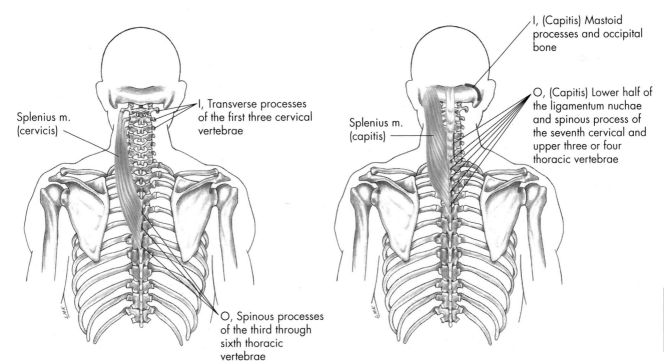

Splenius m. (cervicis)

I, Transverse processes of the first three cervical vertebrae

O, Spinous processes of the third through sixth thoracic vertebrae

I, (Capitis) Mastoid processes and occipital bone

O, (Capitis) Lower half of the ligamentum nuchae and spinous process of the seventh cervical and upper three or four thoracic vertebrae

Splenius m. (capitis)

FIG. 12.12 ● Splenius muscles (cervicis on the left, capitis on the right), posterior view. *O,* Origin; *I,* Insertion.

Muscles of the vertebral column

In the cervical area, the longus colli muscles are located anteriorly and flex the cervical and upper thoracic vertebrae. Posteriorly, the erector spinae group, the transversospinalis group, the interspinal-intertransverse group, and the splenius all run vertically parallel to the spinal column (Fig.12.13, Fig. 12.14, and Table 12.3). This location enables them to extend the spine as well as assist in rotation and lateral flexion. The splenius and erector spinae group are addressed in detail elsewhere in this chapter. The transversospinalis group consists of the semispinalis, multifidus, and rotatores muscles. These muscles all originate on the trans-

verse processes of their respective vertebrae and generally run posteriorly to attach to the spinous processes on the vertebrae just above their vertebrae of origin. All are extensors of the spine and contract to rotate their respective vertebrae to the contralateral side. The interspinal-intertransverse group lies deep to the rotatores and consists of the interspinales and the intertransversarii muscles. As a group, they laterally flex and extend but do not rotate the vertebrae. The interspinales are extensors that connect from the spinous process of one vertebra to the spinous process of the adjacent vertebra. The intertransversarii muscles flex the vertebral column laterally by connecting to the transverse processes of adjacent vertebrae.

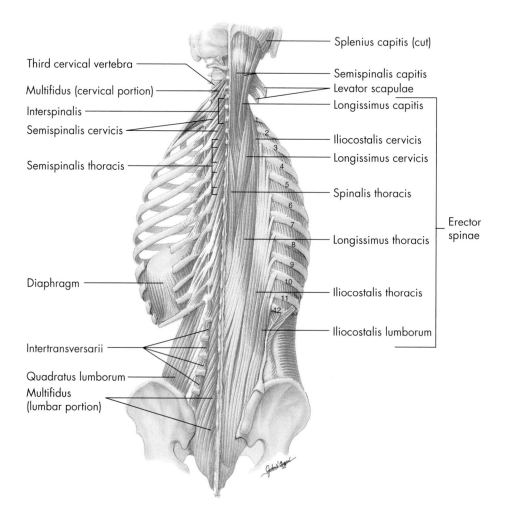

FIG. 12.13 ● Deep back muscles, posterior view. **Right,** The erector spinae group of muscles is demonstrated. **Left,** These muscles have been removed to reveal the deeper back muscles.

Modified from Seeley RR, Stephens TD, Tate P: *Anatomy & physiology,* ed 6, Dubuque, IA, 2003, McGraw-Hill.

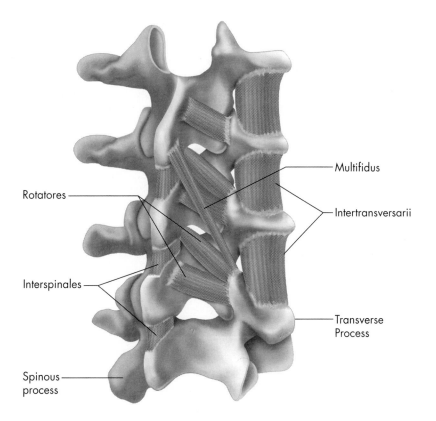

Rotatores

Interspinales

Spinous
process

Multifidus

Intertransversarii

Transverse
Process

Posterolateral view

FIG. 12.14 ● Deep muscles associated with the vertebrae.

From Seeley RR, Stephens TD, Tate P: *Anatomy & physiology,* ed 7, New York, 2006, McGraw-Hill.

Chapter
12

TABLE 12.3 • Muscles of the vertebral column

Muscle	Origin	Insertion	Action	Innervation
Longus colli (superior oblique)	Transverse processes of C3–C5	Anterior arch of atlas	Flexion of cervical spine	C2–C7
Longus colli (inferior oblique)	Bodies of T1–T3	Transverse processes of C5–C6	Flexion of cervical spine	C2–C7
Longus colli (vertical)	Bodies of C5–C7 and T1–T3	Anterior surface of bodies of C2–C4	Flexion of cervical spine	C2–C7
Interspinalis	Spinous process of each vertebra	Spinous process of next vertebra	Extension of spinal column	Posterior primary ramus of spinal nerves
Intertransversarii	Tubercles of transverse processes of each vertebra	Tubercles of transverse processes of next vertebra	Lateral flexion of spinal column	Anterior primary ramus of spinal nerves
Multifidus	Sacrum, iliac spine, transverse processes of lumbar, thoracic, and lower four cervical vertebrae	Spinous processes of 2nd, 3rd, or 4th vertebra above origin	Extension and contralateral rotation of spinal column	Posterior primary ramus of spinal nerves
Rotatores	Transverse process of each vertebra	Base of spinous process of next vertebra above	Extension and contralateral rotation of spinal column	Posterior primary ramus of spinal nerves
Semispinalis cervicis	Transverse processes of T1–T5 or T6	Spinous processes from C2–C5	Extension and contralateral rotation of vertebral column	All divisions, posterior primary ramus of spinal nerves
Semispinalis thoracis	Transverse processes of T6–T10	Spinous processes of C6–C7 and T1–T4	Extension and contralateral rotation of vertebral column	Posterior primary ramus of spinal nerves

Muscles of the thorax

The thoracic muscles are involved almost entirely in respiration (Fig. 12.15). During quiet rest, the diaphragm is responsible for breathing movements. As it contracts and flattens, the thoracic volume is increased and air is inspired to equalize the pressure. When larger amounts of air are needed, such as during exercise, the other thoracic muscles take on a more significant role in inspiration. The scalene muscles elevate the first two ribs to increase the thoracic volume. Further expansion of the chest is accomplished by the external intercostals. Additional muscles of inspiration are the levator costarum and the serratus posterior. Forced expiration occurs with contraction of the internal intercostals, transversus thoracis, and subcostales. All of these muscles are detailed in Table 12.4.

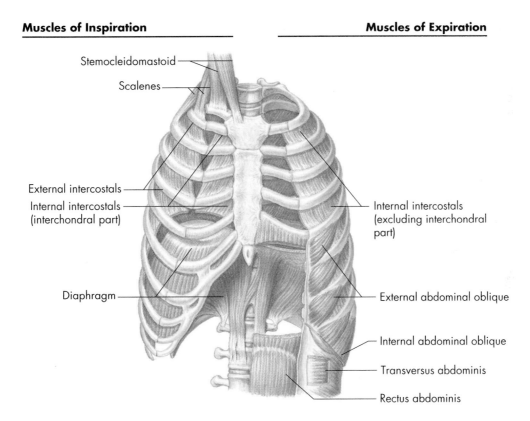

FIG. 12.15 ● Muscles of respiration, anterior view.

Modified from Van De Graaff KM: *Human anatomy,* ed 4, New York, 1995, McGraw-Hill.

Chapter
12

TABLE 12.4 • **Muscles of the thorax**

Muscle	Origin	Insertion	Action	Innervation
Diaphragm	Circumference of thoracic inlet from xiphoid process, costal cartilages 6–12, and lumbar vertebrae	Central tendon of diaphragm	Depresses and draws central tendon forward in inhalation, reduces pressure in thoracic cavity, and increases pressure in abdominal cavity	Phrenic nerve (C3–C5)
Internal intercostals	Longitudinal ridge on inner surface of ribs and costal cartilages	Superior border of next rib below	Elevates costal cartilages of ribs 1–4 during inhalation, depresses all ribs in exhalation	Intercostal branches of T1–T11
External intercostals	Inferior border of ribs	Superior border of next rib below	Elevates ribs	Intercostal branches of T1–T11
Levator costarum	Ends of transverse processes of C7, T2–T12	Outer surface of angle of next rib below origin	Elevates ribs, lateral flexion of thoracic spine	Intercostal nerves
Subcostales	Inner surface of each rib near its angle	Medially on the inner surface of 2nd or 3rd rib below	Draws the ventral part of the ribs downward, decreasing the volume of the thoracic cavity	Intercostal nerves
Scalenus anterior	Transverse processes of C3–C6	Inner border and upper surface of 1st rib	Elevates 1st rib, flexion, lateral flexion, and contralateral rotation of cervical spine	Ventral rami of C5–C6 sometimes C4
Scalenus medius	Transverse processes of C2–C7	Superior surface of 1st rib	Elevates 1st rib, flexion, lateral flexion, and contralateral rotation of cervical spine	Ventral rami of C3–C8
Scalenus posterior	Transverse processes of C5–C7	Outer surface of 2nd rib	Elevates 2nd rib, flexion, lateral flexion, and slight contralateral rotation of cervical spine	Ventral rami of C6–C8
Serratus posterior (superior)	Ligamentum nuchae, spinous processes of C7, T1, and T2 or T3	Superior borders lateral to angles of ribs 2–5	Elevates upper ribs	Branches from anterior primary rami of T1–T4
Serratus posterior (inferior)	Spinous processes of T10–T12 and L1–L3	Inferior borders lateral to angles of ribs 9–12	Counteracts inward pull of diaphragm by drawing last 4 ribs outward and downward	Branches from anterior primary rami of T9–T12
Transversus thoracis	Inner surface of sternum and xiphoid process, sternal ends of costal cartilages of ribs 3–6	Inner surfaces and inferior borders of costal cartilages 3–6	Depresses ribs	Intercostal branches of T3–T6

Chapter

12

Erector spinae muscles*
(sacrospinalis)   FIGS. 12.16, 12.17

(e-rek´tor spi´ne) (sa´kro-spi-na´lis)

Iliocostalis

(il´i-o-kos-ta´lis): lateral layer

Longissimus

(lon-jis´i-mus): middle layer

Spinalis

(spi-na´lis): medial layer

---

*This muscle group includes the iliocostalis, the longissimus dorsi, the spinalis dorsi, and divisions of these muscles in the lumbar, thoracic, and cervical sections of the spinal column.

Origin

Iliocostalis: medial iliac crest, thoracolumbar aponeurosis from sacrum, posterior ribs 3–12

Longissimus: medial iliac crest, thoracolumbar aponeurosis from sacrum, lumbar 1–5 transverse processes and thoracic 1–5 transverse processes, cervical 5–7 articular processes

Spinalis: ligamentum nuchae, seventh cervical spinous process, thoracic 11–12 spinous processes, and lumbar 1–2 spinous processes

Insertion

Iliocostalis: posterior ribs 1–12, cervical 4–7 transverse processes

Longissimus: cervical 2–6 spinous processes, thoracic 1–12 transverse processes, lower nine ribs, mastoid process

Spinalis: second cervical spinous process, thoracic 5–12 spinous processes, occipital bone

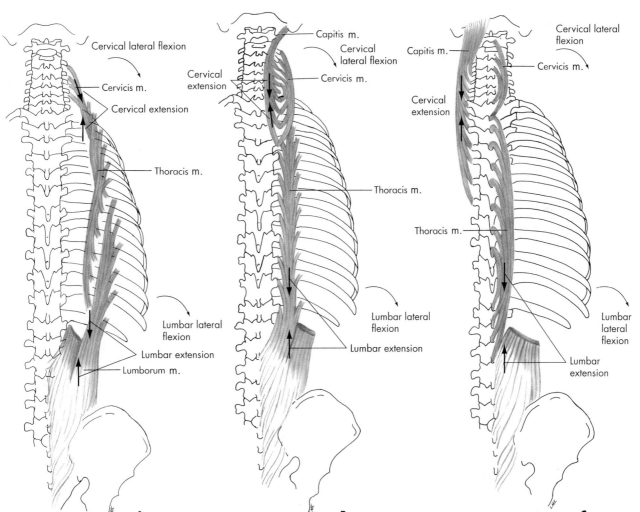

FIG. 12.16 ● Erector spinae (sacrospinalis) muscle, posterior view. **A,** Iliocostalis lumborum, thoracis, and cervicis; **B,** Longissimus thoracis, cervicis, and capitis; **C,** Spinalis thoracis, cervicis, and capitis.

Action

Extension, lateral flexion, and ipsilateral rotation of the spine and head

Anterior pelvic rotation

Lateral pelvic rotation to contralateral side

Palpation

Deep and difficult to distinguish from other muscles in the cervical and thoracic regions; with subject prone, palpate immediately lateral to spinous processes in lumbar region with active extension

Innervation

Posterior branches of the spinal nerves

Application, strengthening, and flexibility

The erector spinae muscle functions best when the pelvis is posteriorly rotated. This lowers the origin of the erector spinae and makes it more effective in keeping the spine straight. As the spine is held straight, the ribs are raised, thus fixing the chest high and consequently making the abdominal muscles more effective in holding the pelvis up in front and flattening the abdominal wall.

An exercise known as the dead lift, employing a barbell, uses the erector spinae in extending the spine. In this exercise, the subject bends over, keeping the arms and legs straight; picks up the barbell; and returns to a standing position. In performing this type of exercise, it is very important to always use correct technique to avoid back injuries. Voluntary static contraction of the erector spinae in the standing position would provide a mild exercise and improve body posture.

The erector spinae and its various divisions may be strengthened through numerous forms of back extension exercises. These are usually done in a prone or face-down position in which the spine is already in some state of flexion. The subject then uses these muscles to move part or all of the spine toward extension against gravity. A weight may be held in the hands behind the head to increase resistance.

Maximal hyperflexion of the entire spine stretches the erector spinae muscle group. Stretch may be isolated to specific segments through specific movements. Maximal flexion of the head and cervical spine stretches the capitis and cervical segments. Flexion combined with lateral flexion to one side accentuates the stretch on the contralateral side. Thoracic and lumbar flexion places the stretch primarily on the thoracis and lumborum segments.

FIG. 12.17 ● Muscles of the back and the neck help move the head (posterior view) and hold the torso erect. The splenius capitis and semispinalis have been removed on the left to show underlying muscles.

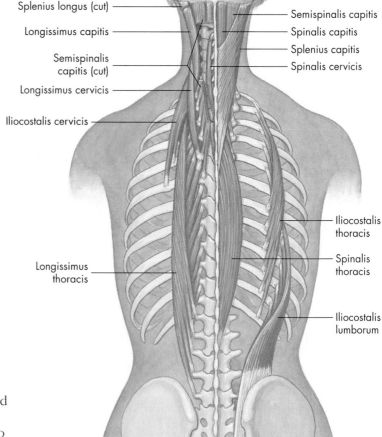

Splenius longus (cut)
Longissimus capitis
Semispinalis capitis (cut)
Longissimus cervicis
Iliocostalis cervicis
Semispinalis capitis
Spinalis capitis
Splenius capitis
Spinalis cervicis
Iliocostalis thoracis
Spinalis thoracis
Iliocostalis lumborum
Longissimus thoracis

Muscles of the abdominal wall

FIGS. 12.18, 12.19, 12.20

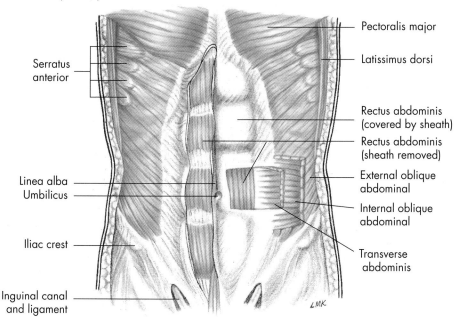

Serratus anterior

Linea alba
Umbilicus

Iliac crest

Inguinal canal and ligament

Pectoralis major

Latissimus dorsi

Rectus abdominis (covered by sheath)

Rectus abdominis (sheath removed)

External oblique abdominal

Internal oblique abdominal

Transverse abdominis

LMK

FIG. 12.18 ● Muscles of the abdomen. External oblique and rectus abdominis. The fibrous sheath around the rectus has been removed on the right side to show the muscle within.

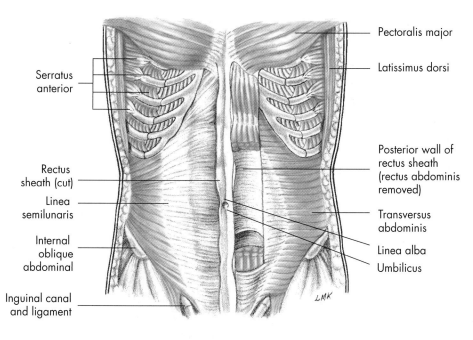

Serratus anterior

Rectus sheath (cut)

Linea semilunaris

Internal oblique abdominal

Inguinal canal and ligament

Pectoralis major

Latissimus dorsi

Posterior wall of rectus sheath (rectus abdominis removed)

Transversus abdominis

Linea alba

Umbilicus

LMK

FIG. 12.19 ● Muscles of the abdomen. The external oblique has been removed on the right to reveal the internal oblique. The external and internal obliques have been removed on the left to reveal the transversus abdominis. The rectus abdominis has been cut to reveal the posterior rectus sheath.

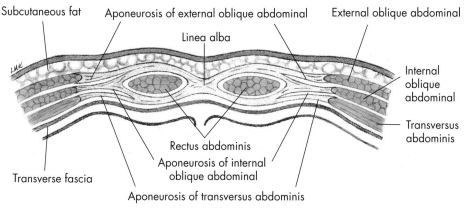

Subcutaneous fat

Aponeurosis of external oblique abdominal

Linea alba

External oblique abdominal

Internal oblique abdominal

Transversus abdominis

Transverse fascia

Rectus abdominis

Aponeurosis of internal oblique abdominal

Aponeurosis of transversus abdominis

FIG. 12.20 ● Abdominal wall. The unique arrangement of the four abdominal muscles with their fascial attachment in and around the rectus abdominis muscle is shown. With no bones for attachments, these muscles can be adequately maintained through exercise.

Chapter
12

Rectus abdominis muscle   FIG. 12.21

(rek´tus ab-dom´i-nis)

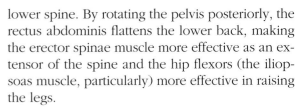

Origin

Crest of the pubis

Insertion

Cartilage of the fifth, sixth, and seventh ribs and the xiphoid process

Action

Both sides: lumbar flexion
Posterior pelvic rotation
Right side: weak lateral flexion to the right
Left side: weak lateral flexion to the left

Palpation

Anteromedial surface of the abdomen, between the rib cage and the pubic bone with isometric trunk flexion

Innervation

Intercostal nerves (T7–T12)

Application, strengthening, and flexibility

The rectus abdominis muscle controls the tilt of the pelvis and the consequent curvature of the lower spine. By rotating the pelvis posteriorly, the rectus abdominis flattens the lower back, making the erector spinae muscle more effective as an extensor of the spine and the hip flexors (the iliopsoas muscle, particularly) more effective in raising the legs.

In a relatively lean person with well-developed abdominals, three distinct sets of lines or depressions may be noted. Each represents an area of tendinous connective tissue connecting or supporting the abdominal arrangement of muscles in lieu of bony attachments. Running vertically from the xiphoid process through the umbilicus to the pubis is the **linea alba**. It divides each rectus abdominis and serves as its medial border. Lateral to each rectus abdominis is the **linea semilunaris**, a crescent, or moon-shaped, line running vertically. This line represents the aponeurosis connecting the lateral border of the rectus abdominis and the medial border of the external and internal abdominal obliques. The **tendinous inscriptions** are horizontal indentations that transect the rectus abdominis at three or more locations, giving the muscle its segmented appearance. Refer to Fig. 12.18.

There are several exercises for the abdominal muscles, such as bent-knee sit-ups, crunches, and isometric contractions. Bent-knee sit-ups with the arms folded across the chest are considered by many to be a safe and efficient exercise. Crunches are also considered to be even more effective for isolating the work to the abdominals. Both of these exercises shorten the iliopsoas muscle and other hip flexors, thus reducing their ability to generate force. Twisting to the left and right brings the oblique muscles into more active contraction. In all of the above exercises, it is important to use proper technique, which involves gradually moving to the up position until the lumbar spine is actively flexed maximally and then slowly returning to the beginning position. Jerking movements using momentum should be avoided. Movement continued beyond full lumbar flexion only exercises the hip flexors, which is not usually an objective. Even though all of these exercises may be helpful in strengthening the abdominals, careful analysis should occur before deciding which are indicated in the presence of various injuries and problems of the low back.

The rectus abdominis is stretched by simultaneously hyperextending both the lumbar and thoracic spine. Extending the hips assists in this process by accentuating the anterior rotation of the pelvis to hyperextend the lumbar spine.

I, Cartilage of fifth, sixth, and seventh ribs, xiphoid process

Lateral flexion

Flexion

Linea alba

Tendinous inscription

Rectus abdominis m.

O, Crest of pubis

FIG. 12.21 ● Rectus abdominis muscle, anterior view. *O*, Origin; *I*, Insertion.

External oblique abdominal muscle   FIG. 12.22

(ek-stur´nel o-bleek´ ab-dom´i-nel)

Origin

Borders of the lower eight ribs at the side of the chest, dovetailing with the serratus anterior muscle*

Insertion

Anterior half of the crest of the ilium, the inguinal ligament, the crest of the pubis, and the fascia of the rectus abdominis muscle at the lower front

Action

Both sides: lumbar flexion
Posterior pelvic rotation
Right side: lumbar lateral flexion to the right and rotation to the left, lateral pelvic rotation to the left
Left side: lumbar lateral flexion to the left and rotation to the right, lateral pelvic rotation to the right

---

*Sometimes the origin and insertion are reversed in anatomy books. This is the result of different interpretations of which bony structure is the more movable. The insertion is considered the most movable part of a muscle.

Palpation

With subject supine palpate lateral to the rectus abdominis between iliac crest and lower ribs with active rotation to the contralateral side

Innervation

Intercostal nerves (T8–T12), iliohypogastric nerve (T12, L1), and ilioinguinal nerve (L1)

Application, strengthening, and flexibility

Working on each side of the abdomen, the external oblique abdominal muscles aid in rotating the trunk when working independently of each other. Working together, they aid the rectus abdominis muscle in its described action. The left external oblique abdominal muscle contracts strongly during sit-ups when the trunk rotates to the right, as in touching the left elbow to the right knee. Rotating to the left brings the right external oblique into action.

Each side of the external oblique must be stretched individually. The right side is stretched by moving into extreme left lateral flexion combined with extension, or by extreme lumbar rotation to the right combined with extension. The opposite movements in extension stretch the left side.

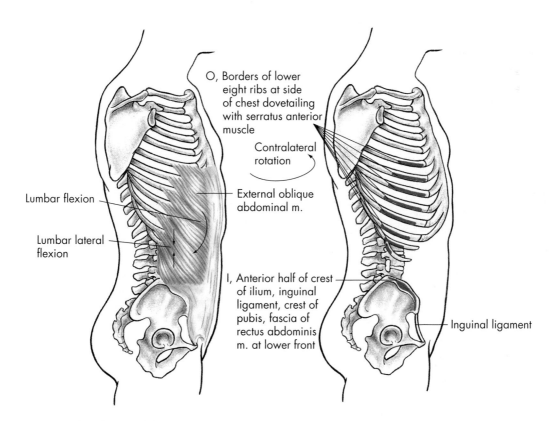

FIG. 12.22 ● External oblique abdominal muscle, lateral view. *O,* Origin; *I,* Insertion.

Internal oblique abdominal muscle

FIG. 12.23

(in-ter´nel o-bleek ab-dom´i-nel)

Origin

Upper half of the inguinal ligament, anterior two-thirds of the crest of the ilium, and lumbar fascia

Insertion

Costal cartilages of the eighth, ninth, and tenth ribs and the linea alba

Action

Both sides: lumbar flexion
Posterior pelvic rotation
Right side: lumbar lateral flexion to the right and rotation to the right, lateral pelvic rotation to the left
Left side: lumbar lateral flexion to the left and rotation to the left, lateral pelvic rotation to the right

Palpation

With subject supine, palpate anterolateral abdomen between iliac crest and lower ribs with active rotation to the ipsilateral side

Innervation

Intercostal nerves (T8–T12), iliohypogastric nerve (T12, L1), and ilioinguinal nerve (L1)

Application, strengthening, and flexibility

The internal oblique abdominal muscles run diagonally in the direction opposite to that of the external obliques. The left internal oblique rotates to the left, and the right internal oblique rotates to the right.

In touching the left elbow to the right knee in crunches, the left external oblique and the right internal oblique abdominal muscles contract at the same time, assisting the rectus abdominis muscle in flexing the trunk to make completion of the movement possible. In rotary movements, the internal oblique and the external oblique on opposite sides from each other always work together.

Like the external oblique, each side of the internal oblique must be stretched individually. The right side is stretched by moving into extreme left lateral flexion and extreme left lumbar rotation combined with extension. The same movements to the right combined with extension stretch the left side.

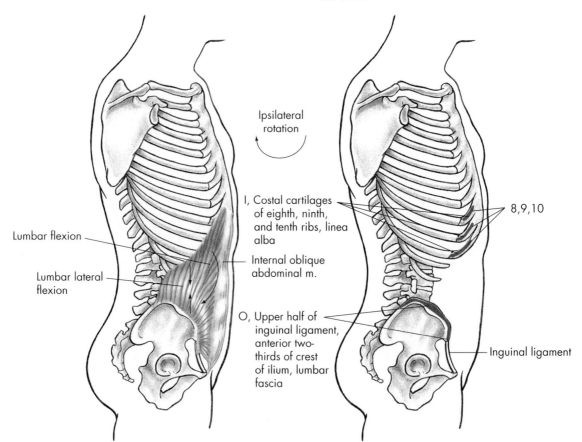

FIG. 12.23 ● Internal oblique abdominal muscle, lateral view. *O*, Origin; *I*, Insertion.

Transversus abdominis muscle

FIG.12.24

(trans-vurs´us ab-dom´i-nis)

Origin

Lateral third of the inguinal ligament, inner rim of the iliac crest, inner surface of the costal cartilages of the lower six ribs, lumbar fascia

Insertion

Crest of the pubis and the iliopectineal line
Abdominal aponeurosis to the linea alba

Action

Forced expiration by pulling the abdominal wall inward

Palpation

With subject supine; anterolateral abdomen between iliac crest and lower ribs during forceful exhalation; very difficult to distinguish from the abdominal obliques

Innervation

Intercostal nerves (T7–T12), iliohypogastric nerve (T12, L1), and ilioinguinal nerve (L1)

Application, strengthening, and flexibility

The transversus abdominis is the chief muscle of forced expiration and is effective—together with the rectus abdominis, the external oblique abdominal, and the internal oblique abdominal muscles—in helping hold the abdomen flat. This abdominal flattening and forced expulsion of the abdominal contents is the only action of this muscle.

The transversus abdominis muscle is exercised effectively by attempting to draw the abdominal contents back toward the spine. This may be done isometrically in the supine position or while standing. A maximal inspiration held in the abdomen applies stretch.

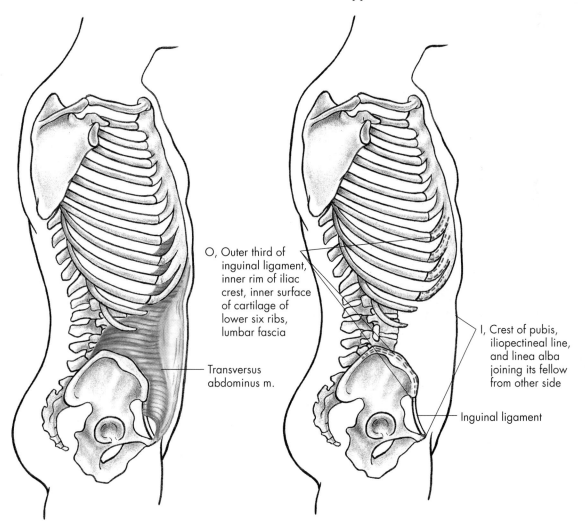

O, Outer third of inguinal ligament, inner rim of iliac crest, inner surface of cartilage of lower six ribs, lumbar fascia

Transversus abdominus m.

I, Crest of pubis, iliopectineal line, and linea alba joining its fellow from other side

Inguinal ligament

FIG. 12.24 ● Transversus abdominis muscle, lateral view. *O*, Origin; *I*, Insertion.

Quadratus lumborum muscle   FIG. 12.25

(kwad-ra´tus lum-bo´rum)

Origin

Posterior inner lip of the iliac crest

Insertion

Approximately one-half the length of the lower border of the twelfth rib and the transverse process of the upper four lumbar vertebrae

Action

Lateral flexion to the ipsilateral side
Stabilizes the pelvis and lumbar spine
Extension of the lumbar spine
Anterior pelvic rotation
Lateral pelvic rotation to contralateral side

Palpation

With subject prone just superior to iliac crest and lateral to lumbar erector spinae with isometric lateral flexion

Innervation

Branches of T12, L1 nerves

Application, strengthening, and flexibility

The quadratus lumborum is important in lumbar lateral flexion and in elevating the pelvis on the same side in the standing position. Trunk rotation and lateral flexion movements against resistance are good exercises for development of this muscle. The position of the body relative to gravity may be changed to increase resistance on this and other trunk and abdominal muscles. Left lumbar lateral flexion while in lumbar flexion stretches the right quadratus lumborum and vice versa.

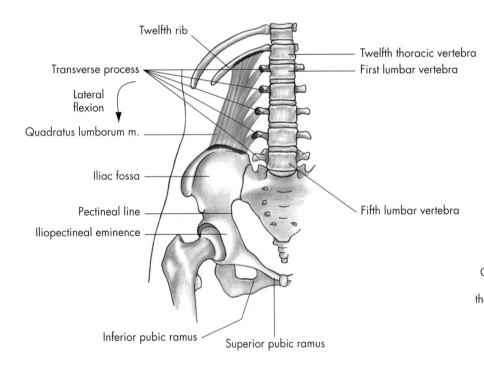

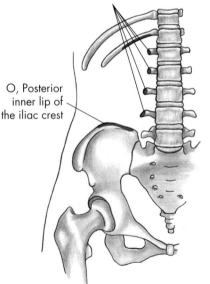

I, Approximately one-half length of lower border of the twelfth rib and transverse process of the upper four lumbar vertebrae

O, Posterior inner lip of the iliac crest

FIG. 12.25 ● Quadratus lumborum muscle. *O,* Origin; *I,* Insertion.

Chapter 12

Web sites

Radiologic Anatomy Browser

http://radlinux1.usuf1.usuhs.mil/rad/iong

Numerous radiological views of the musculoskeletal system.

Loyola University Medical Center: Structure of the Human Body

www.meddean.luc.edu/lumen/meded/grossanatomy/index.htm

Excellent site with many slides, dissections, tutorials, etc., for the study of human anatomy.

University of Arkansas Medical School Gross Anatomy for Medical Students

http://anatomy.uams.edu/anatomyhtml/gross.html

Dissections, anatomy tables, atlas images, links, etc.

Wheeless' Textbook of Orthopaedics

www.wheelessonline.com

Extensive index of links to the fractures, joints, muscles, nerves, trauma, medications, medical topics, and lab tests, as well as links to orthopaedic journals and other orthopaedic and medical news.

Premiere Medical Search Engine

www.medsite.com

Allows the reader to enter any medical condition and it will search the net to find relevant articles.

Virtual Hospital

www.vh.org

Numerous slides, patient information, etc.

Core Stability

www.brianmac.demon.co.uk/corestab.htm

The muscles of the trunk, training techniques, and exercises.

Become Healthy Now.com: The Spine

www.becomehealthynow.com/category/bodyspine

Anatomy and function of spine.

Spine Universe

www.spineuniverse.com

Information on the spine for educating the public about technologies, services, treatments, and research available on spinal disorders.

Stabilization Sensibility

www.calainc.org/Handouts/Participant_handouts/0504%20Stablization.pdf

A discussion on muscles of the abdomen.

Hospital for Joint Disease Spine Center

www.hjdspine.org/hospitals/hjd/hjdspine/musclesandligaments.htm

Muscles and ligaments of the spine.

Worksheet exercise

As an aid to learning, for in-class or out-of-class assignments, or for testing, tear-out worksheets are found at the end of the text (pp. 387 and 388).

Anterior skeletal worksheet (no. 1)

Draw and label the following muscles on the skeletal chart:
a. Rectus abdominis
b. External oblique abdominal
c. Internal oblique abdominal

Posterior skeletal worksheet (no. 2)

Draw and label the following muscles on the skeletal chart:
a. Erector spinae
b. Quadratus lumborum
c. Splenius—cervicis and capitis

LABORATORY AND REVIEW EXERCISES

1. Locate the following parts of the spine on a human skeleton and on a human subject:
   a. Cervical vertebrae
   b. Thoracic vertebrae
   c. Lumbar vertebrae
   d. Spinous processes
   e. Transverse processes
   f. Sacrum
   g. Manubrium
   h. Xiphoid process
   i. Sternum
   j. Rib cage (various ribs)
2. How and where can the following muscles be palpated on a human subject?
   a. Rectus abdominis
   b. External oblique abdominal
   c. Internal oblique abdominal
   d. Erector spinae
   e. Sternocleidomastoid

3. List the planes in which each of the following movements occurs. List the respective axis of rotation for each movement in each plane.
   a. Cervical flexion
   b. Cervical extension
   c. Cervical rotation
   d. Cervical lateral flexion
   e. Lumbar flexion
   f. Lumbar extension
   g. Lumbar rotation
   h. Lumbar lateral flexion
4. Contrast crunches, bent-knee sit-ups, and straight-leg sit-ups. Does having a partner to hold the feet make a difference in the ability to do the bent-knee and straight-leg sit-ups? If so, why?
5. Have a laboratory partner stand and assume a position exhibiting good posture. What motions in each region of the spine does gravity attempt to produce? Which muscles are responsible for counteracting these motions against the pull of gravity?
6. Compare and contrast the spinal curves of a laboratory partner sitting erect with those of one sitting slouched in a chair. Which muscles are responsible for maintaining good sitting posture?
7. Which exercise is better for the development of the abdominal muscles—leg lifts or sit-ups? Analyze each one in regard to the activity of the abdominal muscles. Defend your answer.
8. Why is good abdominal muscular development so important? Why is this area so frequently neglected?
9. Why are weak abdominal muscles frequently blamed for lower back pain?
10. Prepare an oral or a written report on abdominal or back injuries found in the literature.
11. Research common spinal disorders such as brachial plexus neuropraxia, cervical radiculopathy, lumbosacral herniated nucleus pulposus, sciatica, spondylolysis, and spondylolisthesis. Report your findings in class.
12. Fill in the muscle analysis chart below by listing the muscles primarily involved in each movement.
13. Fill in the antagonistic muscle action chart (p. 345) by listing the muscle(s) or parts of muscles that are antagonistic in their actions to the muscles in the left column.
14. After analyzing each of the exercises in the trunk and spine movement analysis chart (p. 345), break each into two primary movement phases such as a lifting phase and lowering phase. For each phase, determine the trunk and spine movements occurring, and then list the trunk and spine muscles primarily responsible for causing/controlling those movements. Beside each muscle in each movement indicate the type of contraction as follows: I-isometric; C-concentric; E-eccentric.

Muscle analysis chart • Cervical and lumbar spine

Cervical spine	
Flexion	Extension
Lateral flexion right	Rotation right
Lateral flexion left	Rotation left

Lumbar spine	
Flexion	Extension
Lateral flexion right	Rotation right
Lateral flexion left	Rotation left

Antagonistic muscle action chart • Cervical and lumbar spine

Agonist	Antagonist
Splenius capitis	
Splenius cervicis	
Sternocleidomastoid	
Erector spinae	
Rectus abdominis	
External oblique abdominal	
Internal oblique abdominal	
Quadratus lumborum	

15. Analyze each skill in the trunk and spine sport skill analysis chart (p. 346) and list the movements of trunk and spine in each phase of the skill. You may prefer to list the initial positions that the trunk and spine are in for the stance phase. After each movement, list the trunk and spine muscle(s) primarily responsible for causing/controlling those movements. Beside each muscle in each movement indicate the type of contraction as follows: I-isometric; C-concentric; E-eccentric. It may be desirable to review the concepts for analysis in Chapter 8 for the various phases.

Trunk and spine movement analysis chart

Exercise	Initial movement phase		Secondary movement phase	
	Movement(s)	Agonist(s)–(contraction type)	Movement(s)	Agonist(s)–(contraction type)
Push-up				
Squat				
Dead lift				
Sit-up, bent knee				
Prone extension				
Rowing exercise				
Leg raises				
Stair machine				

Chapter
12

Trunk and spine sport skill analysis chart

Exercise		Stance phase	Preparatory phase	Movement phase	Follow-through phase
Baseball pitch	(R)				
	(L)				
Football punting	(R)				
	(L)				
Walking	(R)				
	(L)				
Softball pitch	(R)				
	(L)				
Soccer pass	(R)				
	(L)				
Batting	(R)				
	(L)				
Bowling	(R)				
	(L)				
Basketball jump shot	(R)				
	(L)				

References

Clarkson HM, Gilewich GB: *Musculoskeletal assessment: joint range of motion and manual muscle strength,* Baltimore, 1989, Williams & Wilkins.

Day AL: Observation on the treatment of lumbar disc disease in college football players, *American Journal of Sports Medicine* 15:275, January–February 1987.

Field D: *Anatomy: palpation and surface markings,* ed 3, Oxford, 2001, Butterworth-Heinemann.

Gench BE, Hinson MM, Harvey PT: *Anatomical kinesiology,* Dubuque, IA, 1995, Eddie Bowers.

Hamilton N, Luttgens K: *Kinesiology: scientific basis of human motion,* ed 10, Boston, 2002, McGraw-Hill.

Hislop HJ, Montgomery J: *Daniels and Worthingham's muscle testing: techniques of manual examination,* ed 7, Philadelphia, 2002, Saunders.

Holden DL, Jackson DW: Stress fractures of ribs in female rowers, *American Journal of Sports Medicine* 13:277, July–August 1987.

Lindsay DT: *Functional human anatomy,* St. Louis, 1996, Mosby.

Magee DJ: *Orthopedic physical assessment,* ed 4, Philadelphia, 2002, Saunders.

Martens MA, et al: Adductor tendonitis and muscular abdominis tendopathy, *American Journal of Sports Medicine* 15:353, July–August 1987.

Marymont JV: Exercise-related stress reaction of the sacroiliac joint, an unusual cause of low back pain in athletes, *American Journal of Sports Medicine* 14:320, July–August 1986.

Muscolino JE: *The muscular system manual: the skeletal muscles of the human body,* ed 2, St. Louis, 2003, Elsevier Mosby.

National Strength and Conditioning Association; Baechle TR, Earle RW: *Essentials of strength training and conditioning,* ed 2, Champaign, IL, 2000, Human Kinetics.

Oatis CA: *Kinesiology: the mechanics and pathomechanics of human movement,* Philadelphia, 2004, Lippincott Williams & Wilkins.

Perry JF, Rohe DA, Garcia AO: *The kinesiology workbook,* Philadelphia, 1992, Davis.

Prentice WE: *Arnheim's principles of athletic training,* ed 12, New York, 2006, McGraw-Hill.

Rasch PJ: *Kinesiology and applied anatomy,* ed 7, Philadelphia, 1989, Lea & Febiger.

Seeley RR, Stephens TD, Tate P: *Anatomy & physiology,* ed 7, New York, 2006, McGraw-Hill.

Sieg KW, Adams SP: *Illustrated essentials of musculoskeletal anatomy,* ed 2, Gainesville, FL, 1985, Megabooks.

Stone RJ, Stone JA: *Atlas of the skeletal muscles,* New York, 1990, McGraw-Hill.

Thibodeau GA, Patton KT: *Anatomy & physiology,* ed 9, St. Louis, 1993, Mosby.

Van De Graaff KM: *Human anatomy,* ed 6, Dubuque, IA, 2002, McGraw-Hill.

Chapter 8

Muscular Analysis of Trunk and Lower Extremity Exercises

Objectives

● To analyze an exercise to determine the joint movements and the types of contractions occurring in the specific muscles involved in those movements

● To learn to group individual muscles into units that produce certain joint movements

● To begin to think of exercises that increase the strength and endurance of individual muscle groups

● To learn to analyze and prescribe exercises to strengthen major muscle groups

● To apply the concept of the kinetic chain to the lower extremity

---

Online Learning Center Resources

Visit *Manual of Structural Kinesiology's* Online Learning Center at **www.mhhe.com/floyd16e** for additional information and study material for this chapter, including:

- *Self-grading quizzes*
- *Anatomy flashcards*
- *Animations*

Chapter 8 presented an introduction to the analysis of exercise and activities. That chapter included only the analysis of the muscles previously studied in the upper extremity region.

Since that chapter, all the other joints and large muscle groups of the human body have been considered. The exercises and activities found in this chapter concentrate more on the muscles in the trunk and lower extremity.

Strength, endurance, and flexibility of the muscles of the lower extremity, trunk, and abdominal sections are also very important in skillful physical performance and body maintenance.

The type of contraction is determined by whether the muscle is lengthening or shortening during the movement. However, muscles may shorten or lengthen in the absence of a contraction through passive movement caused by other contracting muscles, momentum, gravity, or external forces such as manual assistance and exercise machines.

A concentric contraction is a shortening contraction of the muscles against gravity or resistance, whereas an eccentric contraction is a contraction in which the muscle lengthens under tension to control the joints moving with gravity or resistance.

Contraction against gravity is also quite evident in the lower extremities.

The quadriceps muscle group contracts eccentrically when the body slowly lowers in a weight-bearing movement through lower extremity action. The quadriceps functions as a decelerator to knee joint flexion in weight-bearing movements by contracting eccentrically to prevent too rapid a downward movement. One can easily demonstrate this fact by palpating this muscle group when slowly moving from a standing position to a half-squat. Almost as much work is done in this type of contraction as in concentric contractions.

In this example involving the quadriceps, the slow descent is eccentric, and the ascent from the squatted position is concentric. If the descent were under no muscular control, it would be at the same speed as gravity and the muscle lengthening would be passive. That is, the movement and change in length of the muscle would be both caused and controlled by gravity and not by active muscular contractions.

In recent years, more and more medical and allied health professionals have been emphasizing the development of muscle groups through resistance training and circuit-training activities. Athletes and nonathletes, both male and female, need to have overall muscular development. Even those who do not necessarily desire significant muscle mass are advised to develop and maintain their muscle mass through resistance training. As we age we normally tend to lose muscle mass, and, as a result, our metabolism decreases. This factor combined with improper eating habits results in unhealthy fat accumulation and excessive weight gain. Through increasing our muscle mass, we burn more calories and are less likely to gain excessive fat.

Sport participation does not ensure sufficient development of muscle groups. Also, more and more emphasis has been placed on mechanical kinesiology in physical education and athletic skill teaching. This is desirable and can help bring about more skillful performance. However, one must remember that mechanical principles will be of little or no value to performers without adequate strength and endurance of their muscular system, which is developed through planned exercises and activities. In the fitness and health revolution of recent years, a much greater emphasis has been placed on exercises and activities that improve the physical fitness, strength, endurance, and flexibility of participants. This chapter will continue the practice of analyzing the muscles through simple exercises that began in Chapter 8. When these techniques are mastered, the individual is ready to analyze and prescribe exercises and activities for muscular strength and endurance needed in sport activities and for healthful living.

Sit-up, bent knee   FIG. 13.1

Description

The participant lies on the back, forearms crossed and lying across the chest, with the knees flexed approximately 90 degrees and the feet about hip width apart. The hips and knees are flexed in this manner to reduce the length of the hip flexors, thereby reducing their contribution to the sit-up. This positioning will allow more emphasis on the abdominals, as compared with a straight leg sit-up.

The participant curls up to a sitting position, rotates the trunk to the right, touches the left elbow to the left knee, and then returns to the starting position. On the next repetition, the participant should rotate to the left instead of the right for balanced muscular development.

Analysis

This exercise is divided into four phases for analysis: (1) curling phase to sitting-up position, (2) rotating to right phase, (3) return phase to sitting-up position, and (4) return phase to starting position (Table 13.1).

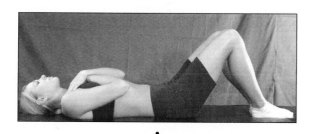
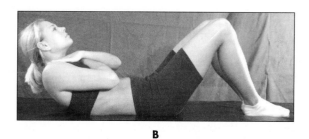
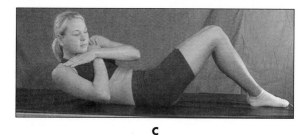

FIG. 13.1 ● Sit-up, bent knee. **A,** Beginning relaxed position; **B,** Sitting-up position; **C,** Rotated to the right position.

Chapter
13

TABLE 13.1 • Sit-up, bent knee

Joint	Curling phase to sitting-up position		Rotating to right phase		Return phase to sitting-up position		Return phase to starting position	
	Action	Agonists	Action	Agonists	Action	Agonists	Action	Agonists
Cervical spine	Flexion	Cervical spine flexors Sternoclei-domastoid	Mainte-nance of cervical flexion	Cervical spine flexors (isometric contraction) Sternoclei-domastoid	Mainte-nance of cervical flexion	Cervical spine flexors (isometric contraction) Sternoclei-domastoid	Extension	Cervical spine flexors (eccentric contraction) Sternoclei-domastoid
Trunk	Flexion	Trunk flexors Rectus abdominis External oblique abdominal Internal oblique abdominal	Right lumbar rotation	Right lumbar rotators (R) Rectus abdominis (L) External oblique abdominal (R) Internal oblique abdominal (R) Erector spinae	Left lumbar rotation to neutral position	Right lumbar rotators (eccentric contraction) (R) Rectus abdominis (L) External oblique abdominal (R) Internal oblique abdominal (R) Erector spinae	Extension	Trunk flexors (eccentric contraction) Rectus abdominis External oblique abdominal Internal oblique abdominal
Hip	Flexion	Hip flexors Iliopsoas Rectus femoris Pectineus	Mainte-nance of hip flexion	Hip flexors (isometric contraction) Iliopsoas Rectus femoris Pectineus	Mainte-nance of hip flexion	Hip flexors (isometric contraction) Iliopsoas Rectus femoris Pectineus	Extension	Hip flexors (eccentric contraction) Iliopsoas Rectus femoris Pectineus

Alternating prone extensions FIG. 13.2

Description

The participant lies in a prone position, face down, with the shoulders fully flexed in a relaxed position lying in front of the body. The head, upper trunk, right upper extremity, and left lower extremity are raised from the floor. The knees are kept in full extension. Then the participant returns to the starting position. On the next repetition, the head, upper trunk, left upper extremity, and right lower extremity are raised from the floor.

Analysis

This exercise is separated into two phases for analysis: (1) lifting phase to raise the right shoulder off the surface and extend the left lower extremity off the floor and (2) lowering phase to relaxed position (Table 13.2).

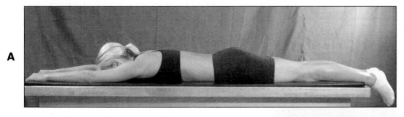

FIG. 13.2 ● Alternating prone extensions. **A,** Beginning relaxed position; **B,** Raised position.

TABLE 13.2 • Alternating prone extensions (Superman exercise)

Joint	Lifting phase to raise upper and lower extremities		Lowering phase to relaxed position	
	Action	Agonists	Action	Agonists
Shoulder	Flexion	Shoulder joint flexors Pectoralis major (clavicular head or upper fibers) Deltoid Coracobrachialis Biceps brachii	Extension	Shoulder joint flexors (eccentric contraction) Pectoralis major (clavicular head or upper fibers) Deltoid Coracobrachialis Biceps brachii
Shoulder girdle	Adduction	Shoulder girdle adductors Trapezius Rhomboids	Abduction	Shoulder girdle adductors (eccentric contraction) Trapezius Rhomboids
Trunk	Extension	Trunk extensors Erector spinae Splenius Quadratus lumborum	Flexion (return to neutral relaxed position)	Trunk and cervical spine extensors (eccentric contraction) Erector spinae Splenius Quadratus lumborum
Hip	Extension	Hip extensors Gluteus maximus Semitendinosus Semimembranosus Biceps femoris	Flexion (return to neutral relaxed position)	Hip extensors (eccentric contraction) Gluteus maximus Semitendinosus Semimembranosus Biceps femoris

Chapter

13

Squat   FIG. 13.3

Description

The participant places a barbell on the shoulders behind the neck and grasps it with the palms-forward position of the hands. The participant squats down, flexing at the hips while keeping the spine in normal alignment, until the thighs are parallel to the floor. The participant then returns to the starting position. This exercise is commonly performed improperly by allowing the knees to move forward beyond the plane of the feet, which greatly increases the risk of injury. Care should be taken to ensure that the shins remain as vertical as possible during this exercise.

The feet should be parallel, with slight external rotation of the lower extremity. The knees should point over the ankles and feet without going in front, between, or outside of the vertical plane of the feet.

Analysis

This exercise is separated into two phases for analysis: (1) lowering phase to the squatted position and (2) lifting phase to the starting position (Table 13.3). NOTE: It is assumed that no movement will take place in the shoulder joint, shoulder girdle, wrists, hands, and back.

FIG. 13.3 ● Squat. **A,** Starting position; **B,** Squatted position.

TABLE 13.3 • Squat

| Joint | Lowering phase to squatted position | | | Lifting phase to starting position | |
	Action	Agonists		Action	Agonists
Hip	Flexion	Hip extensors (eccentric contraction) Gluteus maximus Semimembranosus Semitendinosus Biceps femoris		Extension	Hip extensors Gluteus maximus Semimembranosus Semitendinosus Biceps femoris
Knee	Flexion	Knee extensors (eccentric contraction) Rectus femoris Vastus medialis Vastus intermedius Vastus lateralis		Extension	Knee extensors Rectus femoris Vastus medialis Vastus intermedius Vastus lateralis
Ankle	Dorsiflexion	Plantar flexors (eccentric contraction) Gastrocnemius Soleus		Plantar flexion	Plantar flexors Gastrocnemius Soleus

Dead lift   FIG. 13.4

Description

The participant begins in hip flexed position, keeping the arms, legs, and back straight, and grasps the barbell on the floor. Then a movement to the standing position is made by extending the hips. This exercise, when done improperly by allowing lumbar flexion, may contribute to low back problems. It is essential that the lumbar extensors be used more as isometric stabilizers of the low back while the hip extensors perform the majority of the lift in this exercise.

Analysis

This exercise is divided into two phases for analysis: (1) lifting phase to the hip extended/knee extended position and (2) lowering phase to the hip flexed/knee flexed starting position (Table 13.4).

FIG. 13.4 ● Dead lift. **A,** Beginning hip/knee flexed position; **B,** Ending hip/knee extended position.

TABLE 13.4 • Dead lift

Joint	Lifting phase to hip/knee extended position		Lowering phase to hip/knee flexed position	
	Action	Agonists	Action	Agonists
Wrist and hand	Flexion	Wrist and hand flexors (isometric contraction) Flexor carpi radialis Flexor carpi ulnaris Palmaris longus Flexor digitorum profundus Flexor digitorum superficialis Flexor pollicis longus	Flexion	Wrist and hand flexors (isometric contraction) Flexor carpi radialis Flexor carpi ulnaris Palmaris longus Flexor digitorum profundus Flexor digitorum superficialis Flexor pollicis longus
Trunk	Maintenance of extension	Trunk extensors (isometric contraction) Erector spinae (sacrospinalis) Quadratus lumborum	Maintenance of extension	Trunk extensors (isometric contraction) Erector spinae (sacrospinalis) Quadratus lumborum
Hip	Extension	Hip extensors Gluteus maximus Semimembranosus Semitendinosus Biceps femoris	Flexion	Hip extensors (eccentric contraction) Gluteus maximus Semimembranosus Semitendinosus Biceps femoris
Knee	Extension	Knee extensors (quadriceps) Rectus femoris Vastus medialis Vastus intermedius Vastus lateralis	Flexion	Knee extensors (quadriceps) (eccentric contraction) Rectus femoris Vastus medialis Vastus intermedius Vastus lateralis

Note: Slight movement of the shoulder joint and shoulder girdle is not being analyzed.

Chapter
13

Isometric exercises

An exercise technique called isometrics is a type of muscular activity in which there is contraction of muscle groups with little or no muscle shortening. Although not as productive in terms of overall strength gains as isotonics, isometrics are an effective way to build and maintain muscular strength in a limited range of motion.

A few selected isometric exercises are analyzed muscularly to show how they are designed to develop specific muscle groups. Although there are varying approaches to isometrics, most authorities agree that isometric contractions should be held approximately 7 to 10 seconds for a training effect.

Abdominal contraction   FIG. 13.5

Description

The participant contracts the muscles in the anterior abdominal region as strongly as possible, with no movement of the trunk or hips. This exercise can be performed in sitting, standing, or supine positions. The longer the contraction in seconds, the more valuable the exercise will be, to a degree.

Analysis

Abdomen
  Contraction
    Rectus abdominis
    External oblique abdominal
    Internal oblique abdominal
    Transversus abdominis

FIG. 13.5 ● Abdominal contraction. **A,** Beginning position; **B,** Contracted position.

Leg lifter   FIG. 13.6

Description

The participant sits on a bench or chair with the knees slightly bent and with the left leg over the right. An attempt to raise the right leg while resisting it with the left leg is conducted.

Analysis*

This exercise involves only a single isometric contraction of the muscles as listed in Table 13.5. Upon completion of the exercise, the position of the legs may be reversed so that right leg is over the left in order to exercise the contralateral muscles.

*When the legs are alternated, the muscles used will be the same muscles, but in the other leg.

A   B

FIG. 13.6 ● Leg lifter. **A,** Beginning relaxed position; **B,** Up position.

TABLE 13.5 • Leg lifter

Joint	Right leg—attempting upward movement		Left leg—resisting upward movement	
	Action	Agonists	Action	Agonists
Ankle	Dorsiflexion	Ankle dorsiflexors Tibialis anterior Extensor hallucis longus Extensor digitorum longus Peroneus tertius	Plantar flexion	Plantar flexors Gastrocenemius Soleus
Knee	Extension	Knee extensors (quadriceps) Rectus femoris Vastus medialis Vastus intermedius Vastus lateralis	Flexion	Knee flexors (hamstrings) Biceps femoris Semitendinosus Semimbranosus
Hip	Flexion	Hip flexors Iliopsoas Rectus femoris Pectineus Sartorius Tensor fasciae latae	Extension	Hip extensors Gluteus maximus Biceps femoris Semitendinosus Semimbranosus

Chapter

13

Hip sled FIG. 13.7

Description

The participant lies in a supine position on the floor with the knees and hips flexed in a position close to the chest. The feet are placed on the apparatus plate. The plate is moved upward until the knees and hips are completely extended. Then the participant returns to the starting position.

Analysis

This exercise is divided into two phases for analysis: (1) pushing phase to the straight leg position and (2) lowering phase to the starting position (Table 13.6).

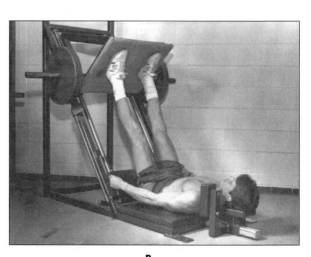

A **B**

FIG. 13.7 ● Hip sled (hip and leg press). **A,** Starting position; **B,** High position.

TABLE 13.6 • Hip sled

| Joint | Pushing phase to straight leg position | | Lowering phase to starting position | |
	Action	Agonists	Action	Agonists
Ankle	Plantar flexion	Ankle plantar flexors Gastrocnemius Soleus	Dorsiflexion	Ankle plantar flexors (eccentric contraction) Gastrocnemius Soleus
Knee	Extension	Knee extensors (quadriceps) Rectus femoris Vastus medialis Vastus intermedius Vastus lateralis	Flexion	Knee extensors (quadriceps) (eccentric contraction) Rectus femoris Vastus medialis Vastus intermedius Vastus lateralis
Hip	Extension	Hip extensors Biceps femoris Semimembranosus Semitendinosus Gluteus maximus	Flexion	Hip extensors (eccentric contraction) Biceps femoris Semimembranosus Semitendinosus Gluteus maximus

Rowing exercise FIG. 13.8

Description

The participant sits on a movable seat with the knees and hips flexed close to the chest. The arms are reaching forward to grasp a horizontal bar. The legs are extended forcibly as the arms are pulled toward the chest. Then the legs and arms are returned to the starting position.

Analysis

This exercise is divided into two movements for analysis: (1) arm pull to chest/leg push to extend knees and hip phase and (2) return phase to the starting position (Table 13.7).

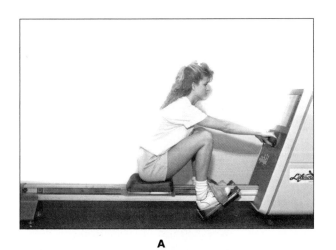
A

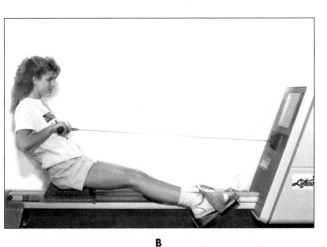
B

FIG. 13.8 ● Rowing exercise machine. **A,** Starting position; **B,** Movement.

TABLE 13.7 • Rowing exercise

Joint	Arm pull/leg push phase		Return phase to starting position	
	Action	Agonists	Action	Agonists
Foot and ankle	Plantar flexion	Ankle plantar flexors Gastrocnemius Soleus	Dorsiflexion	Ankle dorsiflexors Tibialis anterior Extensor hallucis longus Extensor digitorum longus Peroneus tertius
Knee	Extension	Quadriceps (knee extensors) Rectus femoris Vastus medialis Vastus intermedius Vastus lateralis	Flexion	Knee flexors (hamstrings) Biceps femoris Semitendinosus Semimembranosus
Hip	Extension	Hip extensors Gluteus maximus Biceps femoris Semimembranosus Semitendinosus	Flexion	Hip flexors Iliopsoas Rectus femoris Pectineus
Trunk	Extension	Trunk extensors Erector spinae	Flexion	Trunk flexors Rectus abdominis Internal oblique abdominal External oblique abdominal

Chapter

13

TABLE 13.7 (continued) • **Rowing exercise**

| Joint | Arm pull/leg push phase | | Return phase to starting position | |
	Action	Agonists	Action	Agonists

Shoulder girdle	Adduction, downward rotation, and depression	Shoulder girdle adductors, downward rotators, and depressors
		  Trapezius (lower)
		  Rhomboid
		  Pectoralis minor
			Abduction, upward rotation, and elevation	Shoulder girdle adductors, downward rotators, and depressors (eccentric contraction)
				  Trapezius (lower)
				  Rhomboid
				  Pectoralis minor

Shoulder joint	Extension	Shoulder joint extensors
		  Latissimus dorsi
		  Teres major
		  Posterior deltoid
		  Teres minor
		  Infraspinatus
			Flexion	Shoulder joint extensors (eccentric contraction)
				  Latissimus dorsi
				  Teres major
				  Posterior deltoid
				  Teres minor
				  Infraspinatus

Elbow joint	Flexion	Elbow joint flexors
		  Biceps brachii
		  Brachialis
		  Brachioradialis
			Extension	Elbow joint flexors (eccentric contraction)
				  Biceps brachii
				  Brachialis
				  Brachioradialis

Wrist and hand	Flexion	Wrist and hand flexors (isometric contraction)
		  Flexor carpi radialis
		  Flexor carpi ulnaris
		  Palmaris longus
		  Flexor digitorum profundus
		  Flexor digitorum superficialis
		  Flexor pollicis longus
			Flexion	Wrist and hand flexors (isometric contraction)
				  Flexor carpi radialis
				  Flexor carpi ulnaris
				  Palmaris longus
				  Flexor digitorum profundus
				  Flexor digitorum superficialis
				  Flexor pollicis longus

Web sites

American College of Sports Medicine

www.acsm.org

Scientific research, education, and practical applications of sports medicine and exercise science to maintain and enhance physical performance, fitness, health, and quality of life.

Concept II

www.concept2.com/index.html

Information on the technique of rowing and the muscles used.

Fitness World

www.fitnessworld.com

The information at this site is about fitness in general and includes access to *Fitness Management* magazine.

National Council of Strength & Fitness

www.ncsf.org

Personal training certification and continuing education for the fitness professional.

National Strength and Conditioning Association

www.nsca-lift.org

Information on the profession of strength and conditioning specialists and personal trainers.

NSCA Certification Commission

www.nsca-cc.org

The certifying body for the National Strength and Conditioning Association.

President's Council on Physical Fitness and Sports

www.fitness.gov

Information and links from the U.S. government on fitness.

Chapter
13

ExRx.net

www.exrx.net/Lists/Directory.html

A resource for the exercise professional, coach, or fitness enthusiast consisting of over 1,500 pages of exercises and anatomy illustrations.

National Academy of Sports Medicine

www.nasm.org

Offers specific certifications for health and fitness exercise specialists and a valuable resource for continuing education on exercise techniques, etc.

Upper Extremity Conditioning Program

www.eatonhand.com/hw/nirschl.htm

Shows strengthening exercises for the upper body.

Rehab Team Site: Passive Stretching

http://calder.med.miami.edu/pointis/upper.html

Passive range of motion exercises.

Body Map

www.athleticadvisor.com/body_map.html

Describes specific injuries and how to properly rehab with weights.

The Physician and Sports Medicine: Weight Training Injuries

www.physsportsmed.com/issues/1998/03mar/laskow2.htm

Article about upper body injuries and how to strengthen the upper body.

NISMAT Exercise Programs

www.nismat.org/orthocor/programs

Step-by-step instructions for strengthening exercises along with diagrams.

Spine Health.com

www.spine-health.com

Information on the spine including core body strengthening exercises.

#1 Back Pain Site

www.1backpain.com

Information on back pain as well as strengthening and stretching exercises for the back.

Runner Girl.com

www.runnergirl.com

Strengthening and stretching exercises as well as other health and fitness information for women.

Worksheet exercise

As an aid to learning, for in-class and out-of-class assignments, or for testing, a tear-out worksheet is found at the end of the text (pp. 389).

Skill analysis worksheets (no. 1)

Using the techniques taught in this chapter and in Chapter 8, analyze the joint movements and muscles used in each phase of movement for selected exercises from selected sports skills. For each joint, list the initial and subsequent position, the motion and degrees of movement, and the agonists, and denote whether they are contracting concentrically or eccentrically.

LABORATORY AND REVIEW EXERCISES

1. Obtain, describe, and completely analyze five conditioning exercises.
2. Collect, analyze, and evaluate exercises that are found in newspapers, in magazines, on the Internet, or are observed on television.
3. Prepare a set of exercises that will ensure development of all large muscle groups in the body.
4. Select exercises from exercise books for analysis.
5. Bring to class other typical exercises for members to analyze.
6. Analyze the conditioning exercises given by your physical education teachers, coaches, and athletic trainers.
7. Observe children using playground equipment. Analyze muscularly the activities they are performing.
8. Visit the room on your campus where the free weights and specific or multifunction exercise machines are located. Analyze exercises that can be done with each machine. Compare and contrast similar exercises using different exercise machines and free weights.
   NOTE: Manufacturers of all types of exercise apparatus have a complete list of exercises that can be performed with their machines. Secure a copy of recommended exercises and muscularly analyze each exercise.
9. Consider a sport (basketball or any other sport) and develop exercises applying the overload principle that would develop all the large muscle groups used in the sport.
10. Lie supine on a table with the knees flexed and hips flexed 90 degrees and the ankles in the neu-

tral 90-degree position. Extend each joint until your knee is fully extended, your hip is flexed only 10 degrees, and your ankle is plantarflexed 10 degrees by performing each of the following movements before proceeding to the next:

- Full knee extension
- Hip extension to within 10 degrees of neutral
- Plantar flexion to 10 degrees

Analyze the movements and the muscles responsible for each movement at the hip, knee, and ankle.

11. Now stand with your back and buttocks against a smooth wall and place your feet (shoulder width apart) with approximately 12 inches between your heels and the wall. Maintain your feet in position, with the hips and knees each flexed approximately 90 degrees so that the thighs are parallel to the floor. Keeping your feet in place, slowly slide your back and buttocks up the wall until your buttocks are as far away from the floor as possible without removing your feet from their position on the floor. Analyze the movements and the muscles responsible for each movement at the hip, knee, and ankle.

12. What is the difference between the two exercises in questions 10 and 11? Can you perform exercise 11 one step at a time, as you did exercise 10?

13. Analyze each exercise in the exercise analysis chart. Use one row for each joint involved that actively moves during the exercise. Do not include joints where there is no active movement or where the joint is maintained in one position isometrically.

Exercise analysis chart

Exercise	Phase	Joint, movement occurring	Force causing movement (muscle or gravity)	Force resisting movement (muscle or gravity)	Functional muscle group, type of contraction
Sit-up, bent knee	Curling phase to sitting-up position				
	Rotating-to-right phase				
	Return phase to sitting-up position				
	Return phase to starting position				
Alternating prone extensions	Lifting phase				
	Lowering phase				

Exercise analysis chart (continued)

Exercise	Phase	Joint, movement occurring	Force causing movement (muscle or gravity)	Force resisting movement (muscle or gravity)	Functional muscle group, type of contraction
Squat	Lowering phase				
	Lifting phase				
Dead lift	Lifting phase				
	Lowering phase				
Leg press	Pressing phase				
	Return phase				
Hip sled	Pushing phase				
	Lowering phase				

Chapter

13

Exercise analysis chart (continued)

Exercise	Phase	Joint, movement occurring	Force causing movement (muscle or gravity)	Force resisting movement (muscle or gravity)	Functional muscle group, type of contraction
Rowing exercise	Arm pull/leg push phase				
	Return phase to starting position				

References

Adrian M: Isokinetic exercise, *Training and Conditioning* 1:1, June 1991.

Altug Z, Hoffman JL, Martin JL: *Manual of clinical exercise testing, prescription and rehabilitation,* Norwalk, CT, 1993, Appleton & Lange.

Andrews JR, Harrelson GL: *Physical rehabilitation of the injured athlete,* Philadelphia, 1991, Saunders.

Ellenbecker TS, Davies GJ: *Closed kinetic chain exercise: a comprehensive guide to multiple-joint exercise,* Champaign, IL, 2001, Human Kinetics.

Fahey TD: *Athletic training: principles and practices,* Mountain View, CA, 1986, Mayfield.

Logan GA, McKinney WC: *Anatomic kinesiology,* ed 3, New York, 1982, McGraw-Hill.

Matheson O, et al: Stress fractures in athletes, *American Journal of Sports Medicine* 15:46, January-February 1987.

National Strength and Conditioning Association; Baechle TR, Earle RW: *Essentials of strength training and conditioning,* ed 2, Champaign, IL, 2000, Human Kinetics.

Northrip JW, Logan GA, McKinney WC: *Analysis of sport motion: anatomic and biomechanic perspectives,* ed 3, New York, 1983, McGraw-Hill.

Powers SK, Howley ET: *Exercise physiology: theory and application to fitness and performance,* ed 5, New York, 2004, McGraw-Hill.

Prentice WE: *Rehabilitation techniques in sports medicine,* ed 3, New York, 1999, McGraw-Hill.

Steindler A: *Kinesiology of the human body,* Springfield, IL, 1970, Charles C Thomas.

Torg JS, Vegso JJ, Torg E: *Rehabilitation of athletic injuries: an atlas of therapeutic exercise,* Chicago, 1987, Year Book.

Wirhed R: *Athletic ability and the anatomy of motion,* London, 1984, Wolfe.

Appendix

Appendix 1

Range of motion for diarthrodial joints of the upper extremity

Joint	Type	Motion	Range
Sternoclavicular	Arthrodial	Protraction	Moves anteriorly 15°
		Retraction	Moves posteriorly 15°
		Elevation	Moves superiorly 45°
		Depression	Moves inferiorly 5°
		Upward rotation	45°
		Downward rotation	5°
Acromioclavicular	Arthrodial	Protraction-retraction	20°–30° rotational and gliding motion
		Elevation-depression	20°–30° rotational and gliding motion
		Upward rotation–downward rotation	20°–30° rotational and gliding motion
Scapulothoracic	Not a true synovial joint; all movement totally dependent on AC and SC joints	Abduction-adduction	25° total range
		Upward rotation–downward rotation	60° total range
		Elevation-depression	55° total range
Glenohumeral	Enarthrodial	Flexion	90°–100°
		Extension	40°–60°
		Abduction	90°–95°
		Adduction	0° prevented by trunk, 75° anterior to trunk
		Internal rotation	70°–90°
		External rotation	70°–90°
		Horizontal abduction	45°
		Horizontal adduction	135°
Elbow	Ginglymus	Extension	0°
		Flexion	145°–150°
Radioulnar	Trochoid	Supination	80°–90°
		Pronation	70°–90°

Range of motion for diarthrodial joints of the upper extremity *(continued)*

Joint	Type	Motion	Range
Wrist	Condyloid	Flexion	70°–90°
		Extension	65°–85°
		Abduction	15°–25°
		Adduction	25°–40°
Thumb carpometacarpal	Sellar	Flexion	15°–45°
		Extension	0°–20°
		Adduction	0°
		Abduction	50°–70°
Thumb metacarpophalangeal	Ginglymus	Extension	0°
		Flexion	40°–90°
Thumb interphalangeal	Ginglymus	Flexion	80°–90°
		Extension	0°
2nd, 3rd, 4th, and 5th metacarpophalangeal joints	Condyloid	Extension	0°–40°
		Flexion	85°–100°
		Abduction	Variable 10°–40°
		Adduction	Variable 10°–40°
2nd, 3rd, 4th, and 5th proximal interphalangeal joints	Ginglymus	Flexion	90°–120°
		Extension	0°
2nd, 3rd, 4th, and 5th distal interphalangeal joints	Ginglymus	Flexion	80°–90°
		Extension	0°

Appendix 2

Range of motion for diarthrodial joints of the spine and lower extremity

Joint	Type	Motion	Range
Cervical	Arthrodial except atlan-toaxial joint, which is trochoid	Flexion	80°
		Extension	20°–30°
		Lateral flexion	35°
		Rotation unilaterally	45°
Lumbar	Arthrodial	Flexion	45°
		Extension	45°
		Lateral flexion	45°
		Rotation unilaterally	60°
Hip	Enarthrodial	Flexion	130°
		Extension	30°
		Abduction	35°
		Adduction	0°–30°
		External rotation	50°
		Internal rotation	45°
Knee *For internal and external rotation to occur, the knee must be flexed approximately 30° or more.*	Ginglymus (trochoginglymus)	Extension	0°
		Flexion	140°
		Internal rotation	30°
		External rotation	45°
Ankle (talocrural)	Ginglymus	Plantar flexion	50°
		Dorsal flexion	15°–20°
Transverse tarsal and subtalar	Arthrodial	Inversion	20°–30°
		Eversion	5°–15°
Great toe metatarsophalangeal	Condyloid	Flexion	45°
		Extension	70°
		Abduction	Variable 5°–25°
		Adduction	Variable 5°–25°
Great toe interphalangeal	Ginglymus	Flexion	90°
		Extension	0°
2nd, 3rd, 4th, and 5th metatarsophalangeal joints	Condyloid	Flexion	40°
		Extension	40°
		Abduction	Variable 5°–25°
		Adduction	Variable 5°–25°
2nd, 3rd, 4th, and 5th proximal interphalangeal joints	Ginglymus	Flexion	35°
		Extension	0°
2nd, 3rd, 4th, and 5th distal interphalangeal joints	Ginglymus	Flexion	60°
		Extension	30°

Appendix 3

Commonly used exercises for strengthening selected muscles

Some exercises may be more or less specific for certain muscles. In some cases, certain exercises are designed to emphasize specific portions of a particular muscle more than other portions. Some exercises may be modified slightly to further emphasize or deemphasize certain muscles or portions of muscles. In addition to the muscles listed, numerous other muscles in surrounding joints or other parts of the body may be involved by contracting isometrically to maintain appropriate position of the body for the muscles listed to carry out the exercise movement. Appropriate strength and endurance of these stabilizing muscles is essential for correct position and execution of the listed exercises. Finally, these exercises may be documented by different names by different authorities. Illustration and documentation of the proper techniques and indications for these exercises is beyond the scope of this text. Please consult several strength and conditioning texts for further details.

Upper extremity		
Muscle groups	Muscles	Exercise
Scapula abductors	Serratus anterior Pectoralis minor	Front press Dumbbell flys Dumbbell press Bench press Close-grip bench press Incline press Push-ups Incline dumbbell press Pec deck flys Cable crossover flys
Scapula adductors	Rhomboid Trapezius, lower fibers Trapezius, middle fibers	Bent-over lateral raises Pec deck rear delt laterals Seated rows Bent rows T-bar rows Dead lifts Sumo dead lifts
Scapula upward rotators	Serratus anterior Trapezius, lower fibers Trapezius, middle fibers	Dumbbell press Dumbbell flys One-arm dumbbell press Lateral raises Upright rows
Scapula downward rotators	Pectoralis minor Rhomboid	Parallel bar dips Dumbbell pullovers Barbell pullovers Chin-ups Reverse chin-ups Lat pulldowns Back lat pulldowns Close-grip lat pulldowns Straight-arm lat pulldowns One-arm dumbbell rows
Scapula elevators	Rhomboid Levator scapulae Trapezius, upper fibers Trapezius, middle fibers	Upright rows Barbell shrugs Dumbbell shrugs Machine shrugs Dead lifts
Scapula depressors	Pectoralis minor Trapezius, lower fibers	Parallel bar dips

Upper extremity		
Muscle groups	Muscles	Exercise
Shoulder flexors	Deltoid, anterior fibers Deltoid, middle fibers Pectoralis major, clavicular fibers Coracobrachialis	Arm curls Triceps dips One-arm dumbbell press Front raises Low pulley front raises One-dumbbell front raises Barbell front raises
Shoulder extensors	Latissimus dorsi Teres major Triceps brachii, long head Pectoralis major, sternal fibers Deltoid, posterior fibers Infraspinatus Teres minor	Dumbbell pullovers Barbell pullovers Reverse chin-ups Close-grip lat pulldowns Straight-arm lat pulldowns One-arm dumbbell rows
Shoulder abductors	Deltoid, anterior fibers Deltoid, posterior fibers Deltoid, middle fibers Pectoralis major, clavicular fibers Supraspinatus	Back press Front press Dumbbell press One-arm dumbbell press Lateral raises Side-lying lateral raises Low pulley lateral raises Upright rows Nautilus lateral raises Upright rows
Shoulder adductors	Pectoralis major Latissimus dorsi Teres major Subscapularis Coracobrachialis	Triceps dips Parallel bar dips Chin-ups Lat pulldowns Back lat pulldowns
Shoulder internal rotators	Pectoralis major Latissimus dorsi Teres major Subscapularis	Triceps dips Side-lying internal rotations Standing internal rotations at 90 degrees abduction
Shoulder external rotators	Infraspinatus Teres minor	Side-lying external degrees rotations Standing external rotations at 90 degrees abduction

Upper extremity		
Muscle groups	Muscles	Exercise
Shoulder horizontal abductors	Latissimus dorsi Infraspinatus Teres minor Deltoid, middle fibers Deltoid, posterior fibers	Bent-over lateral raises Low pulley bent-over lateral raises Pec deck rear delt laterals Bent rows T-bar rows
Shoulder horizontal adductors	Pectoralis major Coracobrachialis	Dumbbell flys Triceps dips Dumbbell press Bench press Close-grip bench press Incline press Decline press Push-ups Incline dumbbell press Incline dumbbell flys Pec deck flys Cable crossover flys Seated rows
Elbow flexors	Biceps brachii Brachialis Brachioradialis	Arm curls concentration curls Hammer curls Low pulley curls High pulley curls Barbell curls Machine curls Preacher curls Reverse barbell curls Chin-ups Reverse chin-ups Lat pulldowns Back lat pulldowns Close-grip lat pulldowns Seated rows One-arm dumbbell rows Bent rows T-bar rows Upright rows
Elbow extensors	Triceps brachii Triceps brachii, lateral head Triceps brachii, long head Triceps brachii, medial head Anconeus	Barbell pullovers Bench press Close-grip bench press Decline press Dumbbell press Dumbbell pullovers Dumbbell triceps extensions Front press Incline dumbbell press Incline press One-arm dumbbell triceps extensions One-arm reverse pushdowns Parallel bar dips Pushdowns Push-ups Reverse pushdowns Seated dumbbell triceps extensions Seated ez-bar triceps extensions Triceps dips Triceps extensions Triceps kickbacks

Upper extremity		
Muscle groups	Muscles	Exercise
Wrist flexors	Flexor carpi radialis Palmaris longus Flexor carpi ulnaris Flexor digitorum superficialis Flexor digitorum profundus	Wrist curls
Wrist extensors	Extensor carpi radialis longus Extensor carpi radialis brevis Extensor carpi ulnaris	Reverse barbell curls Reverse wrist curls Reverse pushdowns
Finger flexors	Hand intrinsics Flexor digitorum profundus Flexor digitorum superficialis Flexor pollicis longus	Ball squeezes Putty squeezes Rice bucket grips Wrist curls Dead lifts
Finger extensors	Extensor digitorum Extensor digiti minimi Extensor indicis	Reverse barbell curls Reverse wrist curls Rubber band stretches

Lower extremity		
Muscle groups	Muscles	Exercise
Hip flexors	Rectus femoris Iliopsoas Pectineus Tensor fasciae latae	Crunches Sit-ups Gym ladder sit-ups Calves over bench sit-ups Incline bench sit-ups Specific bench sit-ups Machine crunches Incline leg raises Leg raises Hanging leg raises
Hip extensors	Gluteus maximus Biceps femoris, long head Semitendinosus Semi-membranosus	Stiff-legged dead lifts Dead lifts Back extensions Dumbbell squats Squats Front squats Power squats Angled leg press Good mornings Lunges Cable back kicks Machine hip extensions Floor hip extensions Bridging Prone arches
Hip abductors	Gluteus medius Gluteus maximus Tensor fascia latae	Cable hip abductions Standing machine hip abductions Floor hip abductions Seated machine hip abductions

Lower extremity		
Muscle groups	Muscles	Exercise
Hip adductors	Adductor magnus Adductor longus Adductor brevis Gracilis	Sumo dead lifts Power squats Cable adductions Machine adductions
Hip external rotators	Gluteus maximus Piriformis Gamellus superior Gamellus inferior Obturator externus Obturator internus Quadratus femoris	Hip turn-outs Body turn-aways
Knee extensors	Vastus medialis Vastus intermedius Rectus femoris Vastus lateralis	Leg extensions Dead lifts Sumo dead lifts Dumbbell squats Squats Front squats Angled leg press Power squats Hack squats Lunges
Knee flexors	Semitendinosus Biceps femoris, long head Biceps femoris, short head Semi-membranosus Gastrocnemius, lateral head Gastrocnemius, medial head	Standing leg curls Seated leg curls Lying leg curls
Ankle dorsiflexors	Tibialis anterior Extensor hallucis longus Extensor digitorum longus Peroneous tertius	Towel pulls Elastic band pulls
Ankle plantar flexors	Gastrocnemius, lateral head Soleus Gastrocnemius, medial head	Standing calf raises One-leg toe raises Donkey calf raises Seated calf raises Seated barbell calf raises
Transverse tarsal/ subtalar inversion	Tibialis anterior Tibialis posterior Flexor digitorum longus Flexor hallucis longus	Towel drags Elastic band turn-ins
Transverse tarsal/ subtalar eversion	Extensor digitorum longus Peroneus longus Peroneus brevis Peroneus tertius	Towel drags Elastic band turn-outs
Toe extensors	Extensor hallucis longus Extensor digitorum longus	Towel pulls Elastic band pulls
Toe flexors	Flexor digitorum longus Flexor hallucis longus Foot intrinsics	Towel curls Marble pickups Pencil pickups

Cervical spine and trunk		
Muscle groups	Muscles	Exercise
Cervical extensors	Splenius cervicis Splenius capitus Trapezius, upper fibers	Dead lifts Neck extensions
Cervical flexors	Sternocleidomastoid	Chin tucks Sit-ups
Cervical rotators	Sternocleidomastoid Splenius cervicis Splenius capitus	Machine neck rotations
Trunk extensors	Erector spinae	Back extensions Alternating prone extensions Prone arches
Trunk flexors	Rectus abdominis External oblique abdominal Internal oblique abdominal	Dead lifts Crunches Crunch twists Sit-ups Gym ladder sit-ups Calves over bench sit-ups Incline bench sit-ups Specific bench sit-ups High pulley crunches Machine crunches Incline leg raises Leg raises Hanging leg raises
Trunk rotators	External oblique abdominal Internal oblique abdominal	Crunches Crunch twists Sit-up twists Gym ladder sit-ups Calves over bench sit-ups Incline bench sit-ups Specific bench sit-ups High pulley crunches Machine crunches Incline leg raises Leg raises Hanging leg raises Broomstick twists Machine trunk rotations
Trunk lateral flexors	External oblique abdominal Internal oblique abdominal Quadratus lumborum Rectus abdominis	Crunches Crunch twists Sit-ups Gym ladder sit-ups Calves over bench sit-ups Incline bench sit-ups Specific bench sit-ups High pulley crunches Machine crunches Incline leg raises Leg raises Hanging leg raises Broomstick twists Dumbbell side bends Roman chair side bends

Appendix 4

Etymology of commonly used terms in kinesiology

Below are some of the most commonly used terms in naming the muscles, bones, and joints as well as some additional terms utilized in explaining their function. This etymology is provided in order to better understand the origin and historical development of these terms and to provide a more meaningful background as to how these terms came to be used in the study of the body and its movement today.

abdominis Latin: belly

abductor Latin: abducere

acromion Greek: akron, extremity + omus, shoulder

adductor Latin: adducere, adduct-, to bring to, contract

amphiarthrodial Greek: ampho, both + arthron, joint + eidos, form, shape

anconeus Greek: agkon, elbow

antebrachial Latin: ante, before + brachium, arm

antecubital Latin: ante, before + cubitum, elbow

anterior Latin: comparative of ante, before

appendicular Latin: appendere, to hang to

arthrodial Greek: arthron, joint + eidos, form, shape

axillary Latin: axilla

axon Greek: axon, axis

biceps Latin: two-headed, bi-, two; caput, head

brachialis Latin: brachialis, brachial, arm

brachii Latin: bracchium, arm

brachioradialis Latin: bracchium, arm + radialis, radius

brevis Latin: (adj.) short, low, little, shallow

buccal Latin: cheek

bursa Greek: a leather sack

calcaneus Latin: calcaneus, heel, from Latin calcaneum, from calx, calc

cancellous Latin: cancellus, lattice

capitate, capitis Latin: caput, head

carpal Latin: carpalis, from carpus, wrist

carpus, carpi Latin: from Greek karpos, wrist

caudal Latin: caudalis, tail

celiac Greek: koilia, belly

cephalic Greek: kephale, head

cerebellum Latin: little brain

cerebrum Latin: cerebrum, brain

cervicis Latin: cervix, neck

clavicle French: clavicule, collarbone; Latin: clavicula, little key

coccyx Greek: kokkyz, cuckoo

colli Latin: collare, necklace, band or chain for the neck, from collum, the neck

concentric Latin: con, together with + centrum, center

condyle Greek: kondylos, knuckle

condyloidal Greek: kondylos, knuckle + eidos, form, shape

coracobrachialis Greek: from coracoid korax, crow; eidos, form; Greek: brachion + Latin, radial

coracoid Greek: korax, raven + eidos, form, shape

coronal Greek: korone, crown

coronoid Greek: korone, something curved, kind of crown + eidos, form, shape

cortical Latin: rind

costal Latin: costa, rib

coxal Latin: coxa, hip

cranial Latin: cranialis, cranium; Greek: kranion

crest Latin: crista, crest

crural Latin: cruralis, pertaining to leg or thigh

cubital Latin: cubitum, elbow

cuboid Greek: kybos, cube

cuneiforms Latin: cuneus, wedge + forma, form

deltoid, deltoidius Latin: deltoides: Greek deltoeides, triangular: delta, delta + -oeides, -oid

dendrite Greek: dendrites, pertaining to a tree

derma Greek: derma, skin

dermatome Greek: derma, skin + tome, incision

diaphysis Greek: diaphysis, a growing through

diarthrodial Greek: dis, two + arthron, joint + eidos, form, shape

digital, digitorum Latin: digitus, finger or toe

distal Latin: distare, to be distant

dorsal Latin: dorsalis, dorsualis, of the back, from dorsum, back

dorsi Latin: dorsi, genitive of dorsum, back

eccentric Greek: ek, out + kentron, center

enarthrodial Greek: en, in + arthron, joint + eidos, form, shape

endosteum Greek: endon, within + osteon, bone

epiphyseal Greek: epi, above + phyein, to grow

epiphysis Greek: a growing upon

erector Latin: erigere, to erect

extensor Latin: extendere, to stretch out

external, externus Latin: externus, outside, outward

fabella Latin: faba, little bean

facet French: facette, small face

fasciae Latin: fascia, band

femoris Latin: femur, thigh, genitive of femur, thigh

femur Latin: thigh

fibers Latin: fibra, a fiber, filament, of uncertain origin, perhaps related to Latin: filum, thread

fibrous Latin: fibra, fiber

fibula Latin: clasp, brooch

flexor Latin: bender

foramen Latin: hole

fossa Latin: ditch

fovea Latin: pit

frontal   Latin: frontem (nom, frons), forehead, literally that which projects

fusiform   Latin: fusus, spindle + forma, shape

gaster   Greek: gaster, belly

gastrocnemius   Greek: gaster, belly + kneme, leg

gemellus   Latin: twin

ginglymus   Greek: ginglymos, hinge

gluteus   Greek: gloutos, buttock

gomphosis   Greek: bolting together

goniometer   Greek: gonia, angle + metron, measure

gracilis   Latin: graceful

greater   Middle English: grete; Old English: great, thick, coarse; French: grand, which is from Latin: magnus

hallucis   Latin: hallex, large toe

hamate   Latin: hamatus, hooked

head   Latin: from caput

humerus   Latin: a misspelling borrowing umerus, shoulder

hyaline   Greek: hyalos, glass

iliacus   Latin: ilium, flank

iliocostalis   Latin: ilium, flank + costa, rib

ilium   Latin: ilium, groin, flank, variant of Latin: ilia

indicis   Latin: forefinger, pointer, sign, list

inferior   Latin: inferior, lower

infraspinatus, infraspinous   Latin: infra, below + spina, spine

inguinal   Latin: inguinalis, groin

insertion   Latin: in, into + serere, to join

intermediate   Latin: intermediates, lying between; Latin: intermedius, that which is between; from inter, between + medius, in the middle

intermedius   Latin: inter, between, mediare, to divide that which is between

internal, internus   Latin: internus, within

interossei   Latin: inter, between + os, bone

interspinalis   Latin: inter, between + spina, spine

intertransversarii   Latin: inter, between + transverses, cross-direction

ischium   Greek: ischion, hip joint

isokinetic   Greek: isos, equal + kinesis, motion

isometric   Greek: isos, equal + metron, measure

isotonic   Greek: isos, equal + tonus, tone

kinematic   Greek: kinematos, movement

kinesiology   Greek: kinematos, movement + logos, word, reason

kinesthesia   Greek: kinematos, movement + aesthesis, sensation

kyphosis   Greek: humpback

latae   Latin: latus, side

lateral, lateralis   Latin: lateralis, belonging to the side

latissimus   Latin: latissimus, superlative of latus, wide

lesser   Middle English: lesse; Latin: minor

levator   Latin: levator, lifter

lever   Latin: levare, to raise

linea   Latin: linea, line

longissimus   Latin: longest, very long

longus   Latin: long

lordosis   Greek: lordosis, bending

lumbar, lumborum   Latin: lumbus, loin

lumbricales   Latin: lumbus, loin; referred to vermiform; Latin: vermis, worm + forma, form

lunate   Latin: lunatus, past participle of lunare, to bend like a crescent, from luna, moon

magnus   Latin: great

major   Middle English: majour; Latin: major

malleolus   Latin: malleolus, little hammer

mammary   Latin: mamma, breast

mandible   Latin: mandibula, jaw; Latin: mandere, to chew

manubrium   Latin: handle, from manus, hand

margin   Latin: marginalis, border

maxilla   Latin: upper jaw, of mala, jaw, cheekbone

maximus   Latin: greatest

meatus   Latin: meatus, passage

medial, medialis   Latin: medialis, of the middle; Latin: medius, middle

medius   Latin: middle

mental   Latin: mentum, chin

metacarpal   Greek: meta, after, beyond, over + Latin: carpalis, from carpus, wrist

metatarsals   Greek: meta, after, beyond, over + tarsos, flat surface

middle   Old English: middle; Latin: medium

minimus, minimi   Latin: minimum, smallest

minor   Latin: lesser, smaller, junior

multifidus   Latin: multus, many + clefts or segments

muscle   Latin: musculus

myo   Greek: mys, muscle

myotome   Greek: mys, muscle + tome, incision

nasal   Latin: nasus, nose

navicular   Latin: navicula, boat, diminutive of navis, ship

neural   Latin: neuralis, nerve

neuron   Greek: neuron, nerve, sinew

notch   French: noche, indention, depression

nuchal   Latin: nape (back) of the neck

oblique, obliquus   Latin: obliquus, slanted

obturator   Latin: obturare, to close

occiput   Latin: occiput (gen. occipitis), back of the skull, from ob, against, behind + caput, head

olecranon   Greek: elbow

omos   Greek: omos, shoulder

opponens   Latin: opponentem (nom. opponens), prp. of opponere, oppose, object to, set against

oral   Latin: oralis, mouth

orbital   Latin: orbita, track

origin   Latin: origo, beginning

osseous   Latin: osseus, bony

otic   Greek: otikos, ear

palmar   Latin: palma, palm of the hand

palmaris   Latin: palma, palm

patella   Latin: pan, kneecap

pectineus   Latin: pectin, comb

pectoralis Latin: pectoralis, from pectus, pector-, breast; Middle English, French, Latin: pectorale, breastplate, from neuter of pectoralis

pedal Latin: pedalis, foot

pennate Latin: penna, feather

perineal Greek: perinaion, perineum

periosteum Greek: peri, around + osteon, bone

peroneus Greek: perone, brooch

phalanges A plural of phalanx Greek: phalangos, finger or toe bone

phalanx Greek: phalangos, finger or toe bone

piriformis Latin: piriformis, pear shaped

pisiform Latin: pisa, pea + forma, form

plantae Latin: planta, sole of the foot

pollicis Latin: from pollex, thumb, big toe

popliteus Latin: poples, ham of the knee

posterior Latin: comparative of posterus, coming after, from post, afterward

process Latin: processus, going before

profundus Latin: deep, bottomless, vast

pronator Latin: pronare, to bend forward

proximal Latin: proximitatem (nom. proximitas), nearness, vicinity, from proximus, nearest

psoas Greek: psoa, muscle of the loin

pubis Latin: (os) pubis, bone of the groin

quadratus Latin: quadratus, square

quinti Latin: quintus, fifth

radialis Latin: radialis, radius, beam of light

radiate Latin: radiatre, to emit rays

radius Latin: beam of light

ramus Latin: branch

rectus Latin: rectus, straight

rhomboids, rhomboidus From the word rhombus; Latin: flatfish, magician's circle; from Greek: rhombos, rhombus

rotatores Latin: rotare, to rotate

sacrum Latin: os sacrum, sacred bone

sartorius Latin: sartor, tailor

scaphoid Latin: skiff, boat shaped + eidos, form

scapulae, scapula Latin: shoulder; Latin: scapulae, the shoulder blades

scoliosis Greek: scoliosis, crookedness

sellar Latin: Turkish saddle

semimembranosus Latin: semi, half + membrane, membrane

semispinalis Latin: semi, half + spina, spine

semitendinosus Latin: semi, half + tendere, to stretch

serratus Latin: serratus, saw-shaped, from serra, saw

sesamoid Latin: sesamoides, resembling a grain of sesame in size or shape

sinus Latin: curve, hollow

soleus Latin: solea, sandal

somatic Greek: soma, body

sphincter Greek: sphincter, hand

spinae Latin: thorn

spinal Latin: spinalis, spine

splenius Greek: splenion, bandage

sternocleidomastoid Greek: sternon, chest + kleis, key + mastos, breast + eidos, form

sternum Greek: sternon, chest, breast, breastbone

styloid Anglo-Saxon: stigan, to rise + eidos, form

subscapularis Latin: sub, beneath + scapulae, shoulder blades

sulcus Latin: groove

superficialis Latin: superficies, of or pertaining to the surface

superior Latin: superiorem, higher

supinator Latin: reflectere, to bend back

supraspinatus, supraspinous Latin: supra, above + spina, spine

sural Latin: sura, calf

suture Latin: sutura, a seam

symphysis Greek: symphysis, growing together

synarthrodial Greek: syn, together + arthron, joint + eidos, form, shape

synchondrosis Greek: syn, together + chondros, cartilage + osis, condition

syndesmosis Greek: syndesmos, ligament + osis, condition

synovial Latin: synovia, joint fluid

talus Latin: ankle, anklebone, knucklebone

tarsal Greek: tarsalis, ankle

temporal Latin: temporalis, of time, temporary, from tempus (temporis), time, season, proper time or season

tendon Latin: tendo, tendon

tensor Latin: tendere, to stretch

teres Latin: rounded

tertius Latin: third

thoracic Greek: thorax, chest

tibia Latin: shinbone

tibialis Latin: tibia, pipe, shinbone

transverse Latin: transverses, oblique

trapezium Greek: trapezion, a little table

trapezius Latin: trapezium, trapezium, from the shape of the muslces paired

trapezoid Greek: trapezoeides, table shaped

triceps Latin: three-headed; tri-, tri- + caput, head

triquetrum Latin: neuter of triquetrus, three-cornered

trochanter Greek: trokhanter, to run

trochlear Greek: trokhileia, system of pulleys

tubercle Latin: turberculum, a little swelling

tuberosity Latin: tuberositas, tuberosity

ulna Latin: elbow

ulnaris Latin: ulna, elbow

umbilical Latin: umbilicus, naval

vastus Latin: immense, extensive, huge

vertical Latin: verticalis, overhead, vertex, highest point

visceral Latin: viscera, body organs

volar Latin: vola, sole, palm

xiphoid Greek: xiphos, sword + eidos, form, shape

zygoma Latin: zygoma, zygomat-, from Greek zugoma, bolt, from zugoun, to join

Worksheets

Chapter 1

Worksheet no. 1

On the posterior skeletal worksheet, list the names of the bones and all of the prominent features of each bone.

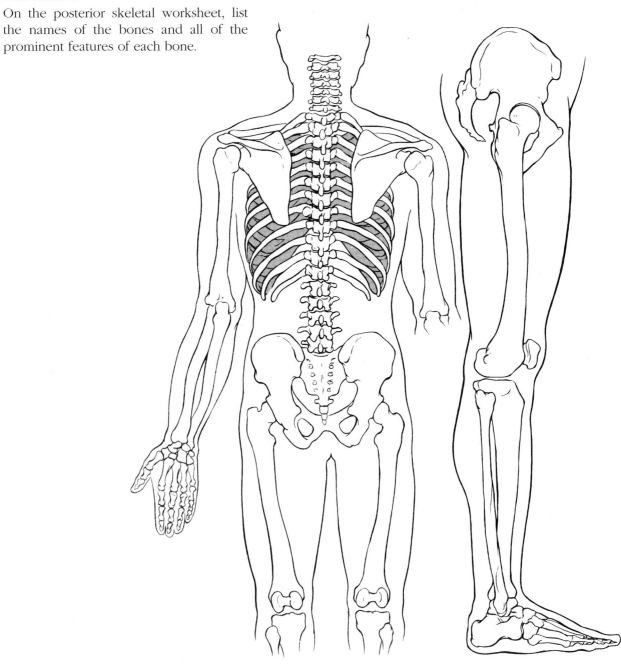

Worksheet no. 2

On the anterior skeletal worksheet, list the names of the bones and all of
the prominent features of each bone.

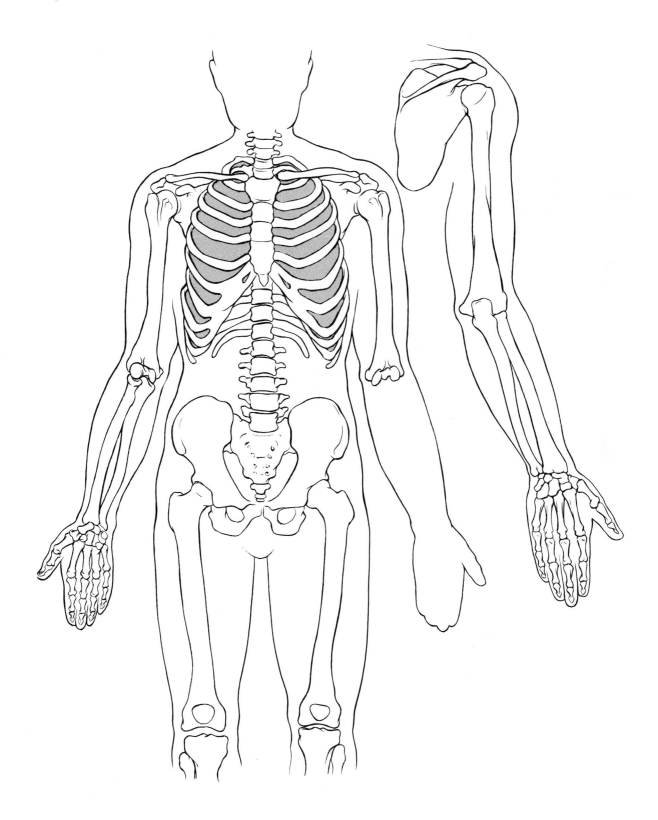

Chapter 4

Worksheet no. 1

Draw and label on the worksheet the following listed muscles. Indicate the origin and insertion of each muscle with an "O" and an "I," respectively.

a. Trapezius

b. Rhomboid

c. Serratus anterior

d. Levator scapulae

e. Pectoralis minor

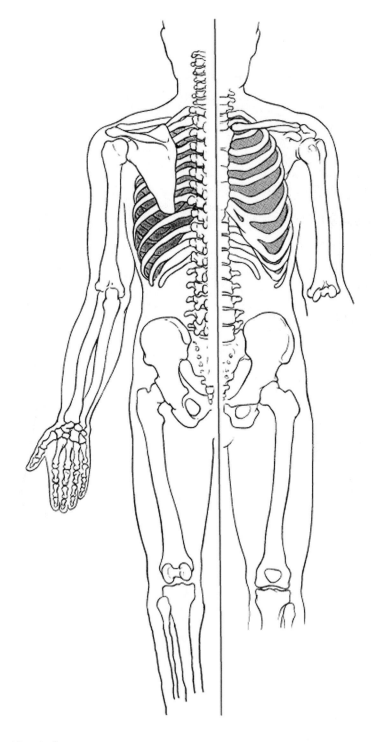

Chapter 4

Worksheet no. 2

Label and indicate by arrows the following movements of the shoulder girdle. For each motion, list the plane in which it occurs and list the axis of rotation.

a. Adduction

b. Abduction

c. Upward rotation

d. Downward rotation

e. Elevation

f. Depression

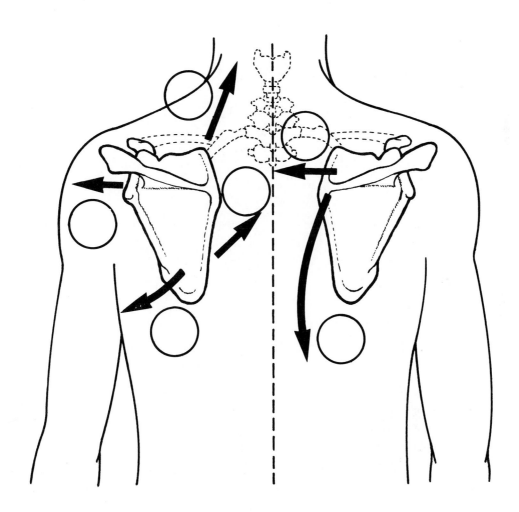

Chapter 5

Worksheet no. 1

Draw and label on the worksheet the following muscles. Indicate the origin and insertion of each muscle with an "O" and an "I," respectively.

a. Deltoid

b. Supraspinatus

c. Subscapularis

d. Teres major

e. Infraspinatus

f. Teres minor

g. Latissimus dorsi

h. Pectoralis major

i. Coracobrachialis

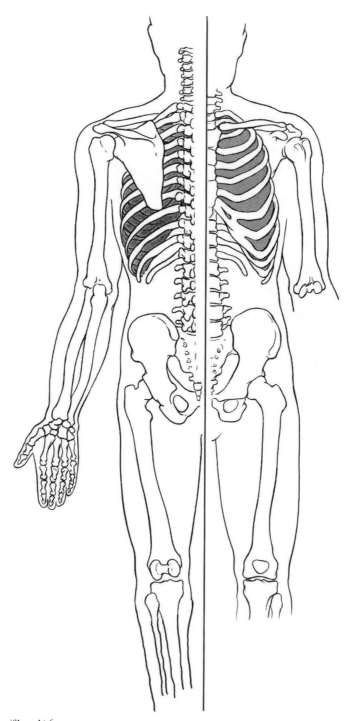

Chapter 5

Worksheet no. 2

Label and indicate by arrows the following listed movements of the shoulder joint. For each motion, list the plane in which it occurs and list the axis of rotation.

a. Abduction
b. Adduction
c. Flexion
d. Extension
e. Horizontal adduction
f. Horizontal abduction

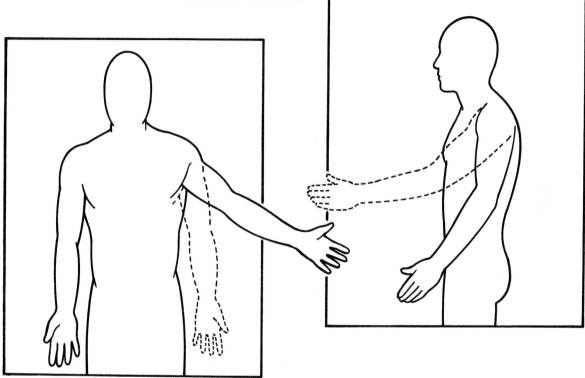

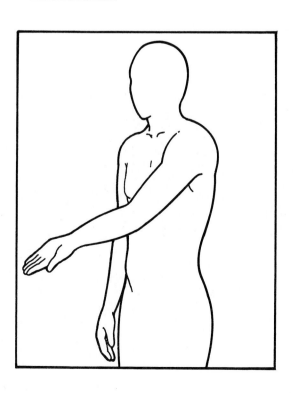

Chapter 6

Worksheet no. 1

Draw and label on the worksheet the following muscles. Indicate the origin and insertion of each muscle with an "O" and an "I," respectively.

a. Biceps brachii   e. Supinator

b. Brachioradialis   f. Triceps brachii

c. Brachialis   g. Anconeus

d. Pronator teres   h. Pronator quadratus

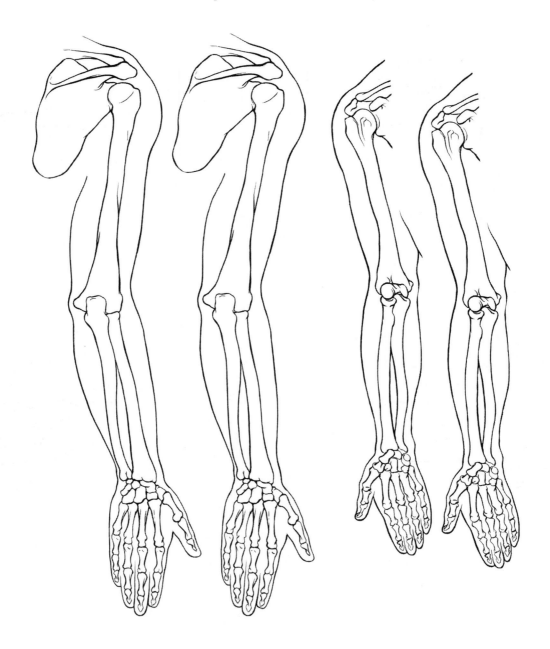

Chapter 6

Worksheet no. 2

Label and indicate by arrows the following movements of the elbow and radioulnar joint. For each motion, list the plane in which it occurs and list the axis of rotation.

Elbow
　　Flexion
　　Extension

Radioulnar joint
　　Pronation
　　Supination

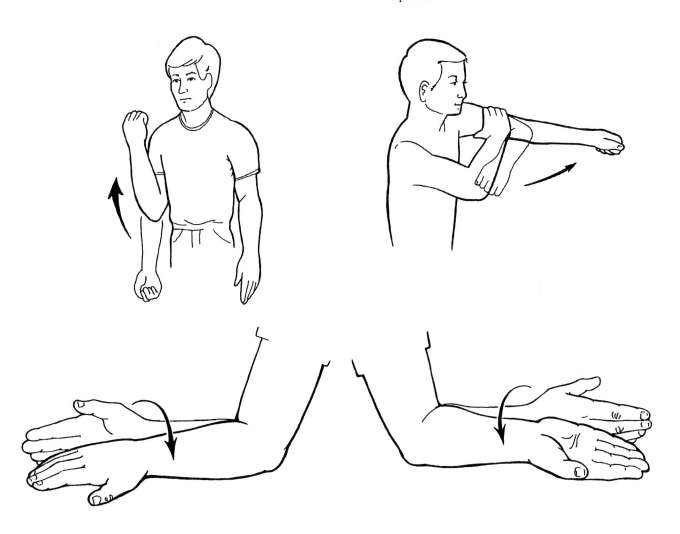

Chapter 7

Worksheet no. 1

Draw and label on the worksheet the following muscles. Indicate the origin and insertion of each muscle with an "O" and an "I," respectively.

a. Flexor pollicis longus
b. Flexor carpi radialis
c. Flexor carpi ulnaris
d. Extensor digitorum
e. Extensor pollicis longus

f. Extensor pollicis brevis
g. Extensor carpi ulnaris
h. Palmaris longus
i. Extensor carpi radialis longus
j. Extensor carpi radialis brevis

k. Extensor digiti minimi
l. Extensor digitorum indicis
m. Flexor digitorum superficialis
n. Flexor digitorum profundus
o. Abductor pollicis longus

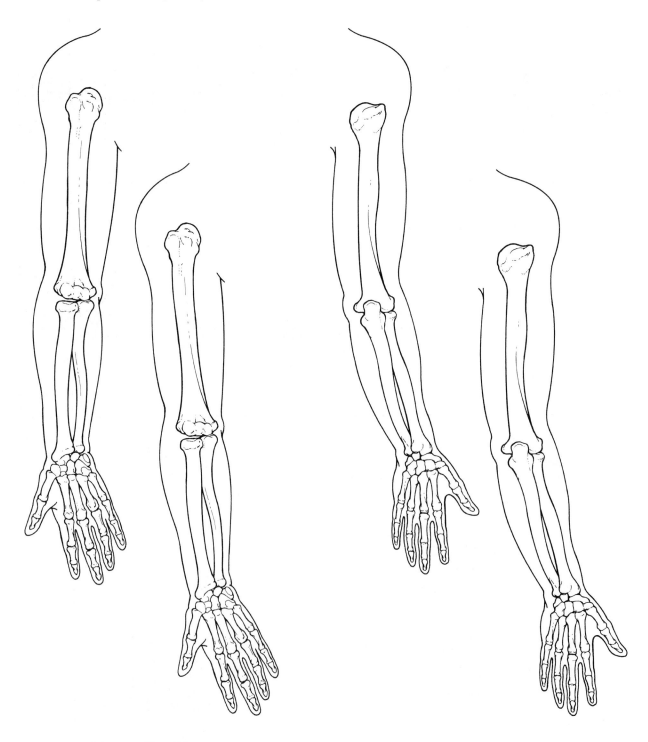

Chapter 7

Worksheet no. 2

Label and indicate by arrows the following movements of the wrist and hands.
For each motion, list the plane in which it occurs and list the axis of rotation.

Wrist and hands
Extension
Flexion
Abduction
Adduction

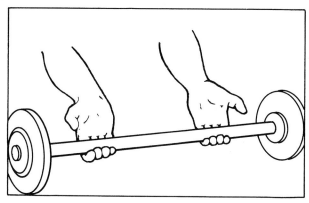

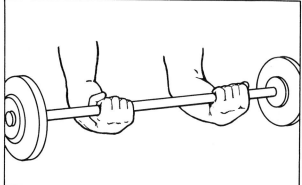

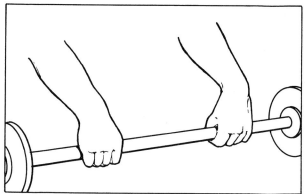

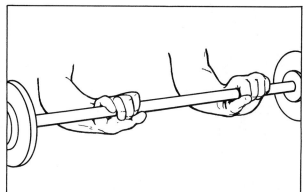

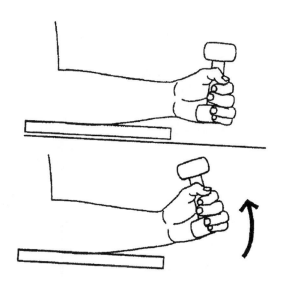

Chapter 8

Worksheet no. 1

Analyze this exercise following the procedures explained in this chapter that include joint movements and muscles that produce these movements.

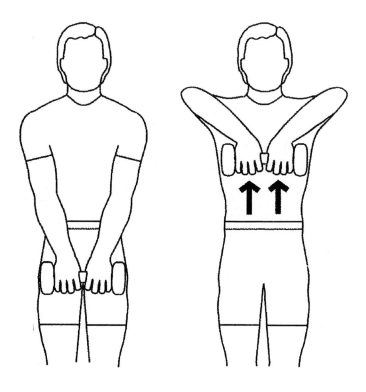

Worksheet no. 2

Analyze this exercise following the procedures explained in this chapter that include joint movements and muscles that produce these movements.

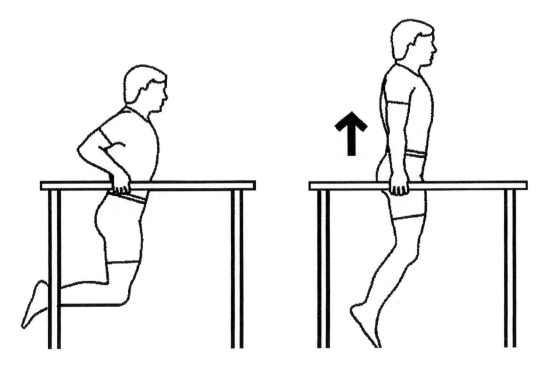

Chapter 9

Worksheet no. 1

Draw and label on the worksheet the anterior hip joint and pelvic girdle muscles. Indicate the origin and insertion of each muscle with an "O" and an "I," respectively.

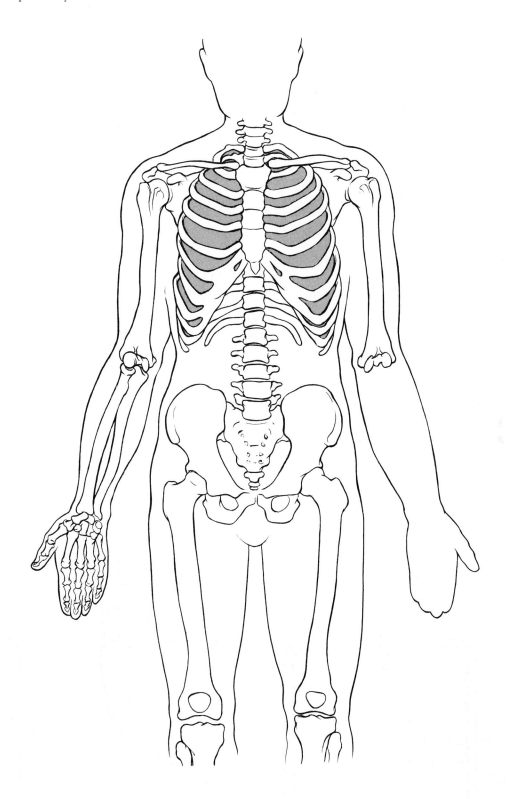

Chapter 9

Worksheet no. 2

Draw and label on the worksheet the posterior hip and pelvic girdle muscles. Indicate the origin and insertion of each muscle with an "O" and an "I," respectively.

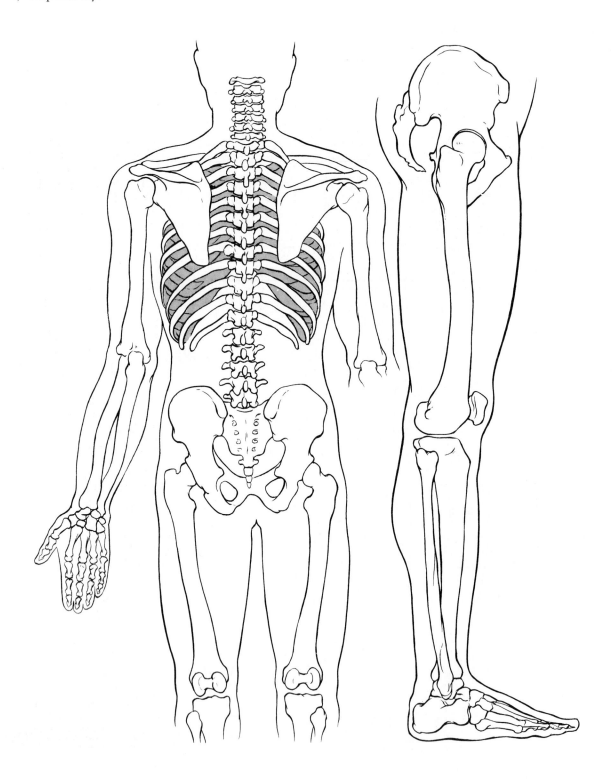

Chapter 10

Worksheet no. 1

Draw and label on the worksheet the knee joint muscles. Indicate the origin and insertion of each muscle with an "O" and an "I," respectively.

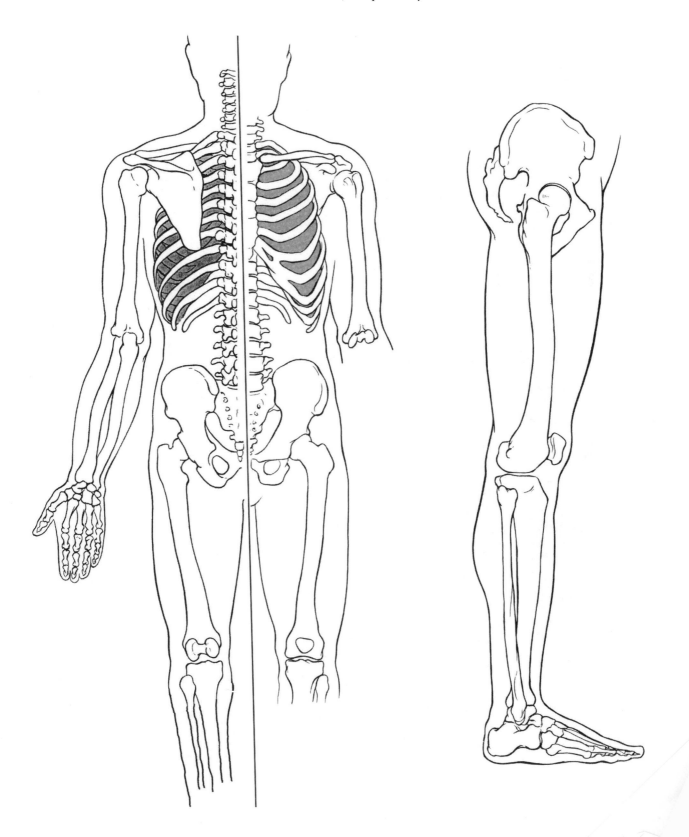

Chapter 11

Worksheet no. 1

Draw and label on the worksheet the following muscles of the ankle and foot. Indicate the origin and insertion of each muscle with an "O" and an "I," respectively.

a. Tibialis anterior
b. Extensor digitorum longus
c. Peroneus longus
d. Peroneus brevis
e. Soleus
f. Peroneus tertius

g. Gastrocnemius
h. Extensor hallucis longus
i. Tibialis posterior
j. Flexor digitorum longus
k. Flexor hallucis longus

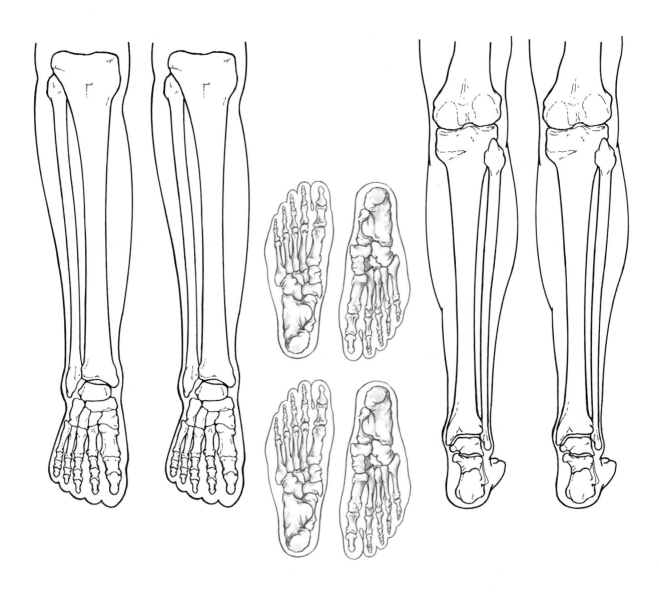

Chapter 12

Worksheet no. 1

Draw and label the following muscles on the skeletal chart. Indicate the origin and insertion of each muscle with an "O" and an "I," respectively.

a. Rectus abdominis

b. External oblique abdominal

c. Internal oblique abdominal

d. Sternocleidomastoid

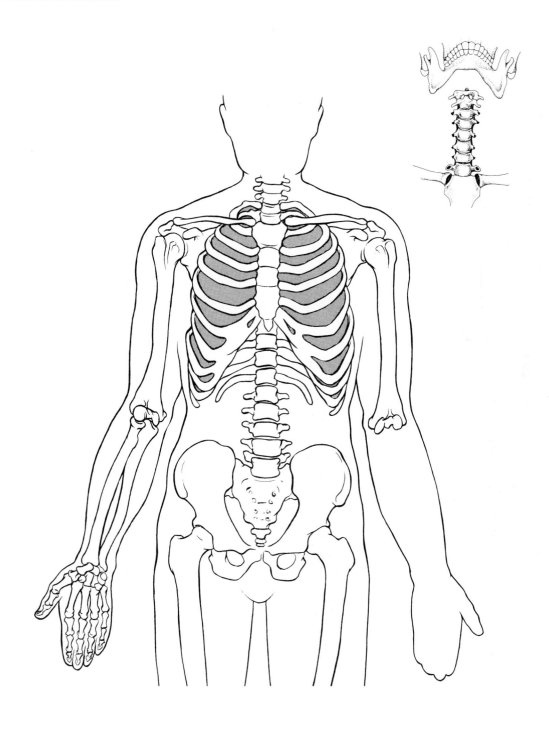

Chapter 12

Worksheet no. 2

Draw and label the following muscles on the skeletal chart. Indicate the origin and insertion of each muscle with an "O" and an "I," respectively.

a. Erector spinae

b. Quadratus lumborum

c. Splenius cervicis and capitis

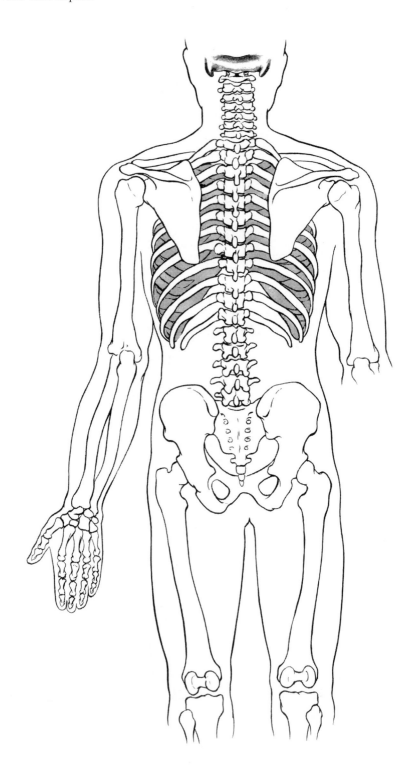

Chapter 13
Worksheet no. 1

Kinesiology skill analysis table

Phase		Toes	TT/ST	Ankle	Knee	Hip	Lumbar	Cervical	Shou G	G-H	Elbow	Wrist	Fingers
	Position												
	Deg/move												
	Agon/con												
	Position												
	Deg/move												
	Agon/con												
	Position												
	Deg/move												
	Agon/con												
	Position												
	Deg/move												
	Agon/con												
	Position												
	Deg/move												
	Agon/con												

Glossary

abduction Lateral movement away from the midline of the trunk, as in raising the arms or legs to the side horizontally.

acceleration The rate of change in velocity.

accessory motion The actual change in relationship between the articular surface of one bone relative to another, characterized as roll, spin, and glide.

action potential Electrical signal transmitted from the brain and spinal cord through axons to the muscle fibers in a particular motor unit providing the stimulus to contract.

active insufficiency Point reached when a muscle becomes shortened to the point that it cannot generate or maintain active tension.

adduction Movement medially toward the midline of the trunk, as in lowering the arms to the side or legs back to the anatomical position.

afferent nerves Nerves that bring impulses from receptors in the skin, joints, muscles, and other peripheral aspects of the body to the central nervous system.

aggregate muscle action Muscles working together in groups rather than independently to achieve given joint motions.

agonist A muscle or muscle group that is described as being primarily responsible for a specific joint movement when contracting.

all or none principle States that regardless of the number involved, the individual muscle fibers within a given motor unit will fire and contract either maximally or not at all.

amphiarthrodial (amphiarthrosis) joints Joints that functionally allow only a very slight amount of movement such as synchondrosis (ex. costochondral joint of the ribs with sternum), syndesmosis (ex. distal tibiofibular), and symphysis (ex. symphysis pubis) joints.

amplitude Range of muscle fiber length between maximal and minimal lengthening.

anatomical position The position of reference in which the subject is in the standing position, with feet together and palms of hands facing forward.

angle Bend or protruding angular projection of a bone such as superior and inferior angle of scapula.

angle of pull The angle between the muscle insertion and the bone on which it inserts.

angular displacement The change in location of a rotating body.

angular motion Motion involving rotation around an axis.

antagonist A muscle or muscle group that counteracts or opposes the contraction of another muscle or muscle group.

anteroposterior axis The axis that has the same directional orientation as the sagittal plane of motion and runs from front to back at a right angle to the frontal plane of motion. Also known as the sagittal or AP axis.

appendicular skeleton The appendages, or the upper and lower extremities, and the shoulder and pelvic girdles.

arthrodial joints Joints in which bones glide on each other in limited movement, as in the bones of the wrist (carpal) or the bones of the foot (tarsal).

arthrokinematics Motion between the actual articular surfaces of the bones at a joint.

arthrosis Joint or articulation between two or more bones.

axial skeleton The skull, vertebral column, ribs, and sternum.

axis of rotation The point in a joint about which a bone moves or turns to accomplish joint motion.

axon An elongated projection that transmits impulses away from the neuron cell body.

balance The ability to control equilibrium, either static or dynamic.

biarticular muscles Those muscles that, from origin to insertion, cross two different joints, allowing them to perform actions at each joint.

bilateral Relating to the right and left sides of the body or of a body structure such as the right and left extremities.

biomechanics The study of mechanics as it relates to the functional and anatomical analysis of biological systems, especially humans.

bipennate A type of pennate muscle with fibers running obliquely on both sides from a central tendon, such as the rectus femoris and flexor hallucis longus.

border or margin Edge or boundary line of a bone such as lateral and medial border of scapula.

brachial plexus Group of spinal nerves composed of cervical nerves 5 through 8, along with thoracic nerve 1; supplies motor and sensory function to the upper extremity and most of the scapula.

cancellous bone Spongy, porous bone that lies under cortical bone.

carpal tunnel syndrome A condition characterized by swelling and inflammation with resultant increased pressure in the carpal tunnel, which interferes with normal function of the median nerve, leading to reduced motor and sensory function of its distribution; particularly common with repetitive use of the hand and wrist in manual labor and clerical work such as typing and keyboarding.

cartilaginous joints Joints joined together by hyaline cartilage or fibrocartilage, allowing very slight movement, such as synchondrosis and symphysis.

center of gravity The point at which all of the body's mass and weight are equally balanced or equally distributed in all directions.

central nervous system (CNS) The cerebral cortex, basal ganglia, cerebellum, brain stem, and spinal cord.

cervical plexus Group of spinal nerves composed of cervical nerves 1 through 4; generally responsible for sensory and motor function from the upper part of the shoulders to the back of the head and front of the neck.

circumduction Circular movement of a bone at the joint, as in movement of the hip, shoulder, or trunk around a fixed point. Combination of flexion, extension, abduction, and adduction.

closed kinetic chain When the distal end of an extremity is fixed, preventing movement of any one joint unless predictable movements of the other joints in the extremity occur.

coefficient of friction The ratio between the force needed to overcome friction over the force holding the surfaces together.

concentric contraction A contraction in which there is a shortening of the muscle that causes motion to occur at the joints it crosses.

condyle Large, rounded projection that usually articulates with another bone, such as the medial or lateral condyle of the femur.

condyloid joint Type of joint in which the bones permit movement in two planes without rotation, as in the wrist between the radius and the proximal row of the carpal bones or the second, third, fourth, and fifth metacarpophalangeal joints.

contractility The ability of muscle to contract and develop tension or internal force against resistance when stimulated.

contraction phase In a single muscle fiber contraction, it is the phase following the latent perion in which the muscle fiber actually begins shortening; lasts about 40 milliseconds.

coronal axis Runs from side to side through the body and is at a right angle to the sagittal plane of motion. Also known as the frontal or lateral axis.

cortex Diaphyseal wall of long bones, formed from hard, dense compact bone.

cortical bone Harder, more compact bone that forms the outer bony surface of the diaphysis.

cranial nerves The group of 12 pairs of nerves originating from the undersurface of the brain and exiting from the cranial cavity through skull openings; they supply specific motor and sensory function to the head and face.

crest Prominent, narrow, ridgelike projection of bone, such as the iliac crest of the pelvis.

curvilinear motion Motion along a curved line.

dendrite One or more branching projections from the neuron cell body that transmit impulses to the neuron and cell body.

depression Inferior movement of the shoulder girdle, as in returning to the normal position from a shoulder shrug.

dermatome A defined area of skin supplied by a specific spinal nerve.

diagonal abduction Movement by a limb through a diagonal plane away from the midline of the body such as in the hip or glenohumeral joint.

diagonal adduction Movement by a limb through a diagonal plane toward and across the midline of the body such as in the hip or glenohumeral joint.

diagonal or oblique axis Axis that runs at a right angle to the diagonal plane. As the glenohumeral joint moves from diagonal abduction to diagonal adduction in overhand throwing, its axis runs perpendicular to the plane through the humeral head.

diagonal plane A combination of more than one plane. Less than parallel or perpendicular to the sagittal, frontal, or transverse plane. Also known as oblique plane.

diaphysis The long cylindrical portion or shaft of long bones.

diarthrodial (diarthrosis) joints Freely movable synovial joints containing a joint capsule and hyaline cartilage and lubricated by synovial fluid.

displacement A change in position or location of an object from its original point of reference.

distal Farthest from the midline or point or reference; the fingertips are the most distal part of the upper extremity.

distance The path of movement; refers to the actual sum length of units of measurement traveled.

dorsal Relating to the back, being or located near, on, or toward the back, posterior part, or upper surface of.

dorsal flexion (dorsiflexion) Flexion movement of the ankle resulting in the top of foot moving toward the anterior tibia.

duration An exercise variable usually referring to the number of minutes per exercise bout.

dynamic equilibrium Occurs when all of the applied and inertial forces acting on the moving body are in balance, resulting in movement with unchanging speed or direction.

dynamics The study of mechanics involving systems in motion with acceleration.

eccentric contraction A contraction in which the muscle lengthens in an attempt to control the motion occurring at the joints that it crosses, characterized by the force of gravity or applied resistance being greater than the contractile force.

eccentric force Force that is applied in a direction not in line with the center of rotation of an object with a fixed axis. In objects without a fixed axis, it is an applied force that is not in line with the object's center of gravity.

efferent nerves Nerves that carry impulses to the outlying regions of the body from the central nervous system.

elasticity The ability of muscle to return to its original length following stretching.

electromyography (EMG) A method utilizing either surface electrodes or fine wire/needle electrodes to detect the action potentials of muscles and provide an electronic readout of the contraction intensity and duration.

elevation Superior movement of the shoulder girdle, as in shrugging the shoulders.

enarthrodial joint Type of joint that permits movement in all planes, as in the shoulder (glenohumeral) and hip joints.

endochondral bones Long bones that develop from hyaline cartilage masses after the embryonic stage.

endosteum Dense, fibrous membrane covering the inside of the cortex of long bones.

epicondyle Projection located above a condyle, such as the medial or lateral epicondyle of the humerus.

epiphyseal plate Thin cartilage plate separating the diaphysis and epiphysis during bony growth; commonly referred to as growth plate.

epiphysis The end of a long bone, usually enlarged and shaped to join the epiphysis of an adjacent bone, formed from cancellous or trabecular bone.

equilibrium State of zero acceleration in which there is no change in the speed or direction of the body.

eversion Turning of the sole of the foot outward or laterally, as in standing with the weight on the inner edge of the foot.

extensibility The ability of muscle to be stretched back to its original length following contraction.

extension Straightening movement resulting in an increase of the angle in a joint by moving bones apart, as when the hand moves away from shoulder during extension of the elbow joint.

external rotation Rotary movement around the longitudinal axis of a bone away from the midline of the body. Also known as rotation laterally, outward rotation, and lateral rotation.

extrinsic muscles Muscles that arise or originate outside of (proximal to) the body part on which they act.

facet Small flat or shallow bony articular surface such as the articular facet of a vertebra.

fascia Fibrous membrane covering, supporting, connecting, and separating muscles.

fibrous joints Joints joined together by connective tissue fibers and generally immovable, such as gomphosis, sutures, and syndesmosis.

first-class lever A lever in which the axis (fulcrum) is between the force and the resistance, as in the extension of the elbow joint.

flat muscles A type of parallel muscle that is usually thin and broad, with fibers originating from broad, fibrous, sheetlike aponeuroses such as the rectus abdominus and external oblique.

flexion Movement of the bones toward each other at a joint by decreasing the angle, as in moving the hand toward the shoulder during elbow flexion.

follow-through phase Phase that begins immediately after the climax of the movement phase, in order to bring about negative acceleration of the involved limb or body segment; often referred to as the deceleration phase. The velocity of the body segment progressively decreases, usually over a wide range of motion.

foramen Rounded hole or opening in bone, such as the foramen magnum in the base of the skull.

force The product of mass times acceleration.

force arm The perpendicular distance between the location of force application and the axis. The shortest distance from the axis of rotation to the line of action of the force. Also known as the moment arm or torque arm.

fossa Hollow, depression, or flattened surface of bone, such as the supraspinatus fossa or iliac fossa.

fovea Very small pit or depression in bone, such as the fovea capitis of the femur.

frequency An exercise variable usually referring to the number of times exercise is conducted per week.

friction Force that results from the resistance between the surfaces of two objects moving upon one another.

frontal plane Plane that bisects the body laterally from side to side, dividing it into front and back halves. Also known as the lateral or coronal plane.

fundamental position Reference position essentially the same as the anatomical position, except that the arms are at the sides and the palms are facing the body.

fusiform muscles A type of parallel muscle with fibers shaped together like a spindle with a central belly that tapers to tendons on each end, such as the brachialis and the biceps brachii.

gaster The central, fleshy, contractile portion of the muscle that generally increases in diameter as the muscle contracts.

ginglymus joint Type of joint that permits a wide range of movement in only one plane, as in the elbow, ankle, and knee joints.

glide (slide) (translation) A type of accessory motion characterized by a specific point on one articulating surface coming in contact with a series of points on another surface.

Golgi tendon organ (GTO) A proprioceptor, sensitive to both muscle tension and active contraction, found in the tendon close to the muscle tendon junction.

gomphosis A type of immovable articulation, as of a tooth inserted into its bony socket.

goniometer Instrument used to measure joint angles or compare the changes in joint angles.

ground reaction force The force of the surface reacting to the force placed on it, as in the reaction force between the body and the ground when running across a surface.

hamstrings A common name given to the group of posterior thigh muscles: biceps femoris, semitendinosus, and semimembranosus.

head Prominent, rounded projection of the proximal end of a bone, usually articulating, such as the humeral or femoral head.

heel-strike First portion of the walking or running stance phase characterized by landing on the heel with the foot in supination and the leg in external rotation.

horizontal abduction Movement of the humerus in the horizontal plane away from the midline of the body.

horizontal adduction Movement of the humerus in the horizontal plane toward the midline of the body.

hyaline cartilage Articular cartilage; covers the end of bones at diarthrodial joints to provide a cushioning effect and reduce friction during movement.

inertia Resistance to action or change; resistance to acceleration or deceleration. Inertia is the tendency for the current state of motion to be maintained, regardless of whether the body segment is moving at a particular velocity or is motionless.

innervation The supplying of a muscle, organ, or body part with nerves.

insertion The distal attachment or point of attachment of a muscle farthest from the midline or center of the body, generally considered the most movable part.

intensity An exercise variable usually referring to a certain percentage of the absolute maximum that a person can sustain.

internal rotation Rotary movement around the longitudinal axis of a bone toward the midline of the body. Also known as rotation medially, inward rotation, and medial rotation.

interneurons Central or connecting neurons that conduct impulses from sensory neurons to motor neurons.

intrinsic muscles Muscles that are entirely contained within a specified body part; usually refers to the small, deep muscles found in the foot and hand.

inversion Turning of the sole of the foot inward or medially, as in standing with the weight on the outer edge of the foot.

irritability The property of muscle being sensitive or responsive to chemical, electrical, or mechanical stimuli.

isokinetic Type of dynamic exercise usually using concentric and/or eccentric muscle contractions in which the speed (or velocity) of movement is constant and muscular contraction (usually maximal contraction) occurs throughout the movement.

isometric contraction A type of contraction with little or no shortening of the muscle resulting in no appreciable change in the joint angle.

isotonic Contraction occurring in which there is either shortening or lengthening in the muscle under tension; also known as a dynamic contraction, and classified as being either concentric or eccentric.

joint capsule Sleevelike covering of ligamentous tissue surrounding diarthrodial joints.

joint cavity The area inside the joint capsule of diarthrodial or synovial joints.

kinematics The description of motion, including consideration of time, displacement, velocity, acceleration, and space factors of a system's motion.

kinesiology The science of movement, which includes anatomical (structural) and biomechanical (mechanical) aspects of movement.

kinesthesia The awareness of the position and movement of the body in space; sense that provides awareness of bodily position, weight, or movement of the muscles, tendons, and joints.

kinetic friction The amount of friction occurring between two objects that are sliding upon one another.

kinetics The study of forces associated with the motion of a body.

Krause's end bulbs A proprioceptor sensitive to touch and thermal changes found in the skin, subcutaneous tissue, lip and eyelid mucosa, and external genitals.

kyphosis Increased anterior concavity of the normal thoracic curve. The lumbar spine may have a reduction of its normal lordotic curve, resulting in a flat-back appearance referred to as lumbar kyphosis.

latent period In a single muscle fiber contraction, it is the brief period of a few milliseconds following the stimulus before the contraction phase begins.

lateral axis Axis that has the same directional orientation as the frontal plane of motion and runs from side to side at a right angle to the sagittal plane of motion. Also known as the frontal or coronal axis.

lateral epicondylitis A common problem quite frequently associated with gripping and lifting activities that usually involves the extensor digitorum muscle near its origin on the lateral epicondyle; commonly known as tennis elbow.

lateral flexion Movement of the head and/or trunk laterally away from the midline; abduction of spine.

lever A rigid bar (bone) that moves about an axis.

ligament A type of tough connective tissue that attaches bone to bone to provide static stability to joints.

line Ridge of bone less prominent than a crest, such as the linea aspera of the femur.

linea alba Tendinous division and medial border of the rectus abdominis running vertically from the xiphoid process through the umbilicus to the pubis.

linear displacement The distance that a system moves in a straight line.

linear motion Motion along a line; also referred to as translatory motion.

linea semilunaris Lateral to the rectus abdominis, a crescent, or moon-shaped, line running vertically that represents the aponeurosis connecting the lateral border of the rectus abdominis and medial border of the external and internal abdominal obliques.

lordosis Increased posterior concavity of the lumbar and cervical curves.

lumbar plexus Group of spinal nerves composed of L1 through L4 and some fibers from T12, generally responsible for motor and sensory function of the lower abdomen and the anterior and medial portions of the lower extremity.

mass The amount of matter in a body.

maximal stimulus A stimulus strong enough to produce action potentials in all of the motor units of a particular muscle.

meatus Tubelike passage within a bone, such as the external auditory meatus of the temporal bone.

mechanical advantage The advantage gained through the uses of machines to increase or multiply the applied force in performing a task; enables a relatively small force to be applied to move a much greater resistance; determined by dividing the load by the effort.

mechanics The study of physical actions of forces; can be subdivided into statics and dynamics.

medial epicondylitis An elbow problem associated with the medial wrist flexor and pronator group near their origin on the medial epicondyle; frequently referred to as golfer's elbow.

median Relating to, located in, or extending toward the middle, situated in the middle, mesial.

medullary cavity   Marrow cavity between the walls of the diaphysis, containing yellow or fatty marrow.

Meissner's corpuscles   A proprioceptor sensitive to fine touch and vibration found in the skin.

midstance   Middle portion of the walking or running stance phase characterized by pronation and internal rotation of the foot and leg, may be divided into loading response, midstance, and terminal stance.

momentum   The quality of motion, which is equal to mass times velocity.

motor neurons   Neurons that transmit impulses away from the brain and spinal cord to muscle and glandular tissue.

motor unit   Consists of a single motor neuron and all of the muscle fibers it innervates.

movement phase   The action part of a skill, sometimes known as the acceleration, action, motion, or contact phase. Phase in which the summation of force is generated directly to the ball, sport object, or opponent, and is usually characterized by near-maximal concentric activity in the involved muscles.

multiarticular muscles   Those muscles that, from origin to insertion, cross three or more different joints, allowing them to perform actions at each joint.

multipennate muscle   A type of pennate muscle that has several tendons with fibers running diagonally between them, such as the deltoid.

muscle spindle   A proprioceptor sensitive to stretch and the rate of stretch that is concentrated primarily in the muscle belly between the fibers.

myotatic or stretch reflex   The reflexive contraction that occurs as a result of the motor neurons of a muscle being activated from the CNS secondarily to a rapid stretch occurring in the same muscle; the knee jerk or patella tendon reflex is an example.

myotome   A muscle or group of muscles supplied by a specific spinal nerve.

neuron   Nerve cell that is the basic functional unit of the nervous system responsible for generating and transmitting impulses.

neutralizers   Muscles that counteract or neutralize the action of other muscles to prevent undesirable movements; referred to as neutralizing, they contract to resist specific actions of other muscles.

notch   Depression in the margin of a bone such as the trochlear and radial notch of the ulna.

open kinetic chain   When the distal end of an extremity is not fixed to any surface, allowing any one joint in the extremity to move or function separately without necessitating movement of other joints in the extremity.

opposition   Diagonal movement of the thumb across the palmar surface of the hand to make contact with the hand and/or fingers.

origin   The proximal attachment or point of attachment of a muscle closest to the midline or center of the body, generally considered the least movable part.

osteoblasts   Specialized cells that form new bone.

osteoclasts   Specialized cells that resorb new bone.

osteokinematic motion   Motion of the bones relative to the three cardinal planes, resulting from physiological movements.

Pacinian corpuscles   A proprioceptor sensitive to pressure and vibration found in the subcutaneous, submucosa, subserous tissues around joints, external genitals, and mammary glands.

palmar flexion   Flexion movement of the wrist in the sagittal plane with the volar or anterior side of the hand moving toward the anterior side of the forearm.

palpation   Using the sense of touch to feel or examine a muscle or other tissue.

parallel muscles   Muscles that have their fibers arranged parallel to the length of the muscle, such as flat, fusiform, strap, radiate, or sphincter muscles.

passive insufficiency   State reached when an opposing muscle becomes stretched to the point where it can no longer lengthen and allow movement.

pennate muscles   Muscles that have their fibers arranged obliquely to their tendons in a manner similar to a feather, such as unipennate, bipennate, and multipennate muscles.

periodization   The intentional variance of overload through a prescriptive reduction or increase in a training program to bring about optimal gains in physical performance.

periosteum   The dense, fibrous membrane covering the outer surface of the diaphysis.

peripheral nervous system (PNS)   Portion of the nervous system containing the sensory and motor divisions of all the nerves throughout the body except those found in the central nervous system.

pes anserinus   Distal tendinous expansion formed by the sartorius, gracilis, and semitendinosus and attaching to the anteromedial aspect of the proximal tibia below the level of the tibial tuberosity.

physiological movement   Normal movements of joints such as flexion, extension, abduction, adduction, and rotation, accomplished by bones moving through planes of motion about an axis of rotation at the joint.

plane of motion   An imaginary two-dimensional surface through which a limb or body segment is moved.

plantar   Relating to the sole or undersurface of the foot.

plantar flexion   Extension movement of the ankle, resulting in the foot and/or toes moving away from the body.

plica   An anatomical variant of synovial tissue folds that may be irritated or inflamed with injuries or overuse of the knee.

preparatory phase   Skill analysis phase, often referred to as the cocking or wind-up phase, used to lengthen the appropriate muscles so that they will be in position to generate more force and momentum as they concentrically contract in the next phase.

process   Prominent projection of a bone, such as the acromion process of the scapula or the olecranon process of the humerus.

pronation   Internally rotating the radius so that it lies diagonally across the ulna, resulting in the palm-down position of the forearm; term also refers to a combination of ankle dorsiflexion, subtalar eversion, and forefoot abduction (toe-out).

proprioception   Feedback relative to the tension, length, and contraction state of muscle, the position of the body and limbs, and movements of the joints provided by internal receptors located in the skin, joints, muscles, and tendons.

protraction Forward movement of the shoulder girdle away from the spine; abduction of the scapula.

proximal Nearest to the midline or point of reference; the forearm is proximal to the hand.

Q angle (quadriceps angle) The angle at the patella formed by the intersection of the line of pull of quadriceps with the line of pull of the patella tendon.

quadriceps A common name given to the four muscles of the anterior aspect of the thigh: rectus femoris, vastus medialis, vastus intermedius, and vastus lateralis.

radial deviation (radial flexion) Abduction movement at the wrist of the thumb side of the hand toward the forearm.

radiate muscles A type of parallel muscle with a combined arrangement of flat and fusiform muscle in that they originate on broad aponeuroses and converge onto a tendon such as the pectoralis major or trapezius. Also described sometimes as being triangular, fan-shaped, or convergent.

ramus Part of an irregularly shaped bone that is thicker than a process and forms an angle with the main body such as the superior and inferior ramus of pubis.

range of motion (ROM) The specific amount of movement possible in a joint.

reciprocal inhibition Activation of the motor units of the agonists, causing a reciprocal neural inhibition of the motor units of the antagonists, which allows them to subsequently lengthen under less tension. Also referred to as reciprocal innervation.

recovery phase Skill analysis phase used after follow-through to regain balance and positioning to be ready for the next sport demand.

rectilinear motion Motion along a straight line.

reduction Return of the spinal column to the anatomic position from lateral flexion; spine adduction.

relaxation phase In a single muscle fiber contraction, it is the phase following the contraction phase in which the muscle fiber begins relaxing; lasts about 50 milliseconds.

reposition Diagonal movement of the thumb as it returns to the anatomical position from opposition with the hand and/or fingers.

resistance arm The distance between the axis and the point of resistance application.

retraction Backward movement of the shoulder girdle toward the spine; adduction of the scapula.

roll (rock) A type of accessory motion characterized by a series of points on one articular surface contacting with a series of points on another articular surface.

rolling friction The resistance to an object rolling across a surface, such as a ball rolling across a court or a tire rolling across the ground.

rotation Movement around the axis of a bone, such as the turning inward, outward, downward, or upward of a bone.

rotator cuff Group of muscles intrinsic to the glenohumeral joint, consisting of the subscapularis, supraspinatus, infraspinatus, and teres minor, that is critical in maintaining dynamic stability of the joint.

Ruffini's corpuscles A proprioceptor sensitive to touch and pressure found in the skin, subcutaneous tissue of fingers, and collagenous fibers of the joint capsule.

sacral plexus Group of spinal nerves composed of L4, L5, and S1 through S4, generally responsible for motor and sensory function of the lower back, pelvis, perineum, posterior surface of the thigh and leg, and dorsal and plantar surfaces of the foot.

sagittal plane Plane that bisects the body from front to back, dividing it into right and left symmetrical halves. Also known as the anteroposterior, or AP plane.

scoliosis Lateral curvatures or sideward deviations of the spine.

second-class lever A lever in which the resistance is between the axis (fulcrum) and the force (effort), as in plantarflexing the foot to raise up on the toes.

sellar joints Type of reciprocal reception that is found only in the thumb at the carpometacarpal joint and permits ball-and-socket movement, with the exception of rotation.

sensory neurons Neurons that transmit impulses to the spinal cord and brain from all parts of the body.

sesamoid bones Small bones embedded within the tendon of a musculotendinous unit that provide protection as well as improve the mechanical advantage of musculotendinous units as in the patella.

sinus Cavity or hollow space within a bone, such as the frontal or maxillary sinus.

somatic nerves (voluntary) Afferent nerves, which are under conscious control and carry impulses to skeletal muscles.

speed How fast an object is moving, or the distance an object travels in a specific amount of time.

sphincter muscle A type of parallel muscle that is a technically endless strap muscle with fibers arranged to surround and close openings upon contraction, such as the orbicularis oris. Also referred to as circular muscles.

spin A type of accessory motion characterized by a single point on one articular surface rotating clockwise or counterclockwise about a single point on another articular surface.

spine (spinous process) Sharp, slender projection of a bone, such as the spinous process of a vertebra or spine of the scapula.

spinal cord The common pathway between the central nervous system and the peripheral nervous system.

spinal nerves The group of 31 pairs of nerves that originate from the spinal cord and exit the spinal column on each side through openings between the vertebrae. They run directly to specific anatomical locations, form different plexuses, and eventually become peripheral nerve branches.

stability The resistance to a change in the body's acceleration; the resistance to a disturbance of the body's equilibrium.

stabilizers Muscles that surround the joint or body part and contract to fixate or stabilize the area to enable another limb or body segment to exert force and move; known as fixators, they are essential in establishing a relatively firm base for the more distal joints to work from when carrying out movements.

stance phase Skill analysis phase that allows the athlete to assume a comfortable and balanced body position from which to initiate the sport skill; emphasis is on setting the various joint angles in the correct positions with respect to one another and to the sport surface.

static equilibrium   The body at complete rest or motionless.

static friction   The amount of friction between two objects that have not yet begun to move.

statics   The study of mechanics involving the study of systems that are in a constant state of motion, whether at rest with no motion or moving at a constant velocity without acceleration. Involves all forces acting on the body being in balance, resulting in the body being in equilibrium.

strap muscles   A type of parallel muscle with fibers uniform in diameter and arranged with essentially all fibers in a long parallel manner, such as the sartorius.

sulcus (groove)   Furrow or groovelike depression on a bone, such as the intertubercular (bicipital) groove of the humerus.

supination   Externally rotating the radius to where it lies parallel to the ulna, resulting in the palm-up position of the forearm; term is also used in referring to the combined movements of inversion, adduction, and internal rotation of the foot and ankle.

suture   Line of union between bones, such as the sagittal suture between the parietal bones of the skull.

syndesmosis joint   Type of joint held together by strong ligamentous structures that allow minimal movement between the bones, such as the coracoclavicular joint and the inferior tibiofibular joint.

synergist   Muscles that assist in the action of the agonists but are not primarily responsible for the action; known as guiding muscles, they assist in refined movement and rule out undesired motions.

synovial joints   Freely movable diarthrodial joints containing a joint capsule and hyaline cartilage and lubricated by synovial fluid.

tendon   Fibrous connective tissue, often cordlike in appearance, that connects muscles to bones and other structures.

tendinous inscriptions   Horizontal indentations that transect the rectus abdominus at three or more locations, giving the muscle its segmented appearance.

tetanus   When stimuli are provided at a frequency high enough that no relaxation can occur between muscle contractions.

third-class lever   A lever in which the force (effort) is between the axis (fulcrum) and the resistance, as in flexion of the elbow joint.

threshold stimulus   When the stimulus is strong enough to produce an action potential in a single motor unit axon and all of the muscle fibers in the motor unit contract.

toe-off   Last portion of the walking or running stance phase characterized by the foot returning to supination and the leg returning to external rotation.

torque   Moment of force. The turning effect of an eccentric force.

transverse plane   Plane that divides the body horizontally into superior and inferior halves; also known as horizontal plane.

treppe   A staircase effect phenomenon of muscle contraction that occurs when rested muscle is stimulated repeatedly with a maximal stimulus at a frequency that allows complete relaxation between stimuli, the second contraction produces a slightly greater tension than the first, and the third contraction produces greater tension than the second.

triceps surae   The gastrocnemius and soleus together; triceps referring to the heads of the medial and lateral gastrocnemius and the soleus; surae referring to the calf.

trochanter   A very large bony projection, such as the greater or lesser trochanter of the femur.

trochoidal joint   Type of joint with a rotational movement around a long axis, as in rotation of the radius at the radioulnar joint.

tubercle   A small, rounded bony projection, such as the greater and lesser tubercles of the humerus.

tuberosity   A large, rounded or roughened bony projection, such as the radial tuberosity or tibial tuberosity.

ulnar deviation (ulnar flexion)   Adduction movement at the wrist of the little finger side of the hand toward the forearm.

uniarticular muscles   Those muscles that, from origin to insertion, cross only one joint, allowing them to perform actions only on the single joint that they cross.

unipennate muscles   A type of pennate muscle with fibers that run obliquely from a tendon on one side only, such as the biceps femoris, extensor digitorum longus, and tibialis posterior.

velocity   Includes the direction and describes the rate of displacement.

ventral   Relating to the belly or abdomen, on or toward the front, anterior part of.

vertical axis   Axis that runs straight down through the top of the head and spinal column and is at a right angle to the transverse plane of motion. Also known as the longitudinal or long axis.

visceral nerves (involuntary)   Nerves that carry impulses to the heart, smooth muscles, and glands; referred to as the autonomic nervous system.

Illustration credits

CHAPTER 1

1.1, 1.4, 1.7, 1.10, Van de Graaff KM: *Human anatomy,* ed 6, New York, 2002, McGraw-Hill; **1.2,** Anthony CP, Kolthoff NJ: *Textbook of anatomy and physiology,* ed 9, St. Louis, 1975, Mosby; **1.3,** Linda Kimbrough; **1.5, 1.9, 1.13, 1.14, 1.16,** Booher JM, Thibodeau GA: *Athletic injury assessment,* ed 4, New York, 2000, McGraw-Hill; **1.8,** Booher JM, Thibodeau GA: *Athletic injury assessment,* ed 4, New York, 2000, McGraw-Hill; Shier D, Butler J, Lewis R: *Hole's Human anatomy & physiology,* ed 9, New York, 2002, McGraw-Hill; Seeley RR, Stephens TD, Tate P: *Anatomy & physiology,* ed 7, New York, 2006, McGraw-Hill; **1.9, 1.11,** Shier D, Butler J, Lewis R: *Hole's Human anatomy & physiology,* ed 9, New York, 2002, McGraw-Hill; **1.12,** Seeley RR, Stephens TD, Tate P: *Anatomy & physiology,* ed 7, New York, 2006, McGraw-Hill; **1.15,** Seeley RR, et al: *Anatomy & physiology,* ed 6, New York, 2000, McGraw-Hill; **1.17,** Prentice WE: *Arnheim's principles of athletic training,* ed 11; New York, 2003, McGraw-Hill; **1.18,** R.T. Floyd; **1.19,** Lisa Floyd; **1.20, 121,** Prentice WE: *Rehabilitation techniques for sports medicine and athletic training,* ed 4; New York, 2004, McGraw-Hill.

CHAPTER 2

2.1, 2.2, Thibodeau GA: *Anatomy and physiology,* St. Louis, 1987, Mosby; **2.3,** Van de Graaff KM: *Human anatomy,* ed 6, Dubuque, IA, 2002, McGraw-Hill; **2.4, 2.5,** Shier D, Butler J, Lewis R: *Hole's human anatomy & physiology,* ed 9, Dubuque, IA, 2002, McGraw-Hill; **2.6,** Booher JM, Thibodeau GA; *Athletic injury assessment,* ed 4, Dubuque, IA, 2000, McGraw-Hill; **2.7, 2.8,** Shier D, Butler J, Lewis R: *Hole's Human anatomy & physiology,* ed 9, New York, 2002, McGraw-Hill; **2.9, 2.12,** Seeley RR, Stephens TD, Tate P: *Anatomy & physiology,* ed 7, New York, 2006, McGraw-Hill; **2.10, 2.11,** Powers SK, Howley ET: *Exercise Physiology: Theory and Application to Fitness and Performance,* ed 4, New York, 2001, McGraw-Hill; **2.13-2.15,** R.T. Floyd; **2.16,** Hall SJ: *Basic biomechanics,* ed 3, Dubuque, IA, 1999, WCB/McGraw-Hill.

CHAPTER 3

3.1, 3.10, Hall SJ: *Basic biomechanics,* ed 3, Dubuque, IA, 1999, WCB/McGraw-Hill; **3.2-3.4,** Booher JM, Thibodeau GA; *Athletic injury assessment,* ed 2, St. Louis, 1989, Mosby, Hall SJ: *Basic biomechanics,* ed 3, Dubuque, IA, 1999, WCB/McGraw-Hill; **3.5-3.9, 3.11, 3.12, 3.14, 3.17,** R.T. Floyd; **3.13, 3.15, 3.16,** Hamilton N, Luttgens K: *Kinesiology: scientific basis of human motion,* ed 10, New York, 2002.

CHAPTER 4

4.1, 4.2A, 4.11, 4.13, Linda Kimbrough; **4.2B,** Shier D, Butler J, Lewis R: *Hole's human anatomy and physiology,* ed 9, New York, 2002, McGraw-Hill; **4.3, 4.4,** Lisa Floyd; **4.5,** Hall SJ: *Basic biomechanics,* ed 3, Dubuque, IA, 1999, WCB/McGraw-Hill; **4.6, 4.7,** Seeley RR, et al: *Anatomy and physiology,* ed 6, Dubuque, IA, 2003, McGraw-Hill; **4.8-4.10, 4.12,** Ernest W. Beck.

CHAPTER 5

5.1-5.3, 5.15, 5.16, 5.19-5.22, Linda Kimbrough; **5.4,** Booher JM, Thibodeau GA; *Athletic injury assessment,* ed 4, Dubuque, IA, 2000, McGraw Hill; **5.5,** Lisa Floyd; **5.6, 5.7,** John Hood; **5.8, 5.9,** Shier D, Butler J, Lewis R: *Hole's essentials of human anatomy and physiology,* ed 9, New York, 2006, McGraw-Hill; **5.12-5.14,** Ernest W. Beck; **5.10, 5.11,** Van de Graaff KM: *Human anatomy,* ed 6, Dubuque, IA. 2002, McGraw-Hill; **5.17,** Seeley RR, et al: *Anatomy and physiology,* ed 6, Dubuque, IA, 2003, McGraw-Hill; **5.18,** Ernest W. Beck with inserts by Linda Kimbrough.

CHAPTER 6

6.1, 6.3A, 6.14-6.21, Linda Kimbrough; **6.2A-B,** Seeley RR, Stephens TD, Tate P: *Anatomy & physiology,* ed 7, New York, 2006, McGraw-Hill; **6.2C,** Shier D, Butler J, Lewis R: *Hole's human anatomy and physiology,* ed 9, New York, 2002, McGraw-Hill; **6.3B, 6.12, 6.13,** Van de Graaff KM: *Human anatomy,* ed 6, Dubuque, IA. 2002, McGraw-Hill; **6.3C,** Jason Alexander; **6.4, 6.5,** Booher JM, Thibodeau GA: *Athletic injury assessment,* ed 4, Dubuque, IA, 2000, McGraw-Hill; **6.6, 6.7, 6.9,** Lisa Floyd; **6.8, 6.10,** Thibodeau GA: *Anatomy and physiology,* St. Louis, 1987, Mosby; **6.11A-B,** Seeley RR, et al: *Anatomy and physiology,* ed 6, Dubuque, IA, 2003, McGraw-Hill.

CHAPTER 7

7.1, Anthony CP, Kolthoff NJ: *Textbook of anatomy and physiology,* ed 9, St. Louis, 1975, Mosby; **7.2, 7.11-7.25,** Linda Kimbrough; **7.3, 7.10, 7.26,** Van de Graaff KM: *Human anatomy,* ed 6, Dubuque, IA, 2002, McGraw-Hill; **7.4-7.6, 7.8,** Booher JM, Thibodeau GA: *Athletic injury assessment,* ed 4, Dubuque, IA, 2000, McGraw-Hill; **7.7,** Lisa Floyd; **7.9A-D,** Seeley RR, et al: *Anatomy and physiology,* ed 6, Dubuque, IA, 2003, McGraw-Hill.

CHAPTER 8

8.1, 8.2, 8.8-8.11, R.T. Floyd; **8.3-8.7,** Lisa Floyd.

CHAPTER 9

9.1, 9.3-9.5, 9.11, 9.21-9.36, Linda Kimbrough; **9.2, 9.9,** Anthony CP, Kolthoff NJ: *Textbook of anatomy and physiology,* ed 9, St. Louis, 1975, Mosby; **9.6,** Booher JM, Thibodeau GA: *Athletic injury assessment,* ed 4, Dubuque, IA, 2000, McGraw-Hill; **9.7A-H, 9.8A-D,** Lisa Floyd; **9.10,** Ernest W. Beck; **9.12, 9.13, 9.15-9.20,** Van de Graaff KM: *Human anatomy,* ed 6, Dubuque, IA, 2002, McGraw-Hill; **9.14,** Shier D, Butler J, Lewis R: *Hole's human anatomy and physiology,* ed 9, Dubuque, IA, 2002, McGraw-Hill

CHAPTER 10

10.1, Prentice WE: *Arnheim's principles of athletic training,* ed 12, New York, 2006, McGraw-Hill; **10.2,** Anthony CP, Kolthoff NJ: *Textbook of anatomy and physiology,* ed 9, St. Louis, 1975, Mosby; **10.3,** Van de Graaff KM: *Human anatomy,* ed 6, Dubuque, IA, 2002, McGraw-Hill; **10.4,** Booher JM, Thibodeau GA: *Athletic injury assessment,* ed 4, Dubuque, IA, 2000, McGraw-Hill; **10.5A-D,** Lisa Floyd; **10.6-10.12,** Linda Kimbrough.

CHAPTER 11

11.1, Prentice WE: *Arnheim's principles of athletic training,* ed 12; New York, 2006, McGraw-Hill; **11.2, 11.3,** Anthony CP, Kolthoff NJ: *Textbook of anatomy and physiology,* ed 9, St. Louis, 1975, Mosby; **11.4, 11.6, 11.8A-D, 11.10, 11.22, 11.23,** Van de Graaff KM: *Human anatomy,* ed 6, Dubuque, IA, 2002, McGraw-Hill; **11.5,** Booher JM, Thibodeau GA; *Athletic injury assessment,* ed 4, Dubuque, IA, 2000, McGraw-Hill; **11.7A-H,** Lisa Floyd; **11.9,** Seeley RR, et al: *Anatomy and physiology,* ed 3, St. Louis, 1995, Mosby; **11.11-11.14, 11.16-11.21,** Ernest W. Beck; **11.15,** Linda Kimbrough.

CHAPTER 12

12.1, 12.13, Seeley RR, et al: *Anatomy and physiology,* ed 6, Dubuque, IA, 2003, McGraw-Hill; **12.2E-F,** Anthony CP, Kolthoff NJ: *Textbook of anatomy and physiology,* ed 9, St. Louis, 1975, Mosby; **12.2A-D, 12.11, 12.12, 12.16, 12.18-12.20, 12.22-12.25,** Linda Kimbrough; **12.3, 12.16,** Shier D, Butler J, Lewis R: *Hole's human anatomy and physiology,* ed 9, Dubuque, IA, 2002, McGraw-Hill; **12.4, 12.9,** Lindsay D: *Functional anatomy,* ed 1, St. Louis, 1996, Mosby; **12.5A-B,** Thibodeau GA, Patton KT: *Anatomy and physiology,* ed 9, St. Louis, 1993, Mosby; **12.5C, 12.14,** Seeley RR, Stephens TD, Tate P: *Anatomy & physiology,* ed 7, New York, 2006, McGraw-Hill; **12.6, 12.7,** Booher JM, Thibodeau GA; *Athletic injury assessment,* ed 4, Dubuque, IA, 2000, McGraw-Hill; **12.8A-H,** Lisa Floyd; **12.10, 12.15,** Van de Graaff KM: *Human anatomy,* ed 6, Dubuque, IA, 2002, McGraw-Hill; **12.21,** Ernest W. Beck.

CHAPTER 13

13.1-13.3, 13.6, Lisa Floyd; **13.4, 13.5,** R.T. Floyd; **13.7, 13.8A-B,** Ron Carlberg.

Index

flexor carpi radialis muscle, *145,* 165, *165,*
169, 171, 172, *173, 203, 204, 205,*
206, 208, 209, 210, 211
flexor carpi ulnaris muscle, 165, *165,* 169,
171, 172, *172, 175, 203, 204, 206,*
208, 209, 210, 211
flexor digiti minimi brevis muscle, 170, 172,
188, 189, *190,* 292, 304, *305, 307*
flexor digitorum brevis muscle, 292, 304,
305, 307
flexor digitorum longus muscle, *236,* 280,
287, 288, 290, 292, *302*
flexor digitorum longus tendon, *305*
flexor digitorum profundus muscle, *161,* 165,
166, 171, 172, *172, 179, 180, 203,*
204, 205, 206, 208, 209, 210, 211
flexor digitorum superficialis muscle, *161,*
165, *166, 171,* 172, *179, 203, 204,*
205, 206, 208, 209, 210, 211
flexor hallucis brevis muscle, 292, 304, *305, 307*
flexor hallucis longus muscle, *236,* 280, *287,*
288, 290, 292, *303*
flexor hallucis longus tendon, 280, *305*
flexor pollicis brevis muscle, 170, 172, *172,*
188, 189
flexor pollicis longus muscle, *145,* 165, *166,*
171, 181, 203, 204, 205, 206, 208,
209, 210, 211
flexor retinaculum muscle, 169, *287*
floating (vertebral) ribs, *316*
follow-through phase, *197*
foot, *281, 305, 306*
  arches, 282, *284*
  bones, 280–81
  eversion, *295, 296, 297*
  inversion, *300, 301, 302, 303*
  joints, *278–324, 282–83*
  ligaments, 282
  movements, *285–86*
  muscles, 287–91, *289, 290*
  nerves, *292*
  terms describing movements, 23
force, 67, 68, *68,* 72, 80, 81
force application, 68, 72
force arm (moment arm, or torque arm), 71,
  72, 73
force magnitude, 71
forearm
  anteromedial aspect, 169
  muscles, *171*
  range-of-motion, *139*
frequency, 200
friction, 79
frontal (lateral, or coronal) plane, 8, *9–10*
fulcrum, 68
fundamental position, 2
fusiform muscles, 39

he.com/floyd16e